Epigenomics in Health and Disease

T0326381

Translational Epigenetics Series

Trygve O. Tollefsbol, Series Editor

Transgenerational Epigenetics
Edited by Trygve O. Tollefsbol, 2014

Personalized Epigenetics
Edited by Trygve O. Tollefsbol, 2015

Epigenetic Technological Applications
Edited by Y. George Zheng, 2015

Epigenetic Cancer Therapy
Edited by Steven G. Gray, 2015

DNA Methylation and Complex Human Disease
By Michel Neidhart, 2015

Epigenomics in Health and Disease

Edited by

Mario F. Fraga
Nanomaterials and Nanotechnology Research Center, Institute of
Oncology of Asturias, Oviedo, Spain

and

Agustín F. Fernández
Cancer Epigenetics Laboratory,
Institute of Oncology of Asturias, Hospital Universitario Central de
Asturias, Universidad de Oviedo, Oviedo, Spain

ELSEVIER

AMSTERDAM • BOSTON • HEIDELBERG • LONDON
NEW YORK • OXFORD • PARIS • SAN DIEGO
SAN FRANCISCO • SINGAPORE • SYDNEY • TOKYO
Academic Press is an imprint of Elsevier

Academic Press is an imprint of Elsevier
125, London Wall, EC2Y 5AS.
525 B Street, Suite 1800, San Diego, CA 92101-4495, USA
225 Wyman Street, Waltham, MA 02451, USA
The Boulevard, Langford Lane, Kidlington, Oxford OX5 1GB, UK

Copyright © 2016 Elsevier Inc. All rights reserved.

No part of this publication may be reproduced or transmitted in any form or by any means, electronic or mechanical, including photocopying, recording, or any information storage and retrieval system, without permission in writing from the Publisher. Details on how to seek permission, further information about the Publisher's permissions policies and our arrangements with organizations such as the Copyright Clearance Center and the Copyright Licensing Agency, can be found at our website: www.elsevier.com/permissions.

This book and the individual contributions contained in it are protected under copyright by the Publisher (other than as may be noted herein).

Notices
Knowledge and best practice in this field are constantly changing. As new research and experience broaden our understanding, changes in research methods, professional practices, or medical treatment may become necessary.

Practitioners and researchers may always rely on their own experience and knowledge in evaluating and using any information, methods, compounds, or experiments described herein. In using such information or methods they should be mindful of their own safety and the safety of others, including parties for whom they have a professional responsibility.

To the fullest extent of the law, neither the Publisher nor the authors, contributors, or editors, assume any liability for any injury and/or damage to persons or property as a matter of products liability, negligence or otherwise, or from any use or operation of any methods, products, instructions, or ideas contained in the material herein.

ISBN: 978-0-12-800140-0

Library of Congress Cataloging-in-Publication Data
A catalog record for this book is available from the Library of Congress.

British Library Cataloguing-in-Publication Data
A catalogue record for this book is available from the British Library.

For information on all Academic Press publications
visit our website at http://store.elsevier.com

Working together
to grow libraries in
developing countries

www.elsevier.com • www.bookaid.org

Publisher: Mica Haley
Acquisition Editor: Catherine Van Der Laan
Editorial Project Manager: Lisa Eppich
Production Project Manager: Melissa Read
Designer: Mark Rogers

Printed and bound in the United States of America

Contents

3. **Ultra-Deep Sequencing of Bisulfite-Modified DNA**

*Tingting Qin, Yongseok Park, Maria E. Figueroa
and Maureen A. Sartor*

4. **Bioinformatics Tools in Epigenomics Studies**

Gustavo F. Bayón, Agustín F. Fernández and Mario F. Fraga

5.   Noncoding RNA Regulation of Health and Disease

Nicolas Léveillé, Carlos A. Melo and Sonia A. Melo

6.   Genome-Wide DNA Methylation Changes During Aging

Kevin C. Johnson and Brock C. Christensen

7.   The Dynamics of Histone Modifications During Aging

Anthony J Bainor and Gregory David

8. Epigenomic Studies in Epidemiology

Valentina Bollati, Valeria Motta, Simona Iodice and Michele Carugno

9. The DNA Methylomes of Cancer

Renée Beekman, Marta Kulis and José Ignacio Martín-Subero

10. Genome-Wide Epigenetic Studies in Neurologic Diseases

Ashwin Woodhoo

11. Epigenetic Deregulation in Autoimmune Disease

Damiana Álvarez-Errico and Esteban Ballestar

List of Contributors

Damiana Álvarez-Errico Chromatin and Disease Group, Cancer Epigenetics and Biology Programme, Bellvitge Biomedical Research Institute, Barcelona, Spain

Anthony J. Bainor Department of Biochemistry and Molecular Pharmacology and NYU Cancer Institute, NYU Langone Medical Center, New York, NY, USA

Esteban Ballestar Chromatin and Disease Group, Cancer Epigenetics and Biology Programme, Bellvitge Biomedical Research Institute, Barcelona, Spain

Gustavo F. Bayón Cancer Epigenetics Laboratory, Institute of Oncology of Asturias, HUCA, Universidad de Oviedo, Oviedo, Spain

Renée Beekman Institut d'Investigacions Biomèdiques August Pi i Sunyer, Barcelona, Departamento de Anatomía Patológica, Farmacología and Microbiología, Universitat de Barcelona, Spain

Marina Bibikova Illumina, Inc., San Diego, CA, USA

Valentina Bollati Center of Molecular and Genetic Epidemiology, Department of Clinical Sciences and Community Health, Università degli Studi di Milano, Milan, Italy; Epidemiology Unit, Fondazione IRCCS Ca' Granda Ospedale Maggiore Policlinico, Milan, Italy

Michele Carugno Center of Molecular and Genetic Epidemiology, Department of Clinical Sciences and Community Health, Università degli Studi di Milano, Milan, Italy

Brock C. Christensen Department of Epidemiology, Geisel School of Medicine at Dartmouth, Hanover, NH, USA; Department of Pharmacology and Toxicology, Geisel School of Medicine at Dartmouth, Hanover, NH, USA; Department of Community and Family Medicine, Geisel School of Medicine at Dartmouth, Hanover, NH, USA

Gregory David Department of Biochemistry and Molecular Pharmacology and NYU Cancer Institute, NYU Langone Medical Center, New York, NY, USA

Agustín F. Fernández Cancer Epigenetics Laboratory, Institute of Oncology of Asturias, HUCA, Universidad de Oviedo, Oviedo, Spain

Maria E. Figueroa Department of Pathology, University of Michigan, Ann Arbor, MI, USA

Mario F. Fraga Nanomaterials and Nanotechnology Research Center, Oviedo, Spain

Holger Heyn Cancer Epigenetics and Biology Program, Bellvitge Biomedical Research Institute, Barcelona, Catalonia, Spain

Simona Iodice Center of Molecular and Genetic Epidemiology, Department of Clinical Sciences and Community Health, Università degli Studi di Milano, Milan, Italy

Kevin C. Johnson Department of Epidemiology, Geisel School of Medicine at Dartmouth, Hanover, NH, USA; Department of Pharmacology and Toxicology, Geisel School of Medicine at Dartmouth, Hanover, NH, USA; Department of Community and Family Medicine, Geisel School of Medicine at Dartmouth, Hanover, NH, USA

Marta Kulis Institut d'Investigacions Biomèdiques August Pi i Sunyer, Barcelona, Departamento de Anatomía Patológica, Farmacología and Microbiología, Universitat de Barcelona, Spain

Nicolas Léveillé Division of Gene Regulation, The Netherlands Cancer Institute, Amsterdam, The Netherlands

Charlotte Ling Department of Clinical Sciences, Epigenetics and Diabetes, Clinical Research Centre, Lund University Diabetes Centre, Malmö, Sweden

José Ignacio Martín-Subero Institut d'Investigacions Biomèdiques August Pi i Sunyer, Barcelona, Departamento de Anatomía Patológica, Farmacología and Microbiología, Universitat de Barcelona, Spain

Michael A. McDevitt Division of Hematology, Johns Hopkins University School of Medicine, Baltimore, MD, USA; Division of Hematological Malignancy, Johns Hopkins University School of Medicine, Baltimore, MD, USA

Carlos A. Melo Division of Gene Regulation, The Netherlands Cancer Institute, Amsterdam, The Netherlands; Doctoral Programme in Biomedicine and Experimental Biology, Centre for Neuroscience and Cell Biology, Coimbra, Portugal

Sonia A. Melo Department of Cancer Biology, Metastasis Research Center, University of Texas MD Anderson Cancer Center, Houston, TX, USA; Instituto de Investigação e Inovação em Saúde, Universidade do Porto, Portugal and Institute of Pathology and Molecular Immunology of the University of Porto, Porto, Portugal

Valeria Motta Center of Molecular and Genetic Epidemiology, Department of Clinical Sciences and Community Health, Università degli Studi di Milano, Milan, Italy

Yongseok Park Department of Biostatistics, University of Pittsburgh, Pittsburgh, PA, USA

Tingting Qin Department of Pathology, University of Michigan, Ann Arbor, MI, USA

Tina Rönn Department of Clinical Sciences, Epigenetics and Diabetes, Clinical Research Centre, Lund University Diabetes Centre, Malmö, Sweden

Maureen A. Sartor Department of Computational Medicine and Bioinformatics, University of Michigan, Ann Arbor, MI, USA

Ashwin Woodhoo CIC bioGUNE, Centro de Investigación Biomédica en Red de Enfermedades Hepáticas y Digestivas, Bizkaia, Spain; IKERBASQUE, Basque Foundation for Science, Bilbao, Spain

Preface

Both genetic and epigenetic mechanisms are involved in the regulation of gene expression, cellular differentiation, and developmental processes. An essential difference is that epigenetic alterations do not involve a change to the DNA sequence, and therefore they can be reverted by the use of therapeutic drugs, such as DNA methylation or histone deacetylases inhibitors. Although many epigenetic modifications are maintained throughout a person's lifetime, at certain genomic loci, epigenetic marks can change over time, and this is thought to depend both on intrinsic, or genetic, and extrinsic, or environmental factors. The impact of genetic variance on epigenetic regulation, especially in pathologic contexts, is covered in Chapter 1 of this volume.

In 1989, only a few years after the first human oncogene (H-RAS) was found to be activated by point mutation, an epigenetic alteration of a tumor suppressor gene (Rb) in human cancer was discovered. The first epigenetic studies analyzed alterations in DNA methylation in tumor suppressor genes by using candidate gene approaches. Later on, in addition to DNA methylation, other epigenetic mechanisms, such as post-translational modifications (PTM) of histones and noncoding RNAs (ncRNAs), have been found altered in association with several human diseases.

During the last few decades, methods and technologies aimed at studying changes in epigenetic marks have evolved from the analysis of a small number of specific candidate genes to analyzing large numbers of genes, but the true qualitative and quantitative leap occurred with the appearance of the techniques of ultra-sequencing complete epigenomes. These new technologies have made it possible to investigate in greater detail the epigenetic alterations associated with human disease. In Chapters 2 and 3, various methodologies for the study and analysis of complete epigenomes, and their usefulness in the identification of new biomarkers associated with diseases are described. The emergence of these new technologies has revolutionized the way of approaching research projects, as much as the way we manage all the information obtained. Chapter 4 explores how to analyze the huge amount of data obtained by these methods, and the software tools currently available for the analysis of epigenomic data are described. Chapters 5–11 of this volume summarize studies where these next-generation technologies have been applied to the search for genome-wide epigenetic changes related to aging, cancer, and other human diseases, including neurologic disorders and autoimmune and metabolic diseases. The focus of Chapter 12 is the importance of factors that should be taken into account in the design of epigenomic epidemiologic studies. And finally, Chapter 13 discusses how epigenomic data can be useful in the clinical management of human disease, particularly in cancer, and the potential applications of epigenetic and genetic biomarkers in optimizing epigenetic therapies are described.

A major future challenge in the field of epigenetics will be to identify epigenetic marks in single cells, since epigenetics, unlike genetics, is cell specific. This epigenomic cell information, along with the genetic information, will provide the basis for defining customized therapies in the treatment of human diseases.

Our most sincere thanks to all the authors for their outstanding contributions.

Agustín F. Fernández
Hospital Universitario Central de Asturias,
Institute of Oncology of Asturias,
Universidad de Oviedo, Spain

Mario F. Fraga
Nanomaterials and Nanotechnology Research
Center, Institute of Oncology of Asturias,
Universidad de Oviedo, Spain

Chapter 1

The Role of the Genetic Code in the DNA Methylation Landscape Formation

Holger Heyn
Cancer Epigenetics and Biology Program, Bellvitge Biomedical Research Institute, Barcelona, Catalonia, Spain

Chapter Outline

1.1 BRINGING THE GENETIC CODE TO LIFE

At the beginning of this millennium, the human genome was decoded [1,2]. The sequencing of the major fraction of the genetic sequence and the resulting first version of an assembled human reference genome displayed an important step toward the understanding of cellular biology and human life. Using the then-current knowledge of genomic organization and sequence characteristics, the genetic sequence was estimated to be roughly 1% coding and the majority of the genome to be noncoding, from which a major part was considered "junk" DNA [3]. However, realizing that the human genome differs only marginally between individuals and is even conserved to around

M. Fraga & A.F. Fernandez (Eds): Epigenomics in Health and Disease.
DOI: http://dx.doi.org/10.1016/B978-0-12-800140-0.00001-7
© 2016 Elsevier Inc. All rights reserved.

1

99% between humans and chimpanzees [4], the noncoding part of the genome was quickly recognized to contribute to phenotype formation and interindividual and interspecies differences.

To illustrate the impact of gene regulation on phenotype formation, the human body presents the best paradigm. Considering the fact that every single cell in the human body carries exactly the same genetic information, it appears remarkable that cells can differ to such a large extent in appearance and function. The difference between a pancreatic cell that produces insulin and a hematopoietic stem cell, giving rise to numerous specialized blood cell types, does not lie within their genetic sequence but is grounded in the activity of the genome. A distinct set of active and repressed genomic regions defines intraindividual differences between cell phenotypes but is also responsible for variation between human individuals (interindividual) and across species (interspecies). Following the sequencing of more than 1000 human genomes [5] and the analyses of variable genetic loci in thousands of individuals in genome-wide association studies (GWAS) [6], it also became apparent that the majority of human phenotypes (traits), ranging from the color of the hair to the susceptibility to develop breast cancer, is encoded in the noncoding fraction of the genome.

Although the protein-coding fraction of the genome, including transcription and translation start sites, was well annotated, the knowledge of the noncoding genome was grounded in annotations based on distance to known genes (promoter regions) or single examples of *cis*-acting regulatory regions (enhancers). As the function of the genome outside the coding context could not be inferred by the sequence itself, efforts were focused on the comprehensive annotation of the noncoding part of the genome that was no longer considered silent or "junk," but rather as important driver for phenotype formation. Here seminal contribution was made by the work performed within the ENCyclopedia Of DNA Elements (ENCODE) project, which, after annotating selected parts of the genome in a pilot phase (1% of the genome) [7], presented a comprehensive catalog of regulatory elements within the human genome and led to different estimates as to the extent of active regions in the human genome [8]. However, although the features tested to determine functionality (ranging from transcription activity to DNA factor binding occupancy and chromosomal conformation) covered a wide range of mechanisms involved in known gene regulatory processes, the tissue specificity of the "regulome" remained a major challenge to drawing a general conclusion. Also, the studies used cell lines from single individuals and hence could not address interindividual differences, and results could have been influenced by biases introduced by cell culture. However, due to its extensive number of performed experiments and width of analyzed features, the resulting annotation of the human genome presented a highly valuable insight into the actual activity of the genetic sequence. Importantly, integrated analysis of the wealth of data sets produced by ENCODE enabled the segmentation of the genome into distinct regulatory elements [9−11]. Specifically, active regulatory regions were annotated as promoter, enhancer, or insulator regions, wherein transcriptional activity was determined as a measure for actively transcribed regions that mark putative genic loci or those for which transcription contributes to the regulatory process (enhancers). In addition, transcriptional silent regions or in those carrying actively repressing marks were annotated in repressed or silent segments.

Segmentation algorithm are extremely powerful to summarize the data produced by multidimensional experiments related to gene regulation and enable an estimate of the activity and function of a given region of interest. However, the diversity of regulatory processes and their individual particularities can only be assessed using in-depth analysis and, importantly, large sample cohorts and particular tailored study designs. This chapter focuses on a regulatory mechanism directly affecting the genetic sequence: DNA methylation. DNA methylation involves the covalent modification of DNA, converting cytosine in 5-methyl-cytosine (5-mC). As part of epigenetic regulatory processes, DNA methylation was shown to be stably inherited throughout cell division and to be crucial in developmental and differentiation processes [12]. However, DNA methylation represents the oldest described epigenetic mechanism and its role in gene regulation is still subject of discussion. In general, high levels of 5-mC at proximal promoters are associated to transcriptional silencing; however, outside these regions, its role is highly dependent on the genomic context [13]. For example, increased 5-mC levels within the gene body were related to an elevated transcriptional activity [14].

Similar to the controversial function of DNA methylation, the question of which elements determine the temporal and spatial distribution of 5-mC within the genome remains unresolved. Although the mechanisms covalently modifying the genetic code are clearly defined, with DNA methyltransferases catalyzing the addition of the methyl-group to cytosine, the driving forces guiding the activity of such enzymes is under intense discussion, and putative causalities are summarized in this chapter.

1.2 INTRINSIC PROPERTIES OF DNA

Taking into account that DNA methylation varies among different cell types within the same individual but is relatively stable in the same tissue of different individuals, the genetic code appears to play an inferior role in the variable nature of DNA methylation profiles. However, there are several examples clearly displaying how the genetic sequence is necessary and sufficient for the formation of the DNA methylation landscape [15]. Here, the density of CpG dinucleotides presents the most prominent example of a genetic feature directly associated to the DNA methylation status of the respective regions [16]. In particular, an elevated CpG density was related to an unmethylated (hypomethylated) status of DNA, whereas CpG-poor regions are mainly highly methylated (hypermethylated). Intriguingly, it was the DNA methylation itself that was shaping the distribution of CpG dinucleotides in the genome throughout evolution. Methylated cytosines tend to be mutated by deamination, which results in the formation of uracil and the incorporation of thymine after cell division (without DNA repair). As these transition mutations mainly affect methylated cytosine and almost exclusively occur in a CpG context, this specific dinucleotide was depleted in the genome during evolution. Consequently, the proportion of CpG dinucleotide within the human genome is far less than would be expected if all dinucleotides would underlie equal evolutionary pressures. Moreover, because hypomethylated CpG sites, a frequent feature of promoter regions, were protected from deamination processes, it led to their enrichment in promoters of constitutively transcribed genes and the formation of CpG clusters or CpG islands. However, CpG islands being located outside proximal promoter regions and a specific set of islands being mainly hypermethylated suggest that additional features shape the human DNA methylome [17].

One of these features was discovered as an intrinsic property of the DNA molecule itself, and such parameters as DNA stiffness or flexibility enlarged our understanding of the code encoded by the one-dimensional succession of nucleotides. Sophisticated analysis and computational modeling of DNA molecule properties resulted in the redefinition of the promoter signature and supported a hypothesis of an ancient regulatory mechanism encoded by the intrinsic physical properties of DNA [18]. Computational strategies, linking the DNA sequence to epigenetic mechanisms, consistently support an intrinsic association between the genome and epigenome. Here the cumulative effects of many weak interactions are thought to guide the epigenetic code and to present the blueprint for subsequent formation processes [19]. In this regard, the epigenetic landscapes could be partially predicted computational models and remarkably could be assembled *in vitro*, suggesting DNA methylation profiles to be partially encoded in the genetic code itself [20].

Interestingly, DNA methylation was described to be capable of altering physical DNA properties and hence modify downstream regulatory processes by altering its flexibility [21]. Specifically, molecular dynamics simulations observed that the high flexibility of the CpG sites is not extrapolated in CpG-dense regions. In contrast, poly-CpG fragments are hardly distinguishable from equally sized dinucleotides in terms of global unwinding and isotropic bending. Importantly, it is the covalent modification of cytosine in CpG-dense regions that dramatically alters the winding properties of DNA, which are particularly important, when DNA is wrapped around nucleosomes, an additional layer in the epigenetic code. Here, DNA methylation changes the winding properties of the DNA by making it stiffer. This was shown to be particularly important in CpG islands, wherein the altered flexibility of the DNA molecule changed the affinity of DNA to assemble into nucleosomes. As nucleosome positioning presents an important feature of transcriptional regulation at transcription start sites, differential DNA methylation can indirectly influence epigenetic regulation through its impact of the physical properties of the genetic sequence.

In conclusion, there is evidence of a direct influence of the DNA sequence on the spatial formation of DNA methylation profiles. These intrinsic properties of the genome on the epigenome present the blueprint for downstream regulatory events, and their understanding will guide our knowledge about cause and consequences in epigenetic regulatory mechanisms. However, physical properties are not sufficient to explain the flexible nature of DNA methylation, which, undoubtedly, involves additional mechanisms. Taking into account that the DNA methylation machinery is embedded in a complex network of regulatory processes, it is likely their interplay that guides the temporal variation of methyl-cytosine distribution in the genome.

1.3   SEQUENCE-PATTERN-DEPENDENT DNA METHYLATION PROFILES

Initial experiments aiming to determine the driving forces behind distinct DNA methylation states identified an important impact of the surrounding genomic environment on the methylation state [22]. However, earlier observations that the binding of transcription factors, such as SP1, can actively contribute to establishing an unmethylated state of targeted region suggested an effect of local regulatory activity and underlying DNA-binding factor sequence motifs on DNA methylation levels [23,24].

Years later, to which extent the genetic sequence acts as determent for the DNA methylation state was still unclear, and crucial evidence that intrinsic properties of the DNA act as major factor for the hypo- and hypermethylated state of DNA came from experiments inserting numerous sequences at a unique genomic locus with subsequent DNA methylation profiling efforts [25]. Here promoter sequences correctly recapitulated the original DNA methylation profiles, even being correctly altered in the transition from pluripotent to differentiated cells. Intriguingly, small methylation-determining regions (MDRs) were shown to control the methylation state of larger regions in cis and were both necessary and sufficient for regulating DNA methylation. Critical features, besides CpG density, that define MDR activity consisted of developmental state and transcription factor-binding motifs, suggesting that a concerted function of the regulatory network mediates the correct establishment of DNA methylation profiles. Subsequent studies further underlined the role of DNA-binding factor for DNA methylation levels [26]. High-throughput engineering of DNA sequence and comprehensive data modeling suggested a principal role for transcription factor binding sites and that CpG density alone is a minor determinant of an unmethylated state. Consistently, sequence variations among individuals at regulatory loci can lead to the differential binding of regulatory factor, with direct impact on the DNA methylation profile in the respective regions [27]. However, it is of note that some transcription factors are sensitive for CpG methylation, also pointing to an active regulatory role of the epigenetic mark in gene regulation [28,29].

In line with the definition of MDRs, a specific set of CpG-dense DNA regions was identified to be exclusively hypermethylated in somatic tissue types and proposed as *cis*-regulatory elements to facilitate methylation at CpG islands containing promoters and to act as the hypermethylation driver sequence [17]. Remarkably, these *cis*-regulatory motifs were utilized to engineer an *in vivo* model system that confirmed the disease-driving potential of epimutations (alteration in epigenetic state) in oncogenesis [30]. Specifically, an MDR that mediates hypermethylation was placed in the promoter region of the tumor suppressor *p16Ink4a*, resulting in epigenetic silencing of the gene. Moreover, the transgenic mice carrying the induced epimutation presented increased incidence of spontaneous cancers. This model perfectly illustrates the impact of genetic sequence on DNA methylation levels and its subsequent mediator function in phenotype formation.

The genetic sequence also provided crucial clues about the widely observed phenomenon of partially methylated domains (PMDs) in cancer samples [31−33]. PMDs span 20−40% of the genome and are thought to contribute to increased regulatory variability within their affected regions [33]. However, the intermediate DNA methylation state was observed by averaging individual CpG levels, and a more detailed analysis at the base-pair resolution level identified an extensive variability along the epigenome of these regions, with DNA methylation levels ranging from 0% to 100% [34]. The heterogeneity of the epigenetic landscape could be traced back to the underlying genetic sequence, wherein sequence features, such as CpG density, presented a major driving feature. Surprisingly, CpG-dense region within PMDs were associated with increased methylation levels, an opposite trend to the previous described hypomethylation in CpG islands. Conclusively, CpG density and sequence motifs present a major factor defining DNA methylation levels. However, as variable epigenetic states were observed for single genetic features, the actual DNA methylation profile highly depends on the underlying context.

1.4 DNA-BINDING FACTORS SHAPING THE EPIGENETIC LANDSCAPE

The binding of transcription factors, such as SP1, to MDRs was established as driving events toward a hypomethylated state of the DNA and confirmed by mutation experiments of respective consensus-binding motifs [25]. It is noteworthy that the study also showed that transcriptional activity is not necessary to maintain hypomethylated states as promoter regions and fragments not related to genic sequence recapitulated the original methylation state when placed in a transcriptional silence context. The scenario that the actual binding of factors to the DNA, and not the sequence itself, determines the methylation state of the region was supported by the fact that bacterial DNA, with comparable sequence composition, did not cause an hypomethylated state of the respective DNA [25]. This closely links DNA methylation to underlying DNA sequence features and suggests that a substantial fraction of methylation changes occur downstream of gene regulation [35].

Whole-genome DNA methylation maps further underpinned the relationship between DNA factor binding and hypomethylation [36]. Here, base-pair resolution profiling of mouse embryonic stem cell (ESC) and derived neuronal progenitor cells revealed the presence of a particular epigenomic feature, represented by small lowly methylated regions (LMRs), which were mostly located distal to promoter regions. LMRs resembled features of *cis*-regulatory elements, such as enhancer regions, being enriched in marker histone modifications (H3K4me1 and H3K27ac) and DNA sensitive to DNaseI digestion, a marker for open chromatin formation. Most notably, LMRs were enriched in DNA-binding factors, such as pluripotency-related factors; in specific cases, they outranged the enrichment observed at promoter regions (e.g., Nanog binding). A specific case relates to the enrichment of the chromatin organization factor CTCF (CCCTC-binding factor), whose binding can not only act to regulate transcription activity but also controls in the spatial organization of chromosomes [37]. Taking into account DNA methylation profiles, the prediction of CTCF binding to the genome could be significantly improved compared with the analysis of sequence motifs alone [36]. Moreover, CTCF was used to confirm the direct relationship between factor binding and differential DNA methylation. The study altered CTCF binding site sequences (in a transgenic approach) to reduce CTCF affinities to the DNA. Consistent with prior results, the absence of DNA occupancy of CTCF conferred a gain of CpG methylation within these regions and directly confirming the active impact of CTCF binding on the DNA methylation signature. Most remarkably, CTCF binding could actively promote a hypomethylated state, as hypermethylated reporter constructs lost methylation following CTCF binding. Interestingly, this effect could even be assessed for genetically variable loci (polymorphisms), wherein a polymorphic genetic sequence can alter CTCF binding affinity. Here, genotypes unfavorable for CTCF binding were associated with increased DNA methylation levels at these loci. However, these results are in conflict with those of previous studies reporting a methylation sensitivity of CTCF binding in imprinted regions [38,39]. Thus, CTCF binding is likely to be able to actively participate in demethylation, however, with DNA methylation preventing CTCF binding in specific regions.

The relationship between transcription factor binding and DNA hypomethylation was further underlined by using genetically engineered knockout mice for the DNA-binding factor REST (REST$^{-/-}$) [36]. In line with previously observed associations, the absence of REST was accompanied by a gain of CpG methylation at loci that were previously hypomethylated and occupied by the transcription factor. Consistently, reintroduction of REST resulted in a *de novo* hypomethylation of the respective loci. These results support the role of transcription factors in the formation of the DNA methylation landscape and predict that the dynamic change of DNA methylation during differentiation is, at least partially, mediated by the function of tissue-specific transcription factors.

1.5 THE DYNAMICS OF TRANSCRIPTION-FACTOR-MEDIATED DNA HYPOMETHYLATION

Further insights into the dynamics of DNA binding and DNA methylation turnover came from ChIP (chromatin immunoprecipitation) experiments enriching for CTCF-bound regions coupled with bisulfite sequencing [40,41]. Herein, the actual molecules occupied by CTCF could be interrogated for their methylation states, and surprisingly, CTCF binding could not be directly related to CpG hypomethylation. As the frequency (ChIP signal intensity) of CTCF binding was significantly associated with the likelihood of a demethylation state and as increased heterogeneity of methylation was detectable in low affinity bound CTCF regions, a dynamic DNA methylation turnover in these regions was hypothesized.

And, indeed, the active involvement of transcription factor binding in DNA hypomethylation mechanisms was elucidated in the ESC and the genome-wide mapping of 5-hydroxy methyl-cytosine (5-hmC). 5-hmC represents a derivate of 5-mC and is considered an intermediate state toward hypomethylation of cytosine. This demethylation cascade is highly active in the ESC as well as neuronal cell types but was also reported, although to a lesser extent, in other somatic tissue types [42,43]. Functionally, the active demethylation is catalyzed by ten-eleven translocation (TET) enzymes, and interestingly, TET1 binding was observed to be enriched at the LMRs in the ESC, suggesting an involvement of the enzyme in the reversible DNA methylation state of these loci [36]. Indeed, profiling the distribution of 5-hmC in the ESC and derived neuronal progenitor cell, determined an enrichment of 5-hmC at the LMRs [41]. Furthermore, comparing both differentiation states revealed differential 5-hmC to be highly frequent at the LMRs. Supporting evidence that transcription factor binding is driving active DNA demethylation came from studies in knockout mice for the DNA-binding factor REST (REST$^{-/-}$), wherein the absence of the transcription factor was associated with decreased levels of 5-hmC at region previously occupied by the factor [41].

In conclusion, these results suggest an active participation of transcription factors in shaping the DNA methylation landscape during differentiation processes. Moreover, the guided turnover of CpG methylation gives valuable mechanistic insights into the dynamics and flexibility of DNA methylation and further underpins a tight interplay between distinct features in the formations in gene regulatory processes that mediate distinct phenotypes, such as cell type states. Although, the intrinsic properties of the DNA sequence act as important determinant for epigenetic states, nongenetic *trans*-acting factors contribute to its final shape. Thus, external stimuli (environmental impacts) and genetic variability contribute to interindividual variability of the epigenetic landscapes.

1.6 TRANSLATIONAL POTENTIAL OF DEMETHYLATION DYNAMICS

Importantly, the knowledge of direct dependencies of DNA methylation and transcription factor binding is of direct translational value, since it can be applied to the identification of causal events that disrupt the epigenetic landscape in diseases. This strategy was applied on comprehensive epigenetic data sets, including several whole-genome DNA methylation landscapes in samples from patients with medulloblastoma [44]. Initially, the study confirmed the relationship between transcription factor binding and DNA hypomethylation, by integrating chromatin—immunoprecipitation data for the transcription factor OXT2 (resulting in a genome-wide map of OXT2 occupancy on DNA) and DNA methylation profiles. Following the validation of a direct relationship between both parameters in the investigated context, the study proceeded to use transcription factor binding motifs for the identification of driving forces for medulloblastoma subtype formation. Particularly, specific hypomethylated regions within previous defined subtypes of patient samples were analyzed for enriched binding motifs, and remarkably, transcription factors that putatively shaped the DNA methylome of the distinct subtypes were determined and were suggested as putative drivers of the different cancer phenotypes. Importantly, this strategy enabled the identification of novel candidates that participated in tumorigenesis as potential disease-driving factors, which could be of translational value for diagnostic or therapeutic approaches.

1.7 DNA METHYLATION QUANTITATIVE TRAIT LOCI

In addition to the relationship between DNA methylation and the directly underlying genomic sequence, wherein changes in genotypes induce a dynamic change in methylation state, genetic variance is also capable of influencing DNA methylation levels over distance. In this regard, genotype variability was described to affect neighboring (*cis*-regulation) or far-reaching (*trans*-regulation) CpG methylation levels [45,46] and led to the definition of DNA methylation quantitative trait loci (meQTLs). As the majority of meQTL-related epigenetic variability occurs in proximity to their associated genotype [47,48], the local regulatory environment and spatial conformation of the chromatin was hypothesized to be the main contributor to the connection. The multidimensional scenario involved the direct alteration of the epigenetic state by genetic variability, which triggers traceable downstream events, including chromatin conformation differences or altered regulatory or transcriptional activity of gene loci [49]. Consistent with this model, sequence polymorphisms were described to influence the binding of transcription factors [50,51], to alter the local profiles of histone modifications [52], or to be associated with distinct chromatin states [53]. Thus, genetic variability confers variant epigenetic states, defines the regulatory potential of the respective regions, and enables a systematic association between genotypes and epitypes [54,55]. Consequently, genetic variability is altering the local regulatory environment and can be detected in downstream events. Besides the triggered variance in DNA methylation (meQTLs), genotypes were shown to directly determine the transcriptional activity of target genes and were defined as expression quantitative trait loci (eQTLs) [56]. Surprisingly, meQTL and eQTL rarely overlap, which suggests the independent nature of both features [57]. Modeling of the relationship between genotypes and epigenetic and gene expression states have consistently revealed a complex relationship among the three features and has questioned an exclusive cause–mediator–consequence relationship [58]. The absence of a direct association between genetically driven variances could also be explained by temporal differences of the events, wherein the expression levels display the status quo of the cell state, and DNA methylation provides information about their history.

1.8 GENOTYPE-DRIVEN VARIANCE IN HUMAN DNA METHYLATION PROFILES

Genetic sequence differs among human individuals [5]. Moreover, human populations present distinct genetic profiles traceable in the course of human evolution [59]. Through comprehensive mapping efforts, in parallel with a systematic integration of phenotype information in GWAS, a plethora of genetic polymorphism could be related to distinct human traits, aiming to explain differences among human individuals. Considering that the majority of the variation is located outside the coding context, and thus unlikely to directly influence protein function or abundance, a regulatory impact of such variation was hypothesized [60–62]. In this regard, DNA methylation variation linked to genetic variability among individuals was suggested to be highly informative to pinpoint downstream effects of distinct genotypes and to explain the relationship between polymorphisms and human traits [63,64]. In this respect, large population studies have proven to be highly valuable in determining the association between genetic and epigenetic variations and the resulting downstream consequences on gene expression and phenotype formation [65].

In an attempt to comprehensively chart the differences of DNA methylation states among human populations, B-cell lines from nearly 300 individuals representing three different populations were analyzed at genome-scale at hundreds of thousands of CpG sites [48]. Several hundred CpG sites were observed to be differentially methylated between the populations and could be linked to differences in phenotypes, including appearance and susceptibility to infections. The integration of DNA methylation profiles of great apes as the outgroup enabled the identification of methylation states putatively driven by local selection pressures, such as infection incidence. Although the results supported a scenario where external stimuli conferred differences in the DNA methylation profile of populations and hence left their fingerprint in the epigenetic landscape of individuals, the majority of population-specific epitypes could be traced back to the underlying genetic sequence. Specifically, more than two-thirds of differences were assigned as meQTLs, which often replicated over tissue type boundaries. In line with the hypothesis of local regulatory states under the control of causative genotypes, meQTLs were found to be in proximity to differentially methylated loci (within 10 kb). These results were consistent with a previous study of an African population (Yoruba) reporting the proximity of the differentially methylated locus and its associated genotype [47].

1.9   EPIGENETIC MEDIATOR FUNCTION FOR HUMAN RISK PHENOTYPE FORMATION

Mining meQTLs for a relationship to human traits, previously identified by GWAS, determined several genotypes associated to both events, the difference in DNA methylation, and phenotypes and suggested the epigenetic regulation to act as mediator between the genetic blueprint and its final product [48]. Particularly, a strong relationship between the risk for chronic hepatitis B virus (HBV) infection and molecular traits that might have established the phenotype could be detected. Here the risk genotypes, more abundant in African and Asian populations, were segregated by hypermethylation of the gene promoter of *HLA-DPA1*, which was strongly associated with chronic HBV infection [66]. Moreover, the gain of DNA methylation of *HLA-DPA1* in individuals at risk was associated with an epigenetic repression of gene expression, being consistent with studies reporting an eQTL for these genotypes [67]. This example illustrates how multidimensional data sets can be integrated to elucidate causalities underlying the associations among the noncoding genetic variations in human traits.

Further studies, implementing genetic and epigenetic data, identified additional relationships, wherein the DNA methylation state guided the interpretation of noncoding human variations. A study comprising an independent cohort of individuals of African and European ancestries determined meQTLs associated with complex traits, such as racial disparities [68]. Specifically, meQTLs suggested an implication of epigenetic regulation in the biology of risk alleles for cardiovascular diseases and high cholesterol levels, conferred by a differential methylation of the *apolipoprotein A-V (APOA5)*.

Moreover, studies with a more directed design to analyze the disease-affected tissues performed meQTL analysis for the identification of causal relationships to explain the disease in question. Brain samples from patients with bipolar disorder were genotyped and epityped, and the integration of both data sets presented previously unsuspected gene candidates to explain the biology of the disease [69]. Moreover, meQTL loci were used to prioritize polymorphisms for genome-wide association studies, significantly reducing the number of tested loci and thus increasing the power to detect functional genetic variations in bipolar disorder.

In order to determine the mechanisms underlying the metabolic disease risk loci, adipose tissue was analyzed for the association of risk variants with differential DNA methylation [57]. Finding a relationship between a risk single nucleotide polymorphism (SNP) for an elevated body mass index and DNA methylation in an enhancer region upstream of the *adenylate cyclase 3* (*ADCY3*), a gene previously linked to obesity, the study suggested a putative novel mechanism causal in the particular human trait. Moreover, the fact that polymorphisms in linkage disequilibrium (LD) overlapped a transcription factor-binding site provided additional insights into the mechanisms underlying the gene deregulation and was in line with previously described dependencies of transcription factor binding and differential DNA methylation [36].

1.10 EPITYPE ASSOCIATION GUIDING THE INTERPRETATION OF CANCER RISK POLYMORPHISMS

The mediator effect of DNA methylation further guided the interpretation of risk variants in studies directly analyzing affected tissue types. Here, cancer risk polymorphisms, in particular, and the use of primary tumor samples revealed a powerful setting to explain the implication of genetic variations in cancer risk. Analyzing around 3000 cancer samples and hundreds of normal tissue specimens, provided by The Cancer Genome Atlas (a publicly available source for multidimensional molecular data sets obtained from primary cancer samples and related normal tissues), enabled the identification of 23 meQTLs, explaining one quarter of the tested cancer risk SNPs [70]. Remarkably, all SNPs were located in a noncoding context, and the application of meQTL analysis directly pointed to the deregulation events of putative cancer genes, supporting the interpretations of the variants. Consistently, differential methylation could be associated with changes in gene expression, further underscoring the functional impact of the meQTLs. Similar studies utilizing changes of expression (eQTL) as a measure of the functional impacts of risk loci revealed eQTL associations in *cis* and *trans*, and both these approaches could complement each other to provide a comprehensive catalog of downstream effects of cancer risk variants [71,72].

Most interestingly, the association between genetic risk variants and differential DNA methylation could also be detected in normal samples of the respective cancer types [70]. These results are in line with the expected biology underlying cancer risk, with the alteration being present in a normal tissue context during lifetime and increasing cancer risk by providing an initial event in the multistep process of tumorigenesis. Previous described oncogenes, such as *MYC* and *COL1A1*, were among the differentially methylated genes identified, suggesting the direct impact of epigenetic alteration in cancer risk biology [70]. Moreover, the identification of meQTLs enabled the identification of putative novel cancer genes, whose impact on cancer formation need to be confirmed in functional studies. Among those, the *FBXO18* was a promising candidate, which, through its function as helicase and its implication in homologous recombination DNA repair mechanisms, could have a direct impact in the destabilization of the genome to favor tumor development [73]. Candidate oncogenic associations between risk genotypes and epitypes also involved the deregulation of the nicotinic acetylcholine receptor subunit *CHRNA4* in lung cancer samples, an interesting candidate to integrate environmental impacts (smoking behavior) in the biology of the disease [74].

Interestingly, molecular association studies also crucially contributed to resolve the search for causative genetic variants of SNPs that segregate in blocks of LD [72]. This presents a major issue in cancer research and shows that, on average, risk loci are correlated with 56 other polymorphisms, making it extremely challenging to pinpoint causative variants [5]. With the assumption that causal variants act as eQTLs and polymorphisms in LD present weaker or absent association with gene expression, a fine-mapped candidate list presented a valuable source to prioritize candidate loci for subsequent validation studies. Although performed using expression as readout for functionality, DNA methylation or additional quantifiable molecular trait are suitable to complement and improve the search for causative genetic risk variants. The value of multidimensional data integration has been further highlighted consistently with the use of chromosomal interactions involving putative cancer risk variants to determine causal SNPs [75]. A sophisticated screening approach, involving the locus-specific capture of cross-linked DNA (conserving the spatial formation of the genome) coupled with deep sequencing strategies, defined putative causal variants through their interaction in *cis* with cancer genes.

Taken together, molecular traits associated with risk polymorphisms present valuable biomarkers for activity and provide an interpretation strategy for variability, which is otherwise challenging to explain. Here, GWAS traits present an interesting example, and larger and more comprehensive data sets from future studies will contribute to explaining the biology of the genome.

1.11 CLOSING REMARKS

The genetic and epigenetic codes are highly interconnected. Epigenetic mechanisms, particularly DNA methylation was crucially involved in shaping the genetic sequence during evolution, and vice versa, the intrinsic properties of the DNA have an important impact on the formation of the DNA methylation landscape [76]. A comprehensive knowledge about dependencies that confer the distribution of methyl-cytosine throughout the genome presents a valuable source to explain the biology of healthy life and diseases. In this respect, profound insights into epigenetic deregulation caused by genetic variation not only guide our understanding of phenotype variance but could also be implemented in future disease management or prevention strategies.

In addition to the contribution of epigenetic regulation to explain genetic variants; genetic alterations, in turn, guide our understanding of variance observed in the epigenetic landscape. For example, highly aberrant DNA methylation signature in solid cancers or leukemia could be explained by causal genetic mutations in the *IDH1* [77] or *TET2* [78] genes, respectively, which are involved in the DNA methylation machinery. Further, DNA methylation differences determined in epigenome-wide association studies (EWAS) could be explained by variations in the genetic information provided by the participants [79].

However, the answer to which extent functional variation of DNA methylation occurs independent of the genetic code still remains elusive. In this respect, transgenerational inheritance of induced epimutations provides a paradigm for the independent and stable transmission of epigenetic variance through generations [80]. Further, variant DNA methylation profiles in genetically identical humans [81] or animals [82] present examples of how epigenetic variance contributes to phenotype formation without being influenced by the genetic sequence.

In conclusion, the integration of genetic, epigenetic, and phenotype information will improve our understanding of the biology of the genome and human variation.

REFERENCES

[1] Lander ES, Linton LM, Birren B, Nusbaum C, Zody MC, Baldwin J, et al. Initial sequencing and analysis of the human genome. Nature 2001;409:860−921.

[2] Venter JC, Adams MD, Myers EW, Li PW, Mural RJ, Sutton GG, et al. The sequence of the human genome. Science 2001;291:1304−51.

[3] Gerstein MB, Bruce C, Rozowsky JS, Zheng D, Du J, Korbel JO, et al. What is a gene, post-ENCODE? History and updated definition. Genome Res 2007;17:669−81.

[4] Chimpanzee Sequencing and Analysis Consortium. Initial sequence of the chimpanzee genome and comparison with the human genome. Nature 2005;437:69−87.

[5] Consortium T 1000 GP. An integrated map of genetic variation from 1,092 human genomes. Nature 2012;491:56−65.

[6] Li MJ, Wang P, Liu X, Lim EL, Wang Z, Yeager M, et al. GWASdb: a database for human genetic variants identified by genome-wide association studies. Nucleic Acids Res 2012;40: D1047−54.

[7] ENCODE Project Consortium. The ENCODE (ENCyclopedia Of DNA Elements) Project. Science 2004;306:636−40.

[8] Consortium TEP. An integrated encyclopedia of DNA elements in the human genome. Nature 2012;489:57−74.

[9] Ernst J, Kellis M. ChromHMM: automating chromatin-state discovery and characterization. Nat Methods 2012;9:215–16.

[10] Ernst J, Kheradpour P, Mikkelsen TS, Shoresh N, Ward LD, Epstein CB, et al. Mapping and analysis of chromatin state dynamics in nine human cell types. Nature 2011;473:43–9.

[11] Hoffman MM, Buske OJ, Wang J, Weng Z, Bilmes JA, Noble WS. Unsupervised pattern discovery in human chromatin structure through genomic segmentation. Nat Methods 2012;9:473–6.

[12] Bird A. DNA methylation patterns and epigenetic memory. Genes Dev 2002;16:6–21.

[13] Jones PA. Functions of DNA methylation: islands, start sites, gene bodies and beyond. Nat Rev Genet 2012;13:484–92.

[14] Kulis M, Heath S, Bibikova M, Queirós AC, Navarro A, Clot G, et al. Epigenomic analysis detects widespread gene-body DNA hypomethylation in chronic lymphocytic leukemia. Nat Genet 2012;44:1236–42.

[15] Jaenisch R, Bird A. Epigenetic regulation of gene expression: how the genome integrates intrinsic and environmental signals. Nat Genet 2003;33 Suppl.:245–54.

[16] Weber M, Hellmann I, Stadler MB, Ramos L, Paabo S, Rebhan M, et al. Distribution, silencing potential and evolutionary impact of promoter DNA methylation in the human genome. Nat Genet 2007;39:457–66.

[17] Shen L, Kondo Y, Guo Y, Zhang J, Zhang L, Ahmed S, et al. Genome-wide profiling of DNA methylation reveals a class of normally methylated CpG island promoters. PLoS Genet 2007;3:2023–36.

[18] Durán E, Djebali S, González S, Flores O, Mercader JM, Guigó R, et al. Unravelling the hidden DNA structural/physical code provides novel insights on promoter location. Nucleic Acids Res 2013;41:7220–30.

[19] Yuan G-C. Linking genome to epigenome. Wiley Interdiscip Rev Syst Biol Med 2012;4:297–309.

[20] Kaplan N, Moore IK, Fondufe-Mittendorf Y, Gossett AJ, Tillo D, Field Y, et al. The DNA-encoded nucleosome organization of a eukaryotic genome. Nature 2009;458:362–6.

[21] Pérez A, Castellazzi CL, Battistini F, Collinet K, Flores O, Deniz O, et al. Impact of methylation on the physical properties of DNA. Biophys J 2012;102:2140–8.

[22] West AG, Gaszner M, Felsenfeld G. Insulators: many functions, many mechanisms. Genes Dev 2002;16:271–88.

[23] Brandeis M, Frank D, Keshet I, Siegfried Z, Mendelsohn M, Nemes A, et al. Sp1 elements protect a CpG island from de novo methylation. Nature 1994;371:435–8.

[24] Macleod D, Charlton J, Mullins J, Bird AP. Sp1 sites in the mouse aprt gene promoter are required to prevent methylation of the CpG island. Genes Dev 1994;8:2282–92.

[25] Lienert F, Wirbelauer C, Som I, Dean A, Mohn F, Schübeler D. Identification of genetic elements that autonomously determine DNA methylation states. Nat Genet 2011;43:1091–7.

[26] Krebs AR, Dessus-Babus S, Burger L, Schübeler D. High-throughput engineering of a mammalian genome reveals building principles of methylation states at CG rich regions. eLife 2014;3:e04094.

[27] Schübeler D. Molecular biology. Epigenetic islands in a genetic ocean. Science 2012;338:756–7.

[28] Kim J, Kollhoff A, Bergmann A, Stubbs L. Methylation-sensitive binding of transcription factor YY1 to an insulator sequence within the paternally expressed imprinted gene, Peg3. Hum Mol Genet 2003;12:233–45.

[29] Perini G, Diolaiti D, Porro A, Della Valle G. *In vivo* transcriptional regulation of N-Myc target genes is controlled by E-box methylation. Proc Natl Acad Sci USA 2005;102:12117–22.

[30] Yu D-H, Waterland RA, Zhang P, Schady D, Chen M-H, Guan Y, et al. Targeted p16Ink4a epimutation causes tumorigenesis and reduces survival in mice. J Clin Invest 2014;124:3708–12.

[31] Berman BP, Weisenberger DJ, Aman JF, Hinoue T, Ramjan Z, Liu Y, et al. Regions of focal DNA hypermethylation and long-range hypomethylation in colorectal cancer coincide with nuclear lamina-associated domains. Nat Genet 2011;44:40–6.

[32] Hon GC, Hawkins RD, Caballero OL, Lo C, Lister R, Pelizzola M, et al. Global DNA hypomethylation coupled to repressive chromatin domain formation and gene silencing in breast cancer. Genome Res 2012;22:246–58.

[33] Hansen KD. Increased methylation variation in epigenetic domains across cancer types. Nat Genet 2011;43:768–75.

[34] Gaidatzis D, Burger L, Murr R, Lerch A, Dessus-Babus S, Schübeler D, et al. DNA sequence explains seemingly disordered methylation levels in partially methylated domains of mammalian genomes. PLoS Genet 2014;10:e1004143.

[35] Baubec T, Schübeler D. Genomic patterns and context specific interpretation of DNA methylation. Curr Opin Genet Dev 2014;25C:85–92.

[36] Stadler MB, Murr R, Burger L, Ivanek R, Lienert F, Schöler A, et al. DNA-binding factors shape the mouse methylome at distal regulatory regions. Nature 2011;480:490–5.

[37] Rao SSP, Huntley MH, Durand NC, Stamenova EK, Bochkov ID, Robinson JT, et al. A 3D map of the human genome at kilobase resolution reveals principles of chromatin looping. Cell 2014;159:1665–80.

[38] Hark AT, Schoenherr CJ, Katz DJ, Ingram RS, Levorse JM, Tilghman SM. CTCF mediates methylation-sensitive enhancer-blocking activity at the H19/Igf2 locus. Nature 2000;405:486–9.

[39] Kanduri C, Pant V, Loukinov D, Pugacheva E, Qi CF, Wolffe A, et al. Functional association of CTCF with the insulator upstream of the H19 gene is parent of origin-specific and methylation-sensitive. Curr Biol 2000;10:853–6.

[40] Brinkman AB, Gu H, Bartels SJJ, Zhang Y, Matarese F, Simmer F, et al. Sequential ChIP-bisulfite sequencing enables direct genome-scale investigation of chromatin and DNA methylation cross-talk. Genome Res 2012;22:1128–38.

[41] Feldmann A, Ivanek R, Murr R, Gaidatzis D, Burger L, Schübeler D. Transcription factor occupancy can mediate active turnover of DNA methylation at regulatory regions. PLoS Genet 2013;9:e1003994.

[42] Kriaucionis S, Heintz N. The nuclear DNA base 5-hydroxymethylcytosine is present in purkinje neurons and the brain. Science 2009;324:929–30.

[43] Tahiliani M, Koh KP, Shen Y, Pastor WA, Bandukwala H, Brudno Y, et al. Conversion of 5-methylcytosine to 5-hydroxymethylcytosine in mammalian DNA by MLL partner TET1. Science 2009;324:930–5.

[44] Hovestadt V, Jones DTW, Picelli S, Wang W, Kool M, Northcott PA, et al. Decoding the regulatory landscape of medulloblastoma using DNA methylation sequencing. Nature 2014;510:537–41.

[45] Kerkel K, Spadola A, Yuan E, Kosek J, Jiang L, Hod E, et al. Genomic surveys by methylation-sensitive SNP analysis identify sequence-dependent allele-specific DNA methylation. Nat Genet 2008;40:904–8.

[46] Shoemaker R, Deng J, Wang W, Zhang K. Allele-specific methylation is prevalent and is contributed by CpG-SNPs in the human genome. Genome Res 2010;20:883–9.

[47] Bell JT, Pai AA, Pickrell JK, Gaffney DJ, Pique-Regi R, Degner JF, et al. DNA methylation patterns associate with genetic and gene expression variation in HapMap cell lines. Genome Biol 2011;12:R10.

[48] Heyn H, Moran S, Hernando-Herraez I, Sayols S, Gomez A, Sandoval J, et al. DNA methylation contributes to natural human variation. Genome Res 2013;23:1363−72.

[49] Heyn H. A symbiotic liaison between the genetic and epigenetic code. Front Genet 2014;5:113.

[50] Kasowski M, Grubert F, Heffelfinger C, Hariharan M, Asabere A, Waszak SM, et al. Variation in transcription factor binding among humans. Science 2010;328:232−5.

[51] Khurana E, Fu Y, Colonna V, Mu XJ, Kang HM, Lappalainen T, et al. Integrative annotation of variants from 1092 humans: application to cancer genomics. Science 2013;342:1235587.

[52] McDaniell R, Lee B-K, Song L, Liu Z, Boyle AP, Erdos MR, et al. Heritable individual-specific and allele-specific chromatin signatures in humans. Science 2010;328:235−9.

[53] Paul DS, Albers CA, Rendon A, Voss K, Stephens J, van der Harst P, et al. Maps of open chromatin highlight cell type−restricted patterns of regulatory sequence variation at hematological trait loci. Genome Res 2013;23:1130−41.

[54] Boyle AP, Hong EL, Hariharan M, Cheng Y, Schaub MA, Kasowski M, et al. Annotation of functional variation in personal genomes using RegulomeDB. Genome Res 2012;22:1790−7.

[55] Schaub MA, Boyle AP, Kundaje A, Batzoglou S, Snyder M. Linking disease associations with regulatory information in the human genome. Genome Res 2012;22:1748−59.

[56] Lappalainen T, Sammeth M, Friedländer MR, 't Hoen PAC, Monlong J, Rivas MA, et al. Transcriptome and genome sequencing uncovers functional variation in humans. Nature 2013;501:506−11.

[57] Grundberg E, Meduri E, Sandling JK, Hedman AK, Keildson S, Buil A, et al. Global analysis of DNA methylation variation in adipose tissue from twins reveals links to disease-associated variants in distal regulatory elements. Am J Hum Genet 2013;93:876−90.

[58] Gutierrez-Arcelus M, Lappalainen T, Montgomery SB, Buil A, Ongen H, Yurovsky A, et al. Passive and active DNA methylation and the interplay with genetic variation in gene regulation. eLife 2013;2:e00523.

[59] Li JZ, Absher DM, Tang H, Southwick AM, Casto AM, Ramachandran S, et al. Worldwide human relationships inferred from genome-wide patterns of variation. Science 2008;319:1100−4.

[60] Freedman ML, Monteiro ANA, Gayther SA, Coetzee GA, Risch A, Plass C, et al. Principles for the post-GWAS functional characterization of cancer risk loci. Nat Genet 2011;43:513−18.

[61] Hindorff LA, Gillanders EM, Manolio TA. Genetic architecture of cancer and other complex diseases: lessons learned and future directions. Carcinogenesis 2011;32:945−54.

[62] Kilpinen H, Dermitzakis ET. Genetic and epigenetic contribution to complex traits. Hum Mol Genet 2012;21:R24−8.

[63] Gibbs JR, van der Brug MP, Hernandez DG, Traynor BJ, Nalls MA, Lai S-L, et al. Abundant quantitative trait loci exist for DNA methylation and gene expression in human brain. PLoS Genet 2010;6:e1000952.

[64] Zhang D, Cheng L, Badner JA, Chen C, Chen Q, Luo W, et al. Genetic control of individual differences in gene-specific methylation in human brain. Am J Hum Genet 2010;86:411−19.

[65] Fraser HB, Lam L, Neumann S, Kobor MS. Population-specificity of human DNA methylation. Genome Biol 2012;13:R8.

[66] Kamatani Y, Wattanapokayakit S, Ochi H, Kawaguchi T, Takahashi A, Hosono N, et al. A genome-wide association study identifies variants in the HLA-DP locus associated with chronic hepatitis B in Asians. Nat Genet 2009;41:591−5.

[67] O'Brien TR, Kohaar I, Pfeiffer RM, Maeder D, Yeager M, Schadt EE, et al. Risk alleles for chronic hepatitis B are associated with decreased mRNA expression of HLA-DPA1 and HLA-DPB1 in normal human liver. Genes Immun 2011;12:428−33.

[68] Moen EL, Zhang X, Mu W, Delaney SM, Wing C, McQuade J, et al. Genome-wide variation of cytosine modifications between european and african populations and the implications for complex traits. Genetics 2013;194(4):987−96.

[69] Gamazon ER, Badner JA, Cheng L, Zhang C, Zhang D, Cox NJ, et al. Enrichment of cis-regulatory gene expression SNPs and methylation quantitative trait loci among bipolar disorder susceptibility variants. Mol Psychiatry 2013;18:340−6.

[70] Heyn H, Sayols S, Moutinho C, Vidal E, Sanchez-Mut JV, Stefansson OA, et al. Linkage of DNA methylation quantitative trait loci to human cancer risk. Cell Rep 2014;7:331−8.

[71] Li Q, Seo J-H, Stranger B, McKenna A, Pe'er I, Laframboise T, et al. Integrative eQTL-based analyses reveal the biology of breast cancer risk loci. Cell 2013;152:633−41.

[72] Li Q, Stram A, Chen C, Kar S, Gayther S, Pharoah P, et al. Expression QTL-based analyses reveal candidate causal genes and loci across five tumor types. Hum Mol Genet 2014;23 (19):5294−302.

[73] Fugger K, Mistrik M, Danielsen JR, Dinant C, Falck J, Bartek J, et al. Human Fbh1 helicase contributes to genome maintenance via pro- and anti-recombinase activities. J Cell Biol 2009;186:655−63.

[74] Scherf DB, Sarkisyan N, Jacobsson H, Claus R, Bermejo JL, Peil B, et al. Epigenetic screen identifies genotype-specific promoter DNA methylation and oncogenic potential of CHRNB4. Oncogene 2013;32:3329−38.

[75] Dryden NH, Broome LR, Dudbridge F, Johnson N, Orr N, Schoenfelder S, et al. Unbiased analysis of potential targets of breast cancer susceptibility loci by Capture Hi-C. Genome Res 2014;24:1854−68.

[76] Feinberg AP, Irizarry RA. Evolution in health and medicine Sackler colloquium: stochastic epigenetic variation as a driving force of development, evolutionary adaptation, and disease. Proc Natl Acad Sci USA 2010;107(Suppl. 1):1757−64.

[77] Turcan S, Rohle D, Goenka A, Walsh LA, Fang F, Yilmaz E, et al. IDH1 mutation is sufficient to establish the glioma hypermethylator phenotype. Nature 2012;483:479−83.

[78] Cancer Genome Atlas Research Network. Genomic and epigenomic landscapes of adult de novo acute myeloid leukemia. N Engl J Med 2013;368:2059−74.

[79] Liu Y, Aryee MJ, Padyukov L, Fallin MD, Hesselberg E, Runarsson A, et al. Epigenome-wide association data implicate DNA methylation as an intermediary of genetic risk in rheumatoid arthritis. Nat Biotechnol 2013;31:142−7.

[80] Radford EJ, Ito M, Shi H, Corish JA, Yamazawa K, Isganaitis E, et al. In utero undernourishment perturbs the adult sperm methylome and intergenerational metabolism. Science 2014;345(6198):1255903.

[81] Fraga MF, Ballestar E, Paz MF, Ropero S, Setien F, Ballestar ML, et al. Epigenetic differences arise during the lifetime of monozygotic twins. Proc Natl Acad Sci USA 2005;102:10604−9.

[82] Herb BR, Wolschin F, Hansen KD, Aryee MJ, Langmead B, Irizarry R, et al. Reversible switching between epigenetic states in honeybee behavioral subcastes. Nat Neurosci 2012;15:1371−3.

Chapter 2

DNA Methylation Microarrays

Marina Bibikova

Illumina, Inc., San Diego, CA, USA

Chapter Outline

2.1 INTRODUCTION

DNA methylation in various regions of the genome is a characteristic of gene activity, tissue type, disease state, and underlying epigenetic regulation of the genome. Precise DNA methylation patterns are established during early embryonic development and are inherited through multiple mitotic cellular divisions [1,2]. DNA methylation is necessary for normal cell development [3,4], X-chromosome inactivation [5,6], control of gene expression patterns [3], and genomic imprinting [4,7] and has effects on cellular growth and genomic stability [8,9]. Knowledge of DNA methylation patterns can greatly advance our ability to understand and diagnose the molecular basis of human diseases. Epigenome alterations, including DNA methylation, have been attracting attention from researchers who are focusing on cancers and neuronal, immune, and metabolic disorders, as well as on cell differentiation and aging.

M. Fraga & A.F. Fernandez (Eds): Epigenomics in Health and Disease.
DOI: http://dx.doi.org/10.1016/B978-0-12-800140-0.00002-9
© 2016 Elsevier Inc. All rights reserved.

Interest in monitoring methylation states of cytosines in CpG dinucleotides has led to the development of various techniques for DNA methylation profiling. These include methylation-specific restriction enzyme digestion [10], bisulfite DNA sequencing [11,12], methylation-specific PCR [13], MethyLight [14], restriction landmark genomic scanning [15,16], pyrosequencing [17,18], and MALDI mass spectrometry [19−21], to name a few. Many novel microarray and next-generation sequencing (NGS)-based technologies have emerged in recent years, attempting to provide high-throughput access to specific sequences in the genome and enable analysis of methylation profiles at high resolution in large sample sets. The microarray-based approaches are largely based on three major techniques: restriction enzyme digest, sodium bisulfite conversion of genomic DNA, and affinity-based assays. The well-known methods based on restriction enzyme digest include differential methylation hybridization (DMH) with various modifications [22−24], *Hpa*II tiny fragment enrichment by ligation-mediated PCR [25], methylated CpG island (CGI) amplification [26,27], methylation-specific fractionation after McrBC digestion coupled with tiling arrays [28,29], and comprehensive high-throughput arrays for relative methylation [30]. To overcome the limitation of enzyme-recognition sites, affinity-based technologies have been developed to enrich the methylated fraction of the genome; these include methylated DNA immunoprecipitation [31] or affinity chromatography over an MBD (methyl-binding domain) [32,33]. The affinity-based enrichment methods can also be combined with NGS readout [34,35].

Bisulfite sequencing of genomic DNA is considered the gold standard for analyzing the methylation state of CpG sites within the genome [36]. Bisulfite treatment of gDNA converts all cytosine bases to uracils, except those protected by methylation [11]. Methylation in the human genome is generally limited to 5-methyl cytosine in the context of CpG sites, although it has been discovered recently that there are several different types of cytosine base modifications, including 5-hydroxymethylcytosine (5hmC) [37], which have previously been indistinguishable. Recent studies have also uncovered non-CpG methylation, primarily in pluripotent cells [38,39]. Although whole-genome bisulfite sequencing (WGBS) can access every CpG site in the genome, the method is still quite expensive and is limited to a low number of samples. A variation of this method, called reduced-representation bisulfite sequencing, uses restriction enzyme digestion to fractionate the genome and targets regions of high CpG density [40].

Although each method referenced above can provide information about methylation states across the genome, they all have limitations. Methods which use methylation-sensitive restriction enzyme digest have to rely on the presence of recognition sites within the region of interest. Affinity-based methods cannot determine DNA methylation level at a unique CpG site and have a bias toward CpG-dense regions. Most NGS platforms still require large amounts of sample input, intensive labor, and complex bioinformatic analyses; in addition, most of these technologies can only analyze a few samples at a time. The above limitations make them difficult to use in large-scale studies where sample materials may be limited. Several reviews compared DNA methylation profiling approaches and discussed the strengths and weaknesses associated with microarray and NGS for DNA methylation profiling [41–46].

Illumina has developed two methylation profiling platforms that provide high sample throughput with single-CpG resolution [47,48]. These assays are based upon the adaptation of the Infinium and GoldenGate single nucleotide polymorphism (SNP) genotyping platforms, using bisulfite-converted DNA. Methylation analysis on DNA arrays can be accomplished by measuring the DNA methylation level, i.e., the fraction of methylated versus unmethylated cytosines, using quantitative genotyping of the "pseudo-SNP" created by bisulfite conversion of gDNA. Well-established Illumina genotyping assays were modified to analyze the methylation state of CpG sites in bisulfite-converted genomic DNA. The GoldenGate genotyping assay and universal arrays were used to assess the methylation level of over 1500 CpG sites in cancer-related genes [48,49]. Similarly, the principles of the Infinium whole-genome genotyping (WGG) assay in combination with bisulfite conversion of DNA were used to enable quantitative methylation analysis on a genome-wide scale [47].

In the following sections, we focus on our approach to genome-wide Infinium DNA methylation array design and some techniques to overcome array design limitations.

2.2   INFINIUM DNA METHYLATION TECHNOLOGY

2.2.1   BeadArray Platform

Illumina DNA methylation profiling technology is based on the well-established Illumina's BeadArray platform. The concept behind BeadArrays is straightforward in principle: Arrays made of oligos linked to silica beads randomly assemble into wells [50]. Among the advantages of this bead-based approach, is that oligo probes are synthesized in solution, using standard phosphoramidite chemistry. Only full-length oligos are attached to beads, and only beads that can hybridize to a target can be decoded (determining which bead is in which well). The necessity of decoding the randomly distributed beads serves as an automatic quality control for the array. The decoding process involves iterative hybridizations with fluorescent oligo pools and has been described in detail by Gunderson et al. [51]. The assignment of fluorescent labels in the oligo pool for each hybridization step allows for the unique identification of each bead by its particular sequence of fluors across a series of hybridization steps. Depending on the array density, there are about 15–30 beads of each type (having the same oligo sequence) present in each array, offering redundancy in analytical measurements. Infinium methylation bead chips are locus-specific arrays consisting of long 50-mer locus-specific sequences concatenated to a decoding sequence. Beads with long target-specific probes are primarily used to build arrays with a high density of features of fixed content, such as Infinium genome-wide methylation arrays [52]. The high density of the Infinium bead chips is achieved with a MEMS-patterned slide substrate that consists of multiple sample sections into which beads are assembled from a pool containing hundreds of thousands of different bead types. Each section receives a same bead pool, thus allowing the analysis of many different samples in parallel. Beads provide a better substrate compared with slides for bulk surface modifications and immobilization of assay oligonucleotides. These bulk processes, in which beads of a particular type are created in one immobilization event, greatly improve array-to-array consistency. This is particularly important for Infinium methylation assay with Infinium I probes that employ a ratiometric comparison between two bead types (see below), with this ratio relatively invariant from one array to another. Bead arrays are also scalable in density. A reduction in bead size and spacing can greatly increase the number of assays per bead chip. The choice of slide substrate has also contributed to the success of the array-based enzymatic Infinium assay.

2.2.2   Infinium Assay

The Infinium assay complexity is limited only by the number of beads that are assembled on the slide section. This assay uses beads with long target-specific probes designed to correspond to individual CpG sites. The Illumina Infinium WGG assay was adapted for measuring CpG methylation using quantitative "genotyping" of bisulfite-converted genomic DNA. The (C/T) polymorphism in the bisulfite-converted DNA resulting from different sensitivity of methylated and unmethylated cytosines to the sodium bisulfite treatment can be queried by using either standard Infinium I methylation-specific assay design consisting of two probes per CpG locus, one "unmethylated" and one "methylated" query probe (Figure 2.1A), or Infinium II design with one probe per locus, where the underlying CpG sites are represented by a "degenerate" R-base, allowing multiple combinations of oligos attached to the bead (Figure 2.1B).

The first Infinium methylation array, HumanMethylation27, was designed by using only Infinium I assays with two "allele-specific" probes for each CpG site [47].

For the next-generation HumanMethylation450 array, we decided to use a degenerate base (R = A or G) in positions complementary to the "C" base in the underlying CpG sites. We first designed probes for over 1000 CpG loci to compare Infinium I (two probes per locus) and Infinium II (one probe per locus with degenerate base) designs. We found that probes with up to three underlying CpG sites can be designed as Infinium II probes, and this allowed us to significantly increase the number of covered CpG sites with the same number of beads in the pool.

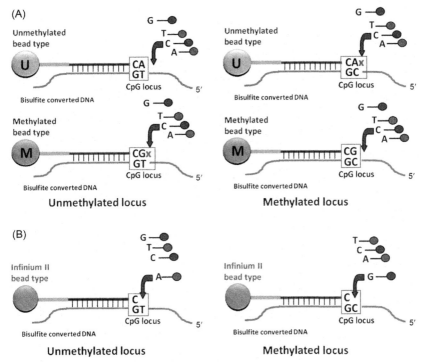

FIGURE 2.1   Infinium assay scheme. (A) Infinium I assay. Two bead types correspond to each CpG locus: one bead type—to methylated (C); another bead type—to unmethylated (T) state of the CpG site. Probe design assumes same methylation status for adjacent CpG sites. Both bead types for the same CpG locus will incorporate the same type of labeled nucleotide, determined by the base preceding the interrogated "C" in the CpG locus, and therefore will be detected in the same color channel. (B) Infinium II assay. One bead type corresponds to each CpG locus. Probe can contain up to three underlying CpG sites, with degenerate R-base corresponding to C in the CpG position. Methylation state is detected by single-base extension. Each locus will be detected in two colors.

The Infinium assay comprises the following major steps. First, bisulfite-converted DNA is whole genome amplified. This amplified DNA is enzymatically fragmented and hybridized to a CpG-specific bead array. After hybridization, the array is processed through a primer extension and an immunohistochemistry staining protocol to allow for the detection of a single-base extension reaction [52–54]. For the Infinium I assays, allele-specific single-base extension is used to determine methylation state of the query locus, requiring two beads for each CpG site—one corresponding to the unmethylated state and another to the methylated state, with underlying CpG sites "in phase" with the queried site. The perfectly matched bead type preferentially extends over the mismatched bead type. For the Infinium II assay, one bead type is used, and the methylation state is determined by single base extension (SBE) using hapten-labeled terminators.

We developed a set of reference standards to assess the quality of Infinium methylation probes [47]. Unmethylated (U), hemimethylated (H), and methylated (M) genomic reference samples were created from genomic DNA that was amplified with the REPLI-g Mini Kit (QIAGEN), according to the manufacturer's recommendations. To minimize representation bias, whole-genome amplification (WGA) reactions were limited to a 100-fold amplification. The WGA amplified DNA was subjected to mung bean nuclease treatment to remove single-stranded DNA. The resultant unmethylated DNA was treated with SssI methylase, which globally methylates all double-stranded CpG sites, to create a nearly completely methylated reference standard. The hemimethylated reference was created by mixing U and M samples in a 1:1 stoichiometric ratio. The standards were used for the development of both HumanMethylation27 and HumanMethylation450 arrays.

2.3 HUMANMETHYLATION450 ARRAY DESIGN AND PERFORMANCE

2.3.1 Design Challenges

There are several challenges to implementing the analysis of bisulfite-converted DNA on a microarray genotyping platform. First of all, most of the cytosines in the genome are converted to uracils, so the uniqueness of any given sequence within the bisulfite-converted genome decreases dramatically. A second issue arises when designing probes to the methylated and unmethylated "alleles" of a particular CpG site in which flanking CpG sites will affect the hybridization and extension of the probes. Furthermore, after bisulfite conversion, opposite strands are no longer complementary, and the WGA step required for the Infinium assay effectively creates four different strands, thus reducing the effective concentration and uniqueness of any given locus.

Illumina DNA methylation arrays are based on "genotyping" of bisulfite-converted genomic DNA. Treatment of a DNA sample with bisulfite results in a change in sequence from C to U for unmethylated cystosines, whereas methylated cytosines remain unchanged. Therefore, the methylation status of a given cytosine base in a CpG site can be interrogated by using a genotyping assay for a C/T polymorphism after bisulfite treatment. We have developed algorithms for designing assay oligos that minimize the impact of the greatly diminished sequence complexity caused by the loss of most cytosines in the genome. Assay probes can be designed for a large fraction of the CpG sites in the genome and to either strand.

One of the challenges with designing probes for bisulfite-converted DNA is dealing with underlying CpG sites. Degenerate bases can be employed, but the concentration of perfectly matched oligos on the beads would be diluted depending on the number of the underlying CpG sites. For the probe design with multiple underlying CpG sites, we assumed methylation is regionally correlated and resolved underlying CpG sites to be in phase with either the "methylated" (C) or "unmethylated" (U) query site. The co-methylation assumption is based on a previous study by Eckhardt et al., in which they performed bisulfite sequencing of chromosomes 6, 20, and 22 [36]. Our probes have a span of 50 bases; within this distance, methylation should be highly correlated. However, there are likely to be exceptions to this regional methylation rule. There are over 28 million CpG sites in the human genome. Using a set of empirical rules, we were able to bioinformatically design Infinium methylation probes for over 16 million of these CpG loci. For the first Infinium methylation array, HumanMethylation27, we designed two "allele-specific" probes for each CpG site. The 3′ terminus of the probe was designed to match either the protected cytosine (methylated design) or the uracil base resulting from bisulfite conversion (unmethylated design). Underlying CpG sites were resolved to their respective C or U phase in the probe design. There are likely exceptions to this local co-methylation rule; nonetheless, this approach allowed us to design probes in CpG-dense regions and inside the CG islands.

To maximize the density of the next-generation HumanMethylation450 array, we tested an Infinium II assay design which requires one probe per locus for CpG sites located in regions of low CpG density. The underlying CpG sites are represented by a "degenerate" R-base, allowing for multiple combinations of oligos attached to the bead. The 3′ terminus of the probe complements the base directly upstream of the query site, whereas a SBE results in the addition of a labeled G or A base, complementary to either the "methylated" C or "unmethylated" T (Figure 2.1B). We demonstrated that Infinium II probes can have up to three underlying CpG sites within the 50-mer probe sequence without compromising data quality. This feature enables the methylation status at a query site to be assessed independently of assumptions on the status of neighboring CpG sites.

2.3.2   Content Selection

HumanMethylation27 array was the first high-throughput platform for DNA methylation profiling. We selected a set of 27,578 CpG sites located within the proximal promoter regions (1 kb upstream and 500 bases downstream of transcription start sites, TSSs) of 14,475 CCDS genes and well-known cancer genes. In addition, we included 254 assays covering 110 miRNA promoters. On average, we designed two assays per CCDS promoter; for a subset of 180 cancer-related genes, we assayed from three to up to 20 CpG sites per promoter region. Assays were preferentially designed to sites within CGIs, whenever possible. We employed a NCBI "relaxed" definition, in which CGIs are identified bioinformatically as DNA sequences (200 base window) with a GC base composition greater than 50% and a CpG ratio of more than 0.6 over what was expected [9]. Using this relaxed definition, 60% of CCDS genes contain one or more CGIs, and 40% contained no CGI. Even though the array content was limited, it allowed the launch of the first epigenome-wide association studies (EWAS) [55] and was selected as a platform of choice by The Cancer Genome Atlas (TCGA) consortium [56].

The development of the next DNA methylation microarray, HumanMethylation450, combined the benefits of Infinium chemistry with substantially expanded genome coverage to provide high quality, genome-wide content with target selection guided by researchers' needs rather than technical limitations [57]. CpG site selection was defined by a set of content categories identified by a consortium of epigenetics researchers. Each category was represented with either publicly available data or experimentally validated sites identified internally or contributed by members of the consortium. Emphasis was placed on gene and CGI regions, for which 99% and 96% coverage, respectively, were achieved. In addition, the format of 12 samples per array provided a throughput capacity for cost-efficient and time-efficient analysis of large sample cohorts. The HumanMethylation450 array covers not only majority of RefSeq genes but also multiple regions outside of the known coding areas. Over 650,000 bead types are designed to interrogate more than 480,000 query sites. We included 485,577 assays (482,421 CpG sites, 3091 non-CpG sites and 65 random SNPs) representing content categories selected with the guidance of a consortium comprising 22 methylation researchers representing 19 institutions worldwide. The consortium identified a series of content categories, including RefSeq genes (http://www.ncbi.nlm.nih.gov/RefSeq/), CGIs, CGI shores [58−60], Hidden Markov Model-defined CGIs [61,62], FANTOM 4 promoters (http:// fantom.gsc.riken.jp/4/) [63,64], MHC regions [65], informatically identified enhancers [66−68], and others, including a small subset of non-CpG loci reported to be differentially methylated in pluripotent cell lines [38,39]. The numbers of sites represented for each content category are listed in Table 2.1.

According to the consortium's recommendations, the highest priority was placed on providing comprehensive coverage across the complete gene and CGI regions. Toward this end, both gene and CGI regions were subdivided according to UCSC classifications [69,70] (Figure 2.2), and each subcategory was targeted individually (Table 2.2). Coverage of CGI regions was further enhanced by including the 2-kb regions flanking CGI shores (referred to here as "CGI shelves") (Figure 2.2B) as well as Hidden Markov Model-defined CGIs [61].

Also included were sites that were shown to be biologically significant or informative based on data that were generated internally or by the members of the consortium. Other categories represented were non-CpG sites [38,39], DNase hypersensitive sites [71,72], and differentially methylated regions (DMRs) [58,73]. Detailed information on this content is available in the HumanMethylation450 manifest (www.illumina.com).

TABLE 2.1 HumanMethylation450 Array Content

Feature Type	Included on Array
Total number of sites	485,577
RefSeq genes	21,231 (99%)
CpG islands (CGIs)	26,658 (96%)
CGI shores (0−2 kb from CGI)	26,249 (92%)
CGI shelves (2−4 kb from CGI)	24,018 (86%)
HMM islands[a]	62,600
FANTOM 4 promoters (high CpG content)[a]	9426
FANTOM 4 promoters (low CpG content)[a]	2328
Differentially methylated regions (DMRs)[a]	16,232
Informatically predicted enhancers[a]	80,538
DNAse hypersensitive sites	59,916
Ensemble regulatory features[a]	47,257
Loci in MHC region	12,334
HumanMethylation27 loci	25,978
Non-CpG loci	3,091

[a]*Features may contain multiple assay probes. One probe may belong to several content categories.*

2.3.3   Gene Coverage

The array provides coverage of a total of 21,231 out of 21,474 UCSC RefGenes (NM and NR) (98.9%) with a global average of 17.2 probes per gene region (Table 2.2). Multiple transcripts of RefSeq genes, as well as additional genes and transcripts not covered by the UCSC database (total of 29,246 transcripts), are included. In order to achieve a comprehensive assessment of gene region methylation, probes covering gene regions were designed across multiple subregions. Promoter regions were divided into two mutually exclusive bins of 200- and 1500-bp blocks upstream of the transcription start site (designated TSS200 and TSS1500, respectively). The 5′ and 3′ UTR, first exon and gene body were independently targeted as well (Figure 2.2A). Details regarding the number of RefSeq genes and subregions, and the average number of CpG sites per gene locus represented on the array are given in Table 2.2.

2.3.4   CGI Coverage

CGIs were defined based on UCSC annotation and according to the criteria previously described [9,74]. We employed a NCBI "strict" definition for

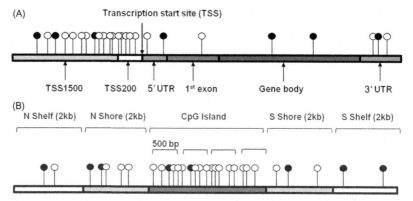

FIGURE 2.2   Infinium methylation probe selection. (A) Coverage of NM and NR transcripts from UCSC database. Each transcript was divided into "functional regions"—TSS200 is the region from transcription start site (TSS) to −200 nt upstream of TSS; TSS1500 covers −200 to −1500 nt upstream of TSS; 5′ UTR, 1st exon, gene body and 3″ UTR were also covered separately. (B) Coverage of CpG islands (CGIs) and adjacent regions. CGIs longer than 500 bp were divided into separate bins. The 2-kb regions immediately upstream and downstream of the CGI boundaries, or "CGI shores," and the 2-kb regions upstream and downstream of the CGI shores, referred to here as "CGI shelves," were also targeted separately.

TABLE 2.2 Coverage of Genes and Transcripts from UCSC Database

Feature Type	Genes Mapped	Percent Genes Covered	Average Number of Loci on Array
NM_TSS200	15,957	84%	3.73
NM_TS1500	18,099	96%	4.31
NM_5′UTR	14,137	79%	4.68
NM_1stExon	15,580	82%	2.54
NM_3′UTR	13,071	72%	1.53
NM_GeneBody	17,117	97%	9.92
NR_TSS200	2140	71%	2.97
NR_TSS1500	2723	90%	3.84
NR_GeneBody	2382	79%	7.15

CGIs as DNA sequences (500 base windows; excluding most repetitive Alu-elements) with a GC base composition greater than 50% and a CpG observed/expected ratio of more than 0.6 [9,74]. As described by Takai and Jones [9], regions of DNA greater than 500 bp with GC composition 55% or greater and observed CpG/expected CpG of 0.65 were more likely to be associated with the 5' regions of genes. Using this definition, 60% of RefSeq genes are found to contain one or more CGI, and 40% contain no CGI. Of the CGIs, 26,658 were covered overall, with an average of 5.63 sites each; 28,249 "north" or upstream and 25,761 "south," or downstream CGI shores, immediately outside of the CGIs, were targeted, with averages of 2.93 and 2.81 sites, respectively. The 2-kb regions upstream and downstream of the CGI shores, referred to here as "CGI shelves," were also targeted with a global average of 2.07 and 2.03 sites each ("North" and "South," respectively) (Figure 2.2B).

Over 94% of loci present on HumanMethylation27 array were included in the HumanMethylation450 array content. All loci which satisfied Infinium II design criteria were redesigned, using one bead per locus. To confirm the accuracy of the methylation measurement across two platforms, we compared the correlation between 450K and 27K arrays, showing an R^2 of >0.95. Good correlation was also confirmed in other studies [75].

To further gauge the technical performance of the HumanMethylation450 array, we assessed data reproducibility between technical replicates by using lymphoblastoid cell lines NA17105 and NA17018, cancer cell line MCF7, and tumor and normal lung tissues. The average correlation R^2 of β-values for technical replicates was 0.992 [57].

2.3.5   Correlation with WGBS Data

High-throughput sequencing technologies enabled generation of single-base-resolution DNA methylation maps for various organisms [76]. The WGBS method provides the most comprehensive measurement of the DNA methylation state at every CpG site, but it is still expensive and allows analysis of only a few samples at a time. We compared DNA methylation β-values measured by the Infinium methylation assay with results from WGBS data generated on a HiSeq2000 (Illumina) using NGS technology. Two comparisons were run—one with a normal lung tissue and the other with a lung tumor sample. WGBS data were filtered to include corresponding HumanMethylation450 loci covered with a minimum of 10 and maximum of 121 aligned reads, resulting in a total of 189,821 and 167,996 loci for comparison in the normal and tumor samples, respectively. The observed β-value correlations were 0.95 for the normal and 0.96 for the tumor samples [57]. These results indicate that the β-values generated by the Infinium HumanMethylation450 array and WGBS are consistent in reporting DNA methylation state across queried CpG loci. Other studies compared DNA methylation values from WGBS and 450K arrays and found that they were highly concordant [77,78].

2.4   HUMANMETHYLATION450 ARRAY ADVANTAGES AND LIMITATIONS

The HumanMethylation450 Infinium assay offers a fast and cost-effective solution for methylation profiling of large numbers of CpG sites in the genome. It provides a true genome-wide scan, at single-CpG site resolution. Data generated from the Infinium Methylation arrays can be easily compared with other data, such as those generated with genomic bisulfite sequencing. The array probe design is very flexible; it is possible to target almost any specific region of interest, such as CGIs and genomic regions with low CpG density. In addition, methylation status at non-CpG sites can be measured without any change to the experimental workflow.

Nevertheless, the array has a number of limitations. There are several challenges in designing Infinium assay probes to query the state of a CpG site in bisulfite-converted DNA. Since most of the cytosines in the genome are converted to uracils, the uniqueness of any given sequence within the bisulfite-converted genome decreases dramatically. This can potentially affect assay specificity, and the fact that opposite strands are no longer complementary reduces the effective concentration of any given locus by a factor of 2. Nonetheless, we found that, in general, the specificity and sensitivity of the Infinium assay was sufficient to read out the requisite loci.

The array design is biased due to loci preselection and inclusion of probes that interrogate certain CpG sites that have been previously identified in methylation-based assays and, therefore, the design is not entirely hypothesis neutral. The probe design assumes that CpG sites located adjacent to those interrogated by the probes will have a similar methylation state, which is known as the "co-methylation assumption" [36]. There is also a concern that there are certain behavioral differences between the Infinium I and Infinium II types of probe design on the array, and a significant number of probes may be affected by SNPs. Underlying SNPs can affect the assay by reducing hybridization efficiency of the probe containing a mismatch nucleotide. By avoiding SNPs in the first 10 bases adjacent to the query site, this effect can be significantly reduced. Some of the concerns listed above are addressed in the data analysis pipelines developed for the Infinium methylation data processing [79,80].

2.5   DNA METHYLATION DATA ANALYSIS

Illumina has developed GenomeStudio software [57,81], which enables basic analysis of DNA methylation data collected by using HumanMethylation450 arrays. Each methylation data point is represented by fluorescent signals from the M and U alleles of each targeted site. The β-value, the ratio of intensities between M and U alleles, is calculated as $\beta = \text{Max}(M,0)/(\text{Max}(U,0) + \text{Max}(M,0) + 100)$ and used to estimate the methylation level of the CpG sites. In Infinium I assays, signals U and M are derived from two different bead types and reported with the same color. In Infinium II assays, U corresponds to the signal in the Red channel and M corresponds to the signal in the Green channel. The β-value calculation results in a number between 0 and 1. In ideal situations, a value of zero indicates that all copies of the CpG site in the sample are completely unmethylated, and a value of one indicates that every copy of the site is methylated.

Background intensity is computed from a set of negative controls and subtracted from each analytical data point. An absolute value is used in the denominator of the formula, as a compensation for any "negative signals" which may arise from oversubtraction (a constant bias of 100 was added to regularize β; this is particularly useful when both U and M values are small) [48].

The software also allows identification of differentially methylated loci. GenomeStudio differential methylation analysis algorithms compare a group of samples (referred to as the condition group) with a reference group using several error models—Mann–Whitney U-test, t-test, or Illumina custom model [81].

However, when large-scale DNA methylation projects and EWAS, aimed at understanding the epigenetic basis of complex diseases, such as cancer, started to emerge, researchers started to encounter roadblocks with the processing and analysis of array-based DNA methylation data. For more in-depth analysis, new methods and tools were developed to overcome problems with data quality control, normalization, batch effects, and identification of differentially methylated sites and regions. Several recent reviews summarize the advantages and shortcomings of various data analysis approaches proposed for handling those issues, focusing on publicly available tools in open-source environments, such as R and Bioconductor [82–84]. The appropriate experimental design and downstream data processing, along with newly developed data analysis tools, facilitate researchers to effectively use the powerful DNA methylation 450K platform, supporting large-scale epigenomic studies and advancing our understanding of human health and disease.

Data quality control is one of the first steps in DNA methylation analysis. The HumanMethylation450 arrays include multiple control probes for determining the data quality, including sample-independent (staining, extension, hybridization, and target removal) controls, and sample-dependent (bisulfite conversion, specificity, nonpolymorphic, and negative) controls [85]. To detect poorly performing samples, visual plots of control probes can be displayed in the diagnostic controls dashboard of GenomeStudio, or generated using the R-package HumMeth27QCReport [85], which was developed for Infinium 27K arrays but is also compatible with 450K array analysis. However, visual inspection is not convenient for large-scale genome-wide DNA methylation studies. To help with the detection of poor-quality samples in large-scale projects, a new visual and interactive Web application, MethylAid, was developed [86]. An alternative approach involves using the raw signal intensities of the control probes and determining whether they are beyond the expected range of the signal intensities across all samples. Other options for quality control of samples, which make use of detection P values, are available in R and Bioconductor packages, such as MethyLumi [87], the preprocessing and analysis pipeline [88], IMA [89], and Minfi [90].

After checking for the sample quality, the next step often involves filtering out probes that have a detection P value above a certain prespecified threshold (i.e., P value <0.05 or <0.01) in a large proportion of samples [57]. In the IMA package [89], probes with missing values, those residing on the X chromosome, and those with a median detection P value >0.05 across samples can be filtered out; other packages allowing such filtering include MethyLumi [87] and other pipelines [88].

A potential issue for the methylation probes' performance can be SNPs within the probe sequence. There may be over 25% probes on the 450K array that can be affected by underlying SNPs [91]. Since methylation levels of a specific locus may be influenced by genotype [92], many investigators want to remove SNP-associated loci from their data, and several R-packages have options for filtering out these loci [88,89]. Genetic effects, however, should not be underestimated in methylation arrays. As was recently demonstrated in a study that investigated population-specific patterns of DNA methylation [93], a large portion of population-specific DNA methylation levels may, in fact, be due to population-specific genetic variants, which are themselves affected by genetic or environmental interactions. Although rare SNPs are unlikely to affect methylation levels to a large extent, somatic mutations (e.g., driver mutations in a tumor) can have a significant impact on methylation levels; therefore, it is important to conduct validation of the methylation data by using orthogonal methods. A number of SNP probes that are unaffected by bisulfite conversion are intentionally included on the Infinium array and can help identify mislabeled samples, as implemented in wateRmelon [80].

One of the recently published studies [94] provided a list of additional probes that researchers may want to remove from their data. The authors observed a number of cross-reactive probes cohybridizing to the sex chromosomes with more than 94% sequence identity and concluded that up to 6% of the array probes can potentially generate spurious signals because of cohybridization to alternative genomic sequences highly homologous to the intended targets.

Background correction is another focus of attention in various data analysis pipelines. It helps to remove the nonspecific background signal from the total signal, and it corrects for possible interarray artifacts. Background correction can be performed by using Illumina's GenomeStudio, or one of several R-packages that contain background correction functions, including lumi [95], MethyLumi [87], and the preprocessing and analysis pipeline for 450K data [88].

Normalization is one of the key steps in array data processing. Normalization concerns the removal of sources of experimental artifacts, random noise, and technical and systematic variation caused by microarray technology, which, if left unaddressed, has the potential to mask true biologic differences [96]. The two main types of normalization are (i) interarray normalization, which removes technical artifacts between samples on different arrays, and (ii) intraarray normalization, which corrects for intensity-related dye biases [97]. GenomeStudio provides an option to normalize data to the internal controls for the 450K assay [81], which is also used in MethyLumi [87] and Minfi [90]; by default, GenomeStudio uses the first sample in the array as the reference and allows the user to reselect the reference sample, as needed, if the original sample has poor quality. There is a lack of consensus regarding the optimal approach for normalization of methylation data, and several reviews discuss the advantages and limitations of the various existing methods [79,80,84,98]. Difficulties in selecting the best normalization algorithm result from an imbalance in methylation levels throughout the genome, creating a skew to the methylation log-ratio distribution, which depends on the levels of methylation in particular samples [97]. This imbalance is caused by the nonrandom distribution of CpG sites throughout the genome and the link between CpG density and DNA methylation; for instance, CGIs are often unmethylated, whereas loci in non-CGI regions are typically highly methylated in normal human cells [99]. Many available normalization methods were designed for gene expression array data and are based on assumptions that may not be appropriate for DNA methylation microarray data [97].

Another potential methodologic concern for HumanMethylation450 data analysis results from the fact that the array uses two different types of probes, Infinium I and Infinium II. Researchers observed that they differ in performance, where Infinium II probes show a reduced dynamic range of measured methylation values compared with the Infinium I probes [83]. The two probe types are generally different in terms of CpG density, with more Infinium I probes containing CpGs mapping to CGIs (57%) compared with type II probes (21%) [57]. The range of β-values obtained from the Infinium II probes is smaller, in general, compared with Infinium I probes; in addition, the Infinium II probes also appear to be less sensitive for the detection of extreme methylation values and display a greater variance between replicates, which leads to the recommendation of data rescaling to make the probe distributions comparable [88,92].

Batch effects represent a significant problem in array analysis that should be accounted for and avoided as much as possible [100]. Understanding and controlling for batch effects requires particular attention in large-scale epigenetic studies. Initially, attention has focused mainly on large differences among batches, such as differences in sample preparation methods, the laboratories in which samples are run, environmental conditions, day of run, reagent lots, equipment used, and laboratory personnel [100,101]. The implications of batch effects due to more subtle differences, such as the location of samples on different chips run concurrently, may not be sufficiently appreciated and taken into account. Several studies illustrate how chip-specific effects can simulate plausible and interesting differential methylation results [102,103]. Normalization has been shown to reduce some components of batch effects but not completely eliminate them [79,83,101,103,104]. Sound study design is critical for proper evaluation of and correction for batch effects: for instance, samples from different study groups should be split randomly or equally into different batches [102,103]. The array literature indicates that array position effects may also exist [105], and therefore new batch correction techniques may be needed to take those into account. By properly correcting for batch effects, one can combine data from multiple batches, which enables greater statistical power to measure epigenetic variations in large datasets [104].

The goal of genome-wide DNA methylation analysis is the identification of differentially methylated CpG loci and DMRs between cases and controls that can explain gene regulation, serve as biomarkers, or assist in diagnostics and treatment of various diseases, including metabolic and neurologic diseases and cancer. Differential methylation can be calculated in GenomeStudio as Delta β [57], where a difference is applied to methylation medians of two groups for each CpG locus; if the absolute value of the difference in medians between loci of each group is higher than 0.2, then that locus is considered differentially methylated. This 0.2 threshold corresponds to the recommended difference in methylation between loci that can be detected with 99% confidence interval [57]. Several methylation data analysis pipelines offer alternative differential methylation analysis algorithms: MethVisual [106] tests whether each CpG site has independent membership between two groups using Fisher's exact test; other packages that allow for the adjustment of potential confounders include CpGassoc [107], MethLAB [108], and IMA [89]. Minfi [90] uses linear regression and an F-test to test for a univariate association between the methylation of individual loci and continuous or categorical phenotypes, respectively.

Differential methylation analysis can also be performed by measuring variability between methylation loci as opposed to using statistical tests on the basis of differences in mean methylation [109]. This is available in the EVORA package, allowing an investigator to use differential variability in methylation of CpGs and to then associate them with a phenotype of interest, such as cancer status [109,110]. ChAMP methylation analysis pipeline is offering a choice of the most popular normalization methods while also introducing new methods for calling DMRs [111].

Investigators also use the M-value, or log ratio of the methylated signal over the unmethylated signal [112]. A normalized M-value near 0 signifies a semimethylated locus, a positive M-value indicates that more molecules are methylated than unmethylated, and negative M-values have the opposite interpretation. The M-value has better statistical properties with approximately constant variance across the methylation range and can be used in many statistical models derived for expression arrays that assume normality [112]. However, β-values are much more biologically interpretable than M-values [113].

A comprehensive summary of available differential methylation analysis methods and pipelines can be found in recent reviews [79,82,84].

2.6   USE OF METHYLATION ARRAYS IN EPIGENETIC STUDIES

2.6.1   DNA Methylation and Cancer

Over the past several decades, an increasing amount of evidence has supported a role of epigenetics in cell biology and physiology. The epigenome of a cell is highly dynamic and depends on a complex interaction between genetic and environmental factors [114]. Normal cellular function relies on the maintenance of epigenomic homeostasis, and epigenetic aberrations, including DNA methylation both at the individual gene level and on a genome-wide scale, result in numerous human diseases, notably cancer [99]. Epigenetic alterations are leading candidates for the development of specific markers for cancer detection, diagnosis, and prognosis. The enzymatic processes that control the epigenome present new opportunities for deriving therapeutic strategies designed to reverse transcriptional abnormalities that are inherent to the cancer epigenome.

Cancer methylome studies are summarized in several reviews, including those by Baylin and Jones, and Stirzaker et al. [99,115]. To date, cancer methylomes for over 20 broad cancer types have been completed, representing over 8000 individual data sets. Infinium DNA methylation arrays became the platform of choice for TCGA consortium [56]. TCGA data portal provides access to DNA methylation results from approximately 7500 samples, which were collected by using the Infinium DNA methylation technology, with majority of data generated by using the 450K array. TCGA findings unveiled similar and unique characteristics of methylomes for various cancer types [116–123]. Likewise, investigators from the International Cancer Genome Consortium (ICGC), which was launched to coordinate large-scale cancer genome studies in tumors from 50 different cancer types and subtypes across the globe, are utilizing 450K arrays along with other DNA methylation profiling technologies for collecting DNA methylation data in a variety of epigenetic studies.

2.6.2   DNA Methylation and Aging

A growing body of research has reported associations between age and the state of the epigenome [124]. The observation that not everyone ages in the same manner has led to the search for molecular markers of age that can be used to predict, monitor, and provide insight into age-associated physiologic decline and disease. Several studies have observed that DNA methylation associates with chronological age over long periods and uncovered a phenomenon named "epigenetic drift," whereby the DNA methylation marks in identical twins increasingly differ as a function of age [125]. Early research focused on age-associated changes affecting a small number of individual CpG loci primarily in high-CpG-density promoters of important cancer genes [126]. Novel DNA methylation analysis technologies, including Infinium methylation arrays, allowed for highly accurate assessment of DNA methylation levels across tens of thousands of CpG sites and demonstrated that age-associated increases in DNA methylation occurs predominantly at bivalent chromatin domain promoters [127]. Two more recent studies [78,128] using HumanMethylation450 arrays have further confirmed that age-associated hypermethylation happens preferentially at high CpG density promoters, which often reside upstream of key developmental genes, such as PolyComb Group Targets [129]. These studies have also confirmed that the majority of changes in the genome involve loss of methylation affecting CpG sites located in low-CpG-density regions, in line with the fact that most of these sites start out being methylated sites [78]. Thus, it would appear that the machinery responsible for maintaining normal DNA methylation patterns becomes gradually deregulated with age, leading to deviations from a normal epigenetic state, or epigenetic drift [130].

The current challenge is to determine whether these changes can be systematically described and modeled to detect different rates of human aging and to tie these rates to related clinical or environmental variables.

2.6.3   Epigenome-Wide Association Studies

New powerful technologies, such as comprehensive DNA methylation microarrays, have accelerated the understanding of human disease−associated epigenetic variations [43]. EWAS have increasingly replaced targeted approaches focusing on individual candidate genes and hold great promise for systematically dissecting the role of epigenetic variation in health and disease. However, good practices and accepted standards for conducting EWAS are only starting to emerge, and several key principles of conducting genome-wide association studies (GWAS) are problematic in the context of EWAS [131].

EWAS examine the epigenetic state at many different loci in a number of individuals and assess whether any of these loci are associated with a trait. EWAS typically focus on the association between DNA methylation in a specific tissue and the presence of disease or other characteristics, such as environmental conditions. These studies use microarray technology or high-throughput sequencing by using a variety of protocols, each of which comes with characteristic strengths and limitations [41,132,133]. Regardless of technology, as the number of loci covered by these methods increases, and as we learn more about interdependence of genetic and epigenetic variations, the statistical analyses required to interpret the large amounts of generated information have rapidly become more complex [82]. Moreover, the dynamic nature of epigenetic signatures, which are variable throughout the life course, renders the design and interpretation of EWAS conceptually different and more difficult than those of GWAS.

The tissue specificity of the epigenomic pattern adds another challenge to the design of EWAS. Most cancer EWAS used primary tumor samples with adjacent tissues as controls. The purity of the tumor material should be confirmed by a pathology review, and laser microdissection may be required to obtain sufficiently pure samples. Cancerous organs may exhibit epigenetic changes even in regions with histologically normal tissues, making organ tissue from healthy control subjects a preferable choice [131]. EWAS for other diseases often measured DNA methylation in blood samples because disease-relevant tissues are difficult to obtain, but the biologic relevance of blood DNA methylation patterns may not be apparent. The use of blood as surrogate tissue requires stringent validation and cautious interpretation of results. To establish a surrogate tissue in a statistically sound manner, it is necessary not only for inter-individual epigenetic differences to correlate between the tissue of interest and the surrogate but also for the exposure to induce similar changes in both tissues. Currently, there is no clear evidence that, in general, epigenetic marks respond to environmental exposures in a similar way across tissues, which makes the identification of a surrogate tissue contentious and problematic.

The Infinium HumanMethylation450 array allows researchers to assess the DNA methylation levels of close to half a million CpG sites distributed across the genome, which corresponds to 2% of all CpG sites of the human genome. As the technology is a high-throughput one, hundreds of samples can be profiled in a short period. Therefore, the Infinium 450K array is currently a good compromise of reagent costs, labor, sample throughput, and coverage. Because microarrays are susceptible to batch effects, they require careful experimental design and appropriate analytical approaches [100,134]. A shortcoming of the chip technology is that in contrast to sequencing-based methods, this approach does not address allele-specific and SNP-specific methylation and does not allow discovery of variations beyond the probed loci. The technical considerations of Infinium data analysis have been discussed above.

EWAS are likely to require large international consortium-based approaches to reach the numbers of subjects, as well as the statistical and scientific rigor, required for robust findings.

2.7 CONCLUSION

The body of literature focused on epigenetics research has rapidly increased over the last several years. This growth has fueled the need for new technologies and, in particular, the capability to run methylation analysis with high-quality, genome-wide coverage on a platform that also offers high-throughput capacity and cost efficiency [44,46]. Infinium HumanMethylation450 DNA Methylation array represents high-throughput and true genome-wide DNA methylation profiling technology, using the Illumina BeadArray platform. With as little as 500 ng of gDNA, the Infinium Methylation assays can measure over 480,000 CpG sites and provide single-CpG site resolution data that can be easily compared with other data, such as those generated by genomic bisulfite sequencing. The array data show strong reproducibility between replicates and high correlation with WGBS data generated on the same samples. The assay can be automated, which greatly increases assay robustness and throughput, thus enabling parallel processing of a large volume of samples.

Unlike the WGBS approach, which has access to every cytosine in the genome, array-based DNA methylation analysis is limited to the targeted CpG loci represented by assay probes immobilized on beads in the Infinium arrays. Nevertheless, the arrays can access individual CpG sites across both CGIs and genomic regions with low CpG density and therefore give a good overview of epigenetic profiles on a genome-wide scale.

The Infinium HumanMethylation450 was designed with the guidance of a consortium comprising methylation researchers to meet the needs for high-throughput methylation profiling. The ability to quickly and affordably run genome-scale methylation analysis aligns with the requirements for large-sample-size studies, such as TCGA and the ICGC initiatives. The growing number of examples of reproducible associations identified through GWAS suggests that similar sample size ranges applied in genome-wide methylation screens could similarly lead to findings that might otherwise be missed. And although important questions pertaining to study design remain, the potential value of EWAS, as well as the integration of genotype and methylation data across sample populations, has already begun to be explored [43,46].

The utility of Infinium HumanMethylation450 was further extended by its applicability to formalin-fixed, paraffin-embedded samples [135] and to 5hmC analysis [136,137].

In summary, the HumanMethylation450 array provides a powerful tool for investigators to fuel the continued, rapid evolution of epigenetic research by offering simple and rapid genome-wide methylation analysis of hundreds of thousands of CpG sites across large numbers of samples. Additional sites of interest will continue to be identified, and this array was designed to provide an efficient, robust, and affordable discovery solution targeting core content of common interest within the epigenetics research community.

Looking forward, the 450K platform is likely to remain useful for some time to come, and it will be the platform of choice for EWAS in the foreseeable future. HumanMethylation450 use and analysis was rapidly growing at the time of writing this chapter.

REFERENCES

[1]   Razin A, Shemer R. DNA methylation in early development. Hum Mol Genet 1995;4 Spec No: 1751–5.

[2]   Reik W, Dean W, Walter J. Epigenetic reprogramming in mammalian development. Science 2001;293(5532):1089–93.

[3]   Bird AP. CpG-rich islands and the function of DNA methylation. Nature 1986;321 (6067):209–13.

[4]   Li E, Beard C, Jaenisch R. Role for DNA methylation in genomic imprinting. Nature 1993;366(6453):362–5.

[5]   Wolf SF, Migeon BR. Studies of X chromosome DNA methylation in normal human cells. Nature 1982;295(5851):667–71.

[6]   Mohandas T, Sparkes RS, Shapiro LJ. Reactivation of an inactive human X chromosome: evidence for X inactivation by DNA methylation. Science 1981;211(4480):393–6.

[7]   Swain JL, Stewart TA, Leder P. Parental legacy determines methylation and expression of an autosomal transgene: a molecular mechanism for parental imprinting. Cell 1987;50(5):719–27.

[8]   Bird A. DNA methylation patterns and epigenetic memory. Genes Dev 2002;16(1):6–21.

[9]   Takai D, Jones PA. Comprehensive analysis of CpG islands in human chromosomes 21 and 22. Proc Natl Acad Sci USA 2002;99(6):3740–5.

[10] Singer-Sam J, LeBon JM, Tanguay RL, Riggs AD. A quantitative HpaII-PCR assay to measure methylation of DNA from a small number of cells. Nucleic Acids Res 1990;18(3):687.

[11] Clark SJ, Harrison J, Paul CL, Frommer M. High sensitivity mapping of methylated cytosines. Nucleic Acids Res 1994;22(15):2990–7.

[12] Frommer M, McDonald LE, Millar DS, Collis CM, Watt F, Grigg GW, et al. A genomic sequencing protocol that yields a positive display of 5-methylcytosine residues in individual DNA strands. Proc Natl Acad Sci USA 1992;89(5):1827–31.

[13] Herman JG, Graff JR, Myohanen S, Nelkin BD, Baylin SB. Methylation-specific PCR: a novel PCR assay for methylation status of CpG islands. Proc Natl Acad Sci USA 1996;93 (18):9821–6.

[14] Eads CA, Danenberg KD, Kawakami K, Saltz LB, Blake C, Shibata D, et al. MethyLight: a high-throughput assay to measure DNA methylation. Nucleic Acids Res 2000;28(8): E32.

[15] Kawai J, Hirotsune S, Hirose K, Fushiki S, Watanabe S, Hayashizaki Y. Methylation profiles of genomic DNA of mouse developmental brain detected by restriction landmark genomic scanning (RLGS) method. Nucleic Acids Res 1993;21(24):5604–8.

[16] Okazaki Y, Hirose K, Hirotsune S, Okuizumi H, Sasaki N, Ohsumi T, et al. Direct detection and isolation of restriction landmark genomic scanning (RLGS) spot DNA markers tightly linked to a specific trait by using the RLGS spot-bombing method. Proc Natl Acad Sci USA 1995;92(12):5610–14.

[17] Colella S, Shen L, Baggerly KA, Issa JP, Krahe R. Sensitive and quantitative universal pyrosequencing methylation analysis of CpG sites. Biotechniques 2003;35(1):146–50.

[18] Dupont JM, Tost J, Jammes H, Gut IG. De novo quantitative bisulfite sequencing using the pyrosequencing technology. Anal Biochem 2004;333(1):119–27.

[19] Ehrich M, Nelson MR, Stanssens P, Zabeau M, Liloglou T, Xinarianos G, et al. Quantitative high-throughput analysis of DNA methylation patterns by base-specific cleavage and mass spectrometry. Proc Natl Acad Sci USA 2005;102(44):15785–90.

[20] Ehrich M, Turner J, Gibbs P, Lipton L, Giovanneti M, Cantor C, et al. Cytosine methyla-
      tion profiling of cancer cell lines. Proc Natl Acad Sci USA 2008;105(12):4844−9.

[21] Tost J, Schatz P, Schuster M, Berlin K, Gut IG. Analysis and accurate quantification of
      CpG methylation by MALDI mass spectrometry. Nucleic Acids Res 2003;31(9):e50.

[22] Gitan RS, Shi H, Chen CM, Yan PS, Huang TH. Methylation-specific oligonucleotide
      microarray: a new potential for high-throughput methylation analysis. Genome Res
      2002;12(1):158−64.

[23] Huang TH, Perry MR, Laux DE. Methylation profiling of CpG islands in human breast
      cancer cells. Hum Mol Genet 1999;8(3):459−70.

[24] Schumacher A, Kapranov P, Kaminsky Z, Flanagan J, Assadzadeh A, Yau P, et al.
      Microarray-based DNA methylation profiling: technology and applications. Nucleic Acids
      Res 2006;34(2):528−42.

[25] Khulan B, Thompson RF, Ye K, Fazzari MJ, Suzuki M, Stasiek E, et al. Comparative
      isoschizomer profiling of cytosine methylation: the HELP assay. Genome Res 2006;16
      (8):1046−55.

[26] Toyota M, Ho C, Ahuja N, Jair KW, Li Q, Ohe-Toyota M, et al. Identification of differen-
      tially methylated sequences in colorectal cancer by methylated CpG island amplification.
      Cancer Res 1999;59(10):2307−12.

[27] Estecio MR, Yan PS, Ibrahim AE, Tellez CS, Shen L, Huang TH, et al. High-throughput
      methylation profiling by MCA coupled to CpG island microarray. Genome Res
      2007;17(10):1529−36.

[28] Ordway JM, Bedell JA, Citek RW, Nunberg A, Garrido A, Kendall R, et al.
      Comprehensive DNA methylation profiling in a human cancer genome identifies novel
      epigenetic targets. Carcinogenesis 2006;27(12):2409−23.

[29] Ordway JM, Budiman MA, Korshunova Y, Maloney RK, Bedell JA, Citek RW, et al.
      Identification of novel high-frequency DNA methylation changes in breast cancer. PLoS
      One 2007;2(12):e1314.

[30] Irizarry RA, Ladd-Acosta C, Carvalho B, Wu H, Brandenburg SA, Jeddeloh JA, et al.
      Comprehensive high-throughput arrays for relative methylation (CHARM). Genome Res
      2008;18(5):780−90.

[31] Weber M, Davies JJ, Wittig D, Oakeley EJ, Haase M, Lam WL, et al. Chromosome-wide
      and promoter-specific analyses identify sites of differential DNA methylation in normal
      and transformed human cells. Nat Genet 2005;37(8):853−62.

[32] Ibrahim AE, Thorne NP, Baird K, Barbosa-Morais NL, Tavare S, Collins VP, et al.
      MMASS: an optimized array-based method for assessing CpG island methylation. Nucleic
      Acids Res 2006;34(20):e136.

[33] Yamada Y, Watanabe H, Miura F, Soejima H, Uchiyama M, Iwasaka T, et al. A compre-
      hensive analysis of allelic methylation status of CpG islands on human chromosome 21q.
      Genome Res 2004;14(2):247−66.

[34] Down TA, Rakyan VK, Turner DJ, Flicek P, Li H, Kulesha E, et al. A Bayesian deconvo-
      lution strategy for immunoprecipitation-based DNA methylome analysis. Nat Biotechnol
      2008;26(7):779−85.

[35] Ruike Y, Imanaka Y, Sato F, Shimizu K, Tsujimoto G. Genome-wide analysis of aberrant
      methylation in human breast cancer cells using methyl-DNA immunoprecipitation com-
      bined with high-throughput sequencing. BMC Genomics 2010;11:137.

[36] Eckhardt F, Lewin J, Cortese R, Rakyan VK, Attwood J, Burger M, et al. DNA methyla-
      tion profiling of human chromosomes 6, 20 and 22. Nat Genet 2006;38(12):1378−85.

[37] Kriaucionis S, Heintz N. The nuclear DNA base 5-hydroxymethylcytosine is present in Purkinje neurons and the brain. Science 2009;324(5929):929−30.

[38] Laurent L, Wong E, Li G, Huynh T, Tsirigos A, Ong CT, et al. Dynamic changes in the human methylome during differentiation. Genome Res 2010;20(3):320−31.

[39] Lister R, Pelizzola M, Dowen RH, Hawkins RD, Hon G, Tonti-Filippini J, et al. Human DNA methylomes at base resolution show widespread epigenomic differences. Nature 2009;462(7271):315−22.

[40] Meissner A, Gnirke A, Bell GW, Ramsahoye B, Lander ES, Jaenisch R. Reduced representation bisulfite sequencing for comparative high-resolution DNA methylation analysis. Nucleic Acids Res 2005;33(18):5868−77.

[41] Plongthongkum N, Diep DH, Zhang K. Advances in the profiling of DNA modifications: cytosine methylation and beyond. Nat Rev 2014;15(10):647−61.

[42] Heyn H, Esteller M. DNA methylation profiling in the clinic: applications and challenges. Nat Rev 2012;13(10):679−92.

[43] Rakyan VK, Down TA, Balding DJ, Beck S. Epigenome-wide association studies for common human diseases. Nat Rev 2011;12(8):529−41.

[44] Beck S. Taking the measure of the methylome. Nat Biotechnol 2010;28(10):1026−8.

[45] Huang YW, Huang TH, Wang LS. Profiling DNA methylomes from microarray to genome-scale sequencing. Technol Cancer Res Treat 2010;9(2):139−47.

[46] Laird PW. Principles and challenges of genome-wide DNA methylation analysis. Nat Rev 2010;11(3):191−203.

[47] Bibikova M, Le J, Barnes B, Saedinia-Melnyk S, Zhou L, Shen R, et al. Genome-wide DNA methylation profiling using Infinium(R) assay. Epigenomics 2009;1 (1):177−200.

[48] Bibikova M, Lin Z, Zhou L, Chudin E, Garcia EW, Wu B, et al. High-throughput DNA methylation profiling using universal bead arrays. Genome Res 2006;16(3):383−93.

[49] Bibikova M, Fan JB. GoldenGate assay for DNA methylation profiling. Methods Mol Biol 2009;507:149−63.

[50] Fan JB, Gunderson KL, Bibikova M, Yeakley JM, Chen J, Wickham Garcia E, et al. Illumina universal bead arrays. Methods Enzymol 2006;410:57−73.

[51] Gunderson KL, Kruglyak S, Graige MS, Garcia F, Kermani BG, Zhao C, et al. Decoding randomly ordered DNA arrays. Genome Res 2004;14(5):870−7.

[52] Steemers FJ, Gunderson KL. Whole genome genotyping technologies on the BeadArray platform. Biotechnol J 2007;2(1):41−9.

[53] Gunderson KL. Whole-genome genotyping on bead arrays. Methods Mol Biol 2009;529:197−213.

[54] Gunderson KL, Steemers FJ, Ren H, Ng P, Zhou L, Tsan C, et al. Whole-genome genotyping. Methods Enzymol 2006;410:359−76.

[55] Flanagan JM. Epigenome-wide association studies (EWAS): past, present, and future. Methods Mol Biol 2015;1238:51−63.

[56] Weisenberger DJ. Characterizing DNA methylation alterations from The Cancer Genome Atlas. J Clin Invest 2014;124(1):17−23.

[57] Bibikova M, Barnes B, Tsan C, Ho V, Klotzle B, Le JM, et al. High density DNA methylation array with single CpG site resolution. Genomics 2011;98(4):288−95.

[58] Doi A, Park IH, Wen B, Murakami P, Aryee MJ, Irizarry R, et al. Differential methylation of tissue- and cancer-specific CpG island shores distinguishes human induced pluripotent stem cells, embryonic stem cells and fibroblasts. Nat Genet 2009;41(12):1350−3.

[59] Irizarry RA, Ladd-Acosta C, Wen B, Wu Z, Montano C, Onyango P, et al. The human colon cancer methylome shows similar hypo- and hypermethylation at conserved tissue-specific CpG island shores. Nat Genet 2009;41(2):178−86.

[60] Rubin AF, Green P. Mutation patterns in cancer genomes. Proc Natl Acad Sci USA 2009;106(51):21766−70.

[61] Irizarry RA, Wu H, Feinberg AP. A species-generalized probabilistic model-based definition of CpG islands. Mamm Genome 2009;20(9-10):674−80.

[62] Hsieh F, Chen SC, Pollard K. A nearly exhaustive search for CpG islands on whole chromosomes. Int J Biostat 2009;5(1):1.

[63] FANTOM4. Functional Annotation of the Mammalian Genome Yokohama city: RIKEN Omics Science Center; 2009 [cited 2010]. Available from: <http://fantom.gsc.riken.jp/4/>.

[64] Severin J, Waterhouse AM, Kawaji H, Lassmann T, van Nimwegen E, Balwierz PJ, et al. FANTOM4 EdgeExpressDB: an integrated database of promoters, genes, microRNAs, expression dynamics and regulatory interactions. Genome Biol 2009;10(4):R39.

[65] Tomazou EM, Rakyan VK, Lefebvre G, Andrews R, Ellis P, Jackson DK, et al. Generation of a genomic tiling array of the human major histocompatibility complex (MHC) and its application for DNA methylation analysis. BMC Med Genomics 2008;1:19.

[66] Birney E, Stamatoyannopoulos JA, Dutta A, Guigo R, Gingeras TR, Margulies EH, et al. Identification and analysis of functional elements in 1% of the human genome by the ENCODE pilot project. Nature 2007;447(7146):799−816.

[67] Heintzman ND, Hon GC, Hawkins RD, Kheradpour P, Stark A, Harp LF, et al. Histone modifications at human enhancers reflect global cell-type-specific gene expression. Nature 2009;459(7243):108−12.

[68] Heintzman ND, Stuart RK, Hon G, Fu Y, Ching CW, Hawkins RD, et al. Distinct and predictive chromatin signatures of transcriptional promoters and enhancers in the human genome. Nat Genet 2007;39(3):311−18.

[69] Fujita PA, Rhead B, Zweig AS, Hinrichs AS, Karolchik D, Cline MS, et al. The UCSC Genome Browser database: update 2011. Nucleic Acids Res 2011;39(Database issue): D876−82.

[70] Rhead B, Karolchik D, Kuhn RM, Hinrichs AS, Zweig AS, Fujita PA, et al. The UCSC Genome Browser database: update 2010. Nucleic Acids Res 2010;38(Database issue): D613−19.

[71] Lian H, Thompson WA, Thurman R, Stamatoyannopoulos JA, Noble WS, Lawrence CE. Automated mapping of large-scale chromatin structure in ENCODE. Bioinformatics 2008;24(17):1911−16.

[72] Xi H, Shulha HP, Lin JM, Vales TR, Fu Y, Bodine DM, et al. Identification and characterization of cell type-specific and ubiquitous chromatin regulatory structures in the human genome. PLoS Genet 2007;3(8):e136.

[73] Rakyan VK, Down TA, Thorne NP, Flicek P, Kulesha E, Graf S, et al. An integrated resource for genome-wide identification and analysis of human tissue-specific differentially methylated regions (tDMRs). Genome Res 2008;18(9):1518−29.

[74] Gardiner-Garden M, Frommer M. CpG islands in vertebrate genomes. J Mol Biol 1987;196(2):261−82.

[75] Sandoval J, Heyn HA, Moran S, Serra-Musach J, Pujana MA, Bibikova M, et al. Validation of a DNA methylation microarray for 450,000 CpG sites in the human genome. Epigenetics 2011;6(6):692−702.

[76] Lister R, Ecker JR. Finding the fifth base: genome-wide sequencing of cytosine methyla-
      tion. Genome Res 2009;19(6):959−66.
[77] Kulis M, Heath S, Bibikova M, Queiros AC, Navarro A, Clot G, et al. Epigenomic analy-
      sis detects widespread gene-body DNA hypomethylation in chronic lymphocytic leuke-
      mia. Nat Genet 2012;44(11):1236−42.
[78] Heyn H, Li N, Ferreira HJ, Moran S, Pisano DG, Gomez A, et al. Distinct DNA
      methylomes of newborns and centenarians. Proc Natl Acad Sci USA 2012;109
      (26):10522−7.
[79] Morris TJ, Beck S. Analysis pipelines and packages for Infinium HumanMethylation450
      BeadChip (450k) data. Methods 2015;72:3−8.
[80] Pidsley R, Wong CCY, Volta M, Lunnon K, Mill J, Schalkwyk LC. A data-driven
      approach to preprocessing Illumina 450K methylation array data. BMC Genomics
      2013;14:293.
[81] Illumina. GenomeStudio methylation module v1.8 user guide 2010. Available from: <http://
      support.illumina.com/content/dam/illumina-support/documents/myillumina/90666eaa-0c66-48
      b4-8199-3be99b2b3ef9/genomestudio_methylation_v1.8_user_guide_11319130_b.pdf>.
[82] Bock C. Analysing and interpreting DNA methylation data. Nat Rev 2012;13
      (10):705−19.
[83] Dedeurwaerder S, Defrance M, Bizet M, Calonne E, Bontempi G, Fuks F. A comprehen-
      sive overview of Infinium HumanMethylation450 data processing. Brief Bioinform
      2014;15(6):929−41.
[84] Wilhelm-Benartzi CS, Koestler DC, Karagas MR, Flanagan JM, Christensen BC, Kelsey
      KT, et al. Review of processing and analysis methods for DNA methylation array data. Br
      J Cancer 2013;109(6):1394−402.
[85] Mancuso FM, Montfort M, Carreras A, Alibes A, Roma G. HumMeth27QCReport: an R
      package for quality control and primary analysis of Illumina Infinium methylation data.
      BMC Res Notes 2011;4:546.
[86] van Iterson M, Tobi EW, Slieker RC, den Hollander W, Luijk R, Slagboom PE, et al.
      MethylAid: visual and interactive quality control of large Illumina 450k datasets.
      Bioinformatics 2014;30(23):3435−7.
[87] Davis S., Du P., Bilke S., Triche T., Bootwalla M. MethyLumi: bioconductor package for
      handling Illumina DNA methylation data 2011. Available from: <http://www.bioconduc-
      tor.org/packages/2.12/bioc/html/methylumi.html>.
[88] Touleimat N, Tost J. Complete pipeline for Infinium((R)) Human Methylation 450K
      BeadChip data processing using subset quantile normalization for accurate DNA methyla-
      tion estimation. Epigenomics 2012;4(3):325−41.
[89] Wang D, Yan L, Hu Q, Sucheston LE, Higgins MJ, Ambrosone CB, et al. IMA: an R
      package for high-throughput analysis of Illumina's 450K Infinium methylation data.
      Bioinformatic 2012;28(5):729−30.
[90] Aryee MJ, Jaffe AE, Corrada-Bravo H, Ladd-Acosta C, Feinberg AP, Hansen KD, et al.
      Minfi: a flexible and comprehensive bioconductor package for the analysis of Infinium
      DNA methylation microarrays. Bioinformatics 2014;30(10):1363−9.
[91] Liu Y, Aryee MJ, Padyukov L, Fallin MD, Hesselberg E, Runarsson A, et al. Epigenome-
      wide association data implicate DNA methylation as an intermediary of genetic risk in
      rheumatoid arthritis. Nat Biotechnol 2013;31(2):142−7.
[92] Dedeurwaerder S, Defrance M, Calonne E, Denis H, Sotiriou C, Fuks F. Evaluation of the
      Infinium Methylation 450K technology. Epigenomics 2011;3(6):771−84.

[93] Fraser HB, Lam LL, Neumann SM, Kobor MS. Population-specificity of human DNA methylation. Genome Biol 2012;13(2):R8.

[94] Chen YA, Lemire M, Choufani S, Butcher DT, Grafodatskaya D, Zanke BW, et al. Discovery of cross-reactive probes and polymorphic CpGs in the Illumina Infinium HumanMethylation450 microarray. Epigenetics 2013;8(2):203−9.

[95] Du P, Kibbe WA, Lin SM. Lumi: a pipeline for processing Illumina microarray. Bioinformatic 2008;24(13):1547−8.

[96] Sun S, Huang YW, Yan PS, Huang TH, Lin S. Preprocessing differential methylation hybridization microarray data. BioData Min 2011;4:13.

[97] Siegmund KD. Statistical approaches for the analysis of DNA methylation microarray data. Hum Genet 2011;129(6):585−95.

[98] Marabita F, Almgren M, Lindholm ME, Ruhrmann S, Fagerstrom-Billai F, Jagodic M, et al. An evaluation of analysis pipelines for DNA methylation profiling using the Illumina HumanMethylation450 BeadChip platform. Epigenetics 2013;8(3):333−46.

[99] Baylin SB, Jones PA. A decade of exploring the cancer epigenome—biological and translational implications. Nat Rev Cancer 2011;11(10):726−34.

[100] Leek JT, Scharpf RB, Bravo HC, Simcha D, Langmead B, Johnson WE, et al. Tackling the widespread and critical impact of batch effects in high-throughput data. Nat Rev 2010;11(10):733−9.

[101] Sun Z, Chai HS, Wu Y, White WM, Donkena KV, Klein CJ, et al. Batch effect correction for genome-wide methylation data with Illumina Infinium platform. BMC Med Genomics 2011;4:84.

[102] Buhule OD, Minster RL, Hawley NL, Medvedovic M, Sun G, Viali S, et al. Stratified randomization controls better for batch effects in 450K methylation analysis: a cautionary tale. Front Genet 2014;5:354.

[103] Harper KN, Peters BA, Gamble MV. Batch effects and pathway analysis: two potential perils in cancer studies involving DNA methylation array analysis. Cancer Epidemiol Biomarkers Prev 2013;22(6):1052−60.

[104] Yousefi P, Huen K, Schall RA, Decker A, Elboudwarej E, Quach H, et al. Considerations for normalization of DNA methylation data by Illumina 450K BeadChip assay in population studies. Epigenetics 2013;8(11):1141−52.

[105] van Eijk KR, de Jong S, Boks MP, Langeveld T, Colas F, Veldink JH, et al. Genetic analysis of DNA methylation and gene expression levels in whole blood of healthy human subjects. BMC Genomics 2012;13:636.

[106] Zackay A, Steinhoff C. MethVisual—visualization and exploratory statistical analysis of DNA methylation profiles from bisulfite sequencing. BMC Res Notes 2010;3:337.

[107] Barfield RT, Kilaru V, Smith AK, Conneely KN. CpGassoc: an R function for analysis of DNA methylation microarray data. Bioinformatic 2012;28(9):1280−1.

[108] Kilaru V, Barfield RT, Schroeder JW, Smith AK, Conneely KN. MethLAB: a graphical user interface package for the analysis of array-based DNA methylation data. Epigenetics 2012;7(3):225−9.

[109] Xu X, Su S, Barnes VA, De Miguel C, Pollock J, Ownby D, et al. A genome-wide methylation study on obesity: differential variability and differential methylation. Epigenetics 2013;8(5):522−33.

[110] Teschendorff AE, Widschwendter M. Differential variability improves the identification of cancer risk markers in DNA methylation studies profiling precursor cancer lesions. Bioinformatics 2012;28(11):1487−94.

[111] Morris TJ, Butcher LM, Feber A, Teschendorff AE, Chakravarthy AR, Wojdacz TK, et al. ChAMP: 450k chip analysis methylation pipeline. Bioinformatics 2014;30 (3):428–30.

[112] Du P, Zhang X, Huang CC, Jafari N, Kibbe WA, Hou L, et al. Comparison of beta-value and M-value methods for quantifying methylation levels by microarray analysis. BMC Bioinformatics 2010;11:587.

[113] Zhuang J, Widschwendter M, Teschendorff AE. A comparison of feature selection and classification methods in DNA methylation studies using the Illumina Infinium platform. BMC Bioinformatics 2012;13:59.

[114] Bernstein BE, Meissner A, Lander ES. The mammalian epigenome. Cell 2007;128 (4):669–81.

[115] Stirzaker C, Taberlay PC, Statham AL, Clark SJ. Mining cancer methylomes: prospects and challenges. Trends Genet 2014;30(2):75–84.

[116] Cancer Genome Atlas N. Comprehensive molecular portraits of human breast tumours. Nature 2012;490(7418):61–70.

[117] Cancer Genome Atlas N. Comprehensive molecular characterization of human colon and rectal cancer. Nature 2012;487(7407):330–7.

[118] Cancer Genome Atlas Research N. Comprehensive genomic characterization defines human glioblastoma genes and core pathways. Nature 2008;455(7216):1061–8.

[119] Cancer Genome Atlas Research N. Integrated genomic analyses of ovarian carcinoma. Nature 2011;474(7353):609–15.

[120] Cancer Genome Atlas Research N. Comprehensive molecular characterization of clear cell renal cell carcinoma. Nature 2013;499(7456):43–9.

[121] Cancer Genome Atlas Research N. Integrated genomic characterization of papillary thyroid carcinoma. Cell 2014;159(3):676–90.

[122] Cancer Genome Atlas Research N. Comprehensive molecular profiling of lung adenocarcinoma. Nature 2014;511(7511):543–50.

[123] Cancer Genome Atlas Research N, Kandoth C, Schultz N, Cherniack AD, Akbani R, Liu Y, et al. Integrated genomic characterization of endometrial carcinoma. Nature 2013;497 (7447):67–73.

[124] Fraga MF, Esteller M. Epigenetics and aging: the targets and the marks. Trends Genet 2007;23(8):413–18.

[125] Fraga MF, Ballestar E, Paz MF, Ropero S, Setien F, Ballestar ML, et al. Epigenetic differences arise during the lifetime of monozygotic twins. Proc Natl Acad Sci USA 2005;102(30):10604–9.

[126] Issa JP, Ottaviano YL, Celano P, Hamilton SR, Davidson NE, Baylin SB. Methylation of the oestrogen receptor CpG island links ageing and neoplasia in human colon. Nat Genet 1994;7(4):536–40.

[127] Rakyan VK, Down TA, Maslau S, Andrew T, Yang TP, Beyan H, et al. Human aging-associated DNA hypermethylation occurs preferentially at bivalent chromatin domains. Genome Res 2010;20(4):434–9.

[128] Hannum G, Guinney J, Zhao L, Zhang L, Hughes G, Sadda S, et al. Genome-wide methylation profiles reveal quantitative views of human aging rates. Mol Cell 2013;49 (2):359–67.

[129] Lee TI, Jenner RG, Boyer LA, Guenther MG, Levine SS, Kumar RM, et al. Control of developmental regulators by polycomb in human embryonic stem cells. Cell 2006;125 (2):301–13.

[130] Teschendorff AE, West J, Beck S. Age-associated epigenetic drift: implications, and a case of epigenetic thrift? Hum Mol Genet 2013;22(R1):R7−15.

[131] Michels KB, Binder AM, Dedeurwaerder S, Epstein CB, Greally JM, Gut I, et al. Recommendations for the design and analysis of epigenome-wide association studies. Nat Methods 2013;10(10):949−55.

[132] Bock C, Tomazou EM, Brinkman AB, Muller F, Simmer F, Gu H, et al. Quantitative comparison of genome-wide DNA methylation mapping technologies. Nat Biotechnol 2010;28(10):1106−14.

[133] Harris RA, Wang T, Coarfa C, Nagarajan RP, Hong C, Downey SL, et al. Comparison of sequencing-based methods to profile DNA methylation and identification of monoallelic epigenetic modifications. Nat Biotechnol 2010;28(10):1097−105.

[134] Teschendorff AE, Zhuang J, Widschwendter M. Independent surrogate variable analysis to deconvolve confounding factors in large-scale microarray profiling studies. Bioinformatics 2011;27(11):1496−505.

[135] Moran S, Vizoso M, Martinez-Cardus A, Gomez A, Matias-Guiu X, Chiavenna SM, et al. Validation of DNA methylation profiling in formalin-fixed paraffin-embedded samples using the Infinium HumanMethylation450 microarray. Epigenetics 2014;9(6):829−33.

[136] Nazor KL, Boland MJ, Bibikova M, Klotzle B, Yu M, Glenn-Pratola VL, et al. Application of a low cost array-based technique—TAB-Array—for quantifying and mapping both 5mC and 5hmC at single base resolution in human pluripotent stem cells. Genomics 2014;104(5):358−67.

[137] Stewart SK, Morris TJ, Guilhamon P, Bulstrode H, Bachman M, Balasubramanian S, et al. oxBS-450K: a method for analysing hydroxymethylation using 450K BeadChips. Methods 2015;72:9−15.

Chapter 3

Ultra-Deep Sequencing of Bisulfite-Modified DNA

Tingting Qin[1], Yongseok Park[2], Maria E. Figueroa[1] and Maureen A. Sartor[3]

[1]*Department of Pathology, University of Michigan, Ann Arbor, MI, USA, [2]Department of Biostatistics, University of Pittsburgh, Pittsburgh, PA, USA, [3]Department of Computational Medicine and Bioinformatics, University of Michigan, Ann Arbor, MI, USA*

Chapter Outline

3.1 INTRODUCTION

Treatment of DNA with sodium bisulfite causes the conversion of unmethylated cytosines to uracil, whereas methylated cytosines remain protected from this conversion. The consequence is that methylated and unmethylated DNA can be distinguished by sequencing. Over the past 5 years, protocols for sodium bisulfite treatment of DNA, coupled with massively parallel sequencing, have become the gold standard approach for assessment of genome-wide DNA methylation. A series of key findings and technologic

M. Fraga & A.F. Fernandez (Eds): Epigenomics in Health and Disease.
DOI: http://dx.doi.org/10.1016/B978-0-12-800140-0.00003-0
© 2016 Elsevier Inc. All rights reserved.

advances over the past years led to our current ability to quantitatively query the methylation status across the whole genome: the development of targeted bisulfite sequencing [1], the discovery that whole-genome amplification of bisulfite-modified DNA does not lead to significant biases [2], and, finally, the adaption of bisulfite conversion to next-generation sequencing for plants [3] and mammals [4]. Whole-genome bisulfite sequencing (WGBS) provides the first, and still the only, method to obtain accurate, quantitative estimates of the percentage of cells in a population that are methylated at each of the millions of CpG sites across the entire genome. The accuracy of the estimates are directly correlated with sequencing depth, and thus, although costly, the possibility exists to obtain estimates at any desired accuracy level with a sufficient number of reads. Unfortunately, since both the CpG sites themselves and their methylation are not evenly distributed across the genome, many of the reads obtained from WGBS do not contain any CpG sites and therefore do not contribute any information to the cells' DNA methylation profiles. In response to this disadvantage, an alternative approach, called reduced representation bisulfite sequencing (RRBS), was introduced to focus on CpG-rich regions [5,6]. This method relies on the use of a restriction enzyme-based approach designed to ensure the capture of at least one CpG site per sequencing read followed by bisulfite treatment of the restriction digested DNA. RRBS allows accurate estimates of percent methylation at CpG resolution across a portion of the genome enriched for CpG islands and promoter regions, at a fraction of the cost of WGBS. As costs have declined and software for aligning these modified DNA sequences to reference genomes were introduced [7,8], both WGBS and RRBS have gained popularity. Although RRBS still captures a fraction of the areas outside of promoters and CpG islands, valuable information present at distal intergenic regions is missed by this approach. In addition, larger-order characteristics of the methylation landscapes spanning megabases in size, such as lamina-associated domains [9] and hypomethylated canyons [10], cannot be adequately captured by this approach and still require the use of WGBS to be characterized.

Sequencing of bisulfite-modified DNA has been the gold standard partially due to its high level of fidelity and reproducibility, with rates well above 99% generally being achieved for conversion of unmethylated cytosine residues to uracil [11]. The data produced by WGBS and RRBS are also attractive because the estimates of percent methylation are free from several biases that can affect alternative approaches. For example, coverage is often affected by guanine−cytosine (GC) content, polymerase chain reaction (PCR) amplification, mapability, and common single nucleotide polymorphisms [12]; although these coverage biases affect the results of antibody pull-down-based approaches to DNA methylation assessment, they do not bias the estimates of percent methylation in WGBS and RRBS experiments due to their equivalent capture of methylated and unmethylated cytosines.

A limitation to both WGBS and RRBS is that bisulfite treatment does not distinguish between DNA methylation based on 5-methylcytosine (5mC) and its related oxidation product 5-hydroxymethylcytosine (5hmC) because 5hmC is similarly resistant to the conversion to uracil [13]. Therefore, the quantification measurements for WGBS or RRBS actually represent 5mC plus 5hmC. This caveat is often overlooked, since 5hmC often accounts for a small minority of the total methylation, although the percentage varies substantially by tissue type [14]. Furthermore, 5hmC is often associated with active gene transcription as opposed to 5mC, which is more likely repressive [15−17]; therefore, in certain circumstances, this limitation can significantly complicate the interpretation of bisulfite sequencing results.

Whether ultra-deep sequencing approaches to genome-wide DNA methylation assessment will continue as the gold standard remains to be seen. The large consortium, The NIH Roadmap Epigenomics, had been using RRBS and now has WGBS data available for several cell types [18]. Possible future directions include direct sequencing and distinction of 5mC, 5hmC, and other alternative DNA modifications. For example, using single-molecular, real-time sequencing, Pacific Biosciences was able to discriminate among several epigenetic modifications by using polymerase kinetics [19]. For numerous reasons, including insufficient throughput, however, this technique is not yet competitive with ultra-deep bisulfite sequencing approaches, especially in mammals.

In this chapter, we introduce the reader to multiple approaches for ultra-deep sequencing of bisulfite-modified DNA. We aim to cover (i) sample preparation and study design, (ii) technical considerations for deep sequencing, (iii) testing for differentially methylated sites or regions, (iv) annotation and visualization, and (v) experimental extensions to bisulfite sequencing. We conclude with specific examples of discoveries that have been made using bisulfite sequencing in cancer research and a brief discussion of future directions.

3.2 SAMPLE PREPARATION AND STUDY DESIGN CONSIDERATIONS FOR ULTRA-DEEP BISULFITE SEQUENCING

Preparation of libraries for WGBS and RRBS requires several steps of sequential processing: (i) genomic DNA isolation, (ii) breakdown of genomic DNA into smaller fragments, (iii) end repair and adapter ligation, (iv) size selection, and (v) bisulfite conversion and amplification of the final library [20] (Figure 3.1). In each step, certain critical parameters must be selected carefully. First, the method for genomic DNA extraction should be chosen carefully because it can critically affect the quality of WGBS and RRBS libraries. For example, the length of sample incubation and the amount of proteinase K, which affect the dissociation of genomic DNA from any contaminating nucleosomal or DNA-binding proteins, correspond directly to the quality of libraries [5,21]. This is because the nucleosomal contamination

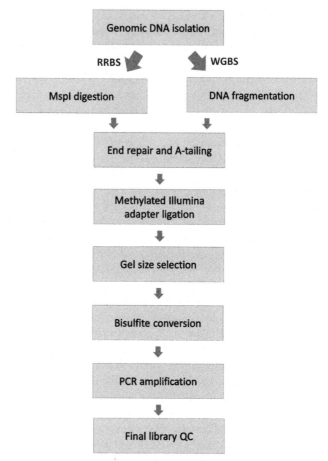

FIGURE 3.1 Comparison of the reduced representation bisulfite sequencing (RRBS) and WGBS library construction workflow. The flowchart highlights the key differences between the two protocols at the level of genomic DNA fragmentation.

can alter the properties of DNA denaturation and prevent suitable C-to-T conversion because the bisulfite reaction requires single-stranded DNA molecules. Genomic DNA isolation is followed by the fragmentation of DNA into smaller-molecular-weight fragments and varies depending on whether the approach of choice is WGBS or RRBS. WGBS relies on sonication of naked DNA to generate fragments less than 500 bp in length. Reduced representation approaches, on the other hand, rely on methylation-insensitive restriction enzyme digestion that cut DNA at CpG sites to generate DNA fragments containing at least one CpG dinucleotide per sequencing read. MspI, which targets 3'-CCGG-5' sequences and cleaves the phosphodiester bonds upstream of the CpG dinucleotide, is commonly used [22].

The subsequent end-repair, A-tailing, and adapter ligation reactions can follow standard next-generation protocols, but special considerations should be taken in certain circumstances. For example, in RRBS, the use of only C and G nucleotides for end repair ensures that only sequences with at least one CpG site at the end will be included in the library. For WGBS, the exclusion of C nucleotides from end-repair reactions prevents the introduction of falsely unmethylated Cs at that specific location [23]. However, newer computational approaches also allow for the removal of this inappropriately unmethylated cytosine after sequencing using a bioinformatics method, so that four nucleotides can also be used in WGBS for end repair [24]. Next, both WGBS and RRBS approaches go through a size selection step to ensure that only appropriately sized fragments are sequenced. In RRBS, the inclusion of fragments in which MspI cuts at higher frequency ensures the sequencing of fragments that are usually CpG rich. Size selection depends on the genomic regions of interest and the desired trade-off between breadth and depth of coverage; a size selection of 40−220 bp are enriched for CpG islands and promoter regions [22,25]. Moreover, in order to reduce possible size-related amplification and/or sequencing biases, generating two separate libraries in parallel rather than a single 40−220 bp library is suggested. Therefore, two gel bands ranging between 40−120 and 120−220 bp should be cut from each sample, and the fragments need to be amplified separately, resulting in short- and long-fragment libraries.

Finally, the efficiency of the bisulfite conversion reaction is the most critical parameter affecting the quality of sequenced reads, and an overall conversion efficiency greater than 99% for unmethylated cytosines is required. The library preparation protocol has to be optimized to provide the greatest possible conversion efficiency while minimizing overconversion (conversion of methylated cytosines) and sample degradation. In addition, before final amplification of converted libraries, it is essential to determine the optimal PCR cycle number for each independent sample by performing a semi-quantitative PCR set across a reasonable cycle range. In the PCR amplification step, the choice of polymerase is also essential and must be uracil insensitive to avoid uracil poisoning that would otherwise dramatically reduce the library quality and yield. Prior to sequencing, library validation by visual inspection of the final diagnostic gel is a critical quality control step. Individual molecular weight bands of a single sample are not expected to dramatically overlap and both in WGBS generated by sonication and in RRBS libraries, repetitive fragments can be expected to be present at a greater intensity than that observed for standard DNA sequencing.

Enhanced reduced representation bisulfite sequencing (ERRBS) is a modification of the original RRBS protocol for expanded coverage of genomic CpG methylation [26]. The considerations of library preparation as described above are also applicable to ERRBS samples, although some modifications and additional care should be taken. The major differences in library preparation between RRBS and ERRBS lie in the end-repair and size-selection steps. End repair during ERRBS library preparation is carried out using all four nucleotides. Although this leads to the inclusion of some sheared DNA sequences that may not contain any CpG overhangs nor internal CpG sites, it does lead to the recovery of a significant fraction of reads with internal CpG dinucleotides that lack a CpG overhang (likely due to sheared DNA or overdigestion of the DNA) thus making it a worthy trade-off [27]. In terms of size selection, RRBS libraries are inherently biased toward representing CpG islands, since only short MspI digested fragments (40−220 bp) are selected before bisulfite conversion [22,25]. In contrast, MspI fragments ranging from 70 to 320 bp are selected in ERRBS libraries, which increases the capture of regions beyond CpG islands [26].

Following library preparation, the bisulfite-converted DNA fragments are sequenced using ultra-deep sequencing technology, such as the Illumina HiSeq 2000 or 2500 sequencing instruments. Certain technical considerations are necessary to ensure a high quality of the sequenced reads, including sequencing depth, read length, and sequencing mode (single- or paired-end), and quality controls should be carried out throughout the entire process.

3.3   TECHNICAL CONSIDERATIONS FOR ULTRA-DEEP BISULFITE SEQUENCING APPROACHES

3.3.1   Sequencing Depth

Sequencing depth, the average number of times that a particular nucleotide is represented in a collection of raw sequences, is one of the key considerations in genomic analysis [12]. It is commonly used interchangeably with depth of coverage or coverage. Ultra-deep sequencing approaches are characterized by short-read length ($\leq$250 bp), and reads may contain errors that are indistinguishable from a sequence variant, leading to ambiguous mapping to the reference genome. Increasing sequence depth can overcome this problem, since the combination of identical reads that cover the location of the variant can produce a strongly supported variant call [28]. Likewise, when inferring methylation information through sequencing of bisulfite-treated fragments, the combination of reads that cover the location of a cytosine can produce an accurate estimate of the methylation level at that CpG site. Therefore, in order to make a reliable estimation of the CpG methylation status across a population of cells, a sufficient number of sequence reads is required for each CpG site: $10\times$ coverage will produce 11 possible combinations of C/T calling, which is a larger representative sample of the complete population compared with the only 3 possible combinations that $2\times$ coverage can produce. Moreover, sequencing depth varies across a genome due to the biases introduced in sample preparation, sequencing, and genomic alignment, resulting in some genomic regions lacking any coverage, while others have substantially higher coverage than theoretically expected. For example, methylated CpG regions that are resistant to C-to-T conversion by bisulfite treatment, especially GC-dense regions, such as CpG islands, may have lower coverage due to poor amplification efficiency at these sites [29]; however, some repetitive regions are prone to be oversequenced due to their overrepresentation in the genome and thereby overamplification by PCR. Therefore, it is critical to apply appropriate cutoffs of coverage before analyzing bisulfite sequencing data to ensure that sufficient reads are collected for methylation estimation and that the methylation information is not skewed due to the amplification bias. Generally, a minimum of 10× average coverage is recommended to give a realible estimate of percent methylation [30]. However, if WGBS is required and costs need to be controlled, lower coverage (e.g., ∼3× average coverage) in combination with data smoothing or tiling may also provide reliable regional information at the expense of sacrifing base-pair resolution [31].

3.3.2   Read Length

Read length is another critical parameter, since it can directly affect the performance of sequence mapping and assembly. Mapping short reads from bisulfite-treated DNA to a reference genome often results in gaps (regions with no aligned reads) that are caused by repetitive regions of length that either approaches or exceeds that of the reads. It also results in a high probability of multiple, equally likely mapping locations for each read. This is due to the low sequence complexity of bisulfite-treated DNA, since most of it is reduced to three possible base-pairs. To overcome these problems, the read length can be increased to 75 or 100+ bp when sequencing the genome, in which case the genomic alignment gaps will be reduced because not only will the chance of reads overlapping increase, but also the mapping accuracy will be improved, since the sequence complexity increases with read length. However, increasing the read length does increase the sequencing cost and the chance of sequencing into the ligated adapters for shorter fragments [32].

As an alternative to increasing the read length of single-end sequencing, paired-end sequencing is applied in some circumstances [33,34]. In contrast to single-end sequencing, in which the sequence fragments are sequenced from one direction only, in paired-end sequencing, a single fragment is sequenced from both ends, giving rise to forward and reverse strand reads. The paired-end reads can be separated by a certain number of bases, or they can be overlapping and produce a contiguous longer single fragment after merging, but this should generally be avoided. When a paired-end read is mapped to the reference genome, the mapping information about one end can be used to infer the likely position for the other, leading to a higher probability of being uniquely mapped to the genome and increasing the chances of capturing data for repetitive sequences, which are frequently epigenetically regulated, and may be of interest to scientists performing DNA methylation analyses. However, as with longer reads, the use of paired-end reads likewise results in an increase of the sequencing costs and increased chances of sequencing into the adapters of shorter fragments.

3.3.3   Quality Controls

To ensure high-quality bisulfite sequencing data, several technical concerns require careful attention during library preparation and sequencing [35]. First, it is crucial that the DNA methylation status of each fragment is not artificially modified before treatment with bisulfite; any prior PCR amplification step will erase any methylation markers that were present. The end-repair step in most.protocols introduces constitutively unmethylated cytosines at both ends of the DNA fragments, which will align perfectly against the reference genome but will not maintain the original methylation status and thus results in a globally biased (underestimated) DNA methylation level. Therefore, care must be taken to bioinformatically trim these locations within the biased sequencing reads prior to methylation calling being performed; this can be done, for example, using the R package BSeQC [24].

Second, the bisulfite conversion rate is one of the most important factors that can directly influence the estimation accuracy of the methylation levels. In a bisulfite sequencing experiment, we implicitly assume that all ummethylated cytosines have been converted into uracils (which are then changed to and sequenced as thymines); however, this conversion may not be totally complete, resulting in an overestimation of methylation. Extended or repeated bisulfite treatment can increase the conversion rate but at the cost of increased DNA degradation and overconversion of some methylated cytosines [36]. Thus, the sensitivity and specificity of the bisulfite conversion should always be monitored, ideally by spike-in control DNA with known methylation levels added before bisulfite treatment. Lambda-gDNA is a common control for unmethylated DNA, and some commercial kits contain known methylated controls.

Third, a large proportion of DNA fragments is often shorter than the read length, resulting in sequencing into the adapters on the 3' end; this is a concern both for RRBS and WGBS applications. The presence of an adapter sequence in a read dramatically decreases its likelihood of mapping correctly, resulting in lower coverage and less accurate methylation calls. Therefore, it is critical to trim the adapter sequences before the sequence alignment is performed.

Finally, any duplicate reads introduced by PCR amplification should be discarded to prevent bias by retaining only one of multiple reads having the exact same sequence. For data sets with high coverage, two or three reads may be kept. However, in reduced representation approaches, such as (E)RRBS in which the DNA is digested at specific restriction sites, identifying duplicate reads introduced by PCR bias becomes difficult, since all reads should align at MspI restriction sites. In this case, routinely removing very high coverage reads (e.g., >400×) is recommended. Routine validation of sites having a wide range of coverage levels using a targeted approach is important in order to assess the accuracy of methylation calls being performed.

Overall, rigorous quality controls should be carried out throughout library preparation, sequencing, and mapping. Several software programs have been developed for these purposes: FastQC (http://www.bioinformatics.babraham. ac.uk/projects/fastqc/), PRINSEQ [37] (http://prinseq.sourceforge.net/), and SolexaQA [38] (http://solexaqa.sourceforge.net/) can be applied on raw sequencing data to report sequence quality; Cutadapt (http://code.google.com/ p/cutadapt/), the FASTX toolkit (http://hannonlab.cshl.edu/fastx_toolkit/ index.html), BSeQC [24] (http://code.google.com/p/bseqc/), and Trimmomatic [39] (http://www.usadellab.org/cms/index.php?page=trimmomatic) can be used to perform adaptive quality trimming and adapter trimming of sequences.

3.3.4   Approaches to Mapping to a Reference Genome

After the raw sequences have been generated and appropriately trimmed, the short reads are ready to be analyzed for methylation inference. The first step in the analysis of bisulfite sequencing data is alignment to the reference genome, which needs to account for the fact that most cytosines (all unmethylated cytosines) will have been coverted to thymines in the sequencing reads. When performing an alignment, it is important to discriminate between the different types of bisulfite-treated DNA libraries [40]: (i) In directional libraries, only the original top or bottom strands of DNA fragments are sequenced; (ii) in nondirectional libraries, all four DNA strands that arise through bisulfite treatment and subsequent PCR amplification can be sequenced with the same frequency, leading to up to four different strand alignments for each sequence. Due to the complexity of bisulfite sequencing alignment, standard short-read aligners cannot be used directly, and thus alternative approaches have been developed (Table 3.1).

The first type of aligners, which includes BSMAP [8], GSNAP [43], Last [44], Pash [45], RMAP [46], RRBSMAP [47], and segemehl [48], are methylation-"aware" alignment tools that consider both cytosine and thymine as potential matches to a genomic cytosine. They either replace cytosine in the genomic DNA sequence by the wild-card letter Y, which matches both cytosine and thymine in the read sequence, or they exempt mismatches between cytosine in the genomic DNA sequence and thymine in the read sequence from penalization in the alignment scoring matrix [35]. Because these

TABLE 3.1 Comparison of Different Software Packages for BS-seq Alignment

	Bismark	BRAT	BS-Seeker	BSMAP	MethylCoder	RMAP-BS
Aligner	Bowtie [41]	Reference hashing and wild-card matching	Bowtie [41]	SOAP [42]	Bowtie [41]/ GSNAP [43]	Wild-card/position-weight-matrix matching
Year	2011	2010	2010	2009	2011	2008
Version	0.5.0	1.2.2	N/A	0.2.1	0.14.1	2.05
Language	Perl	C++	Python	C++	C/C++/Python	C++
Library type	D and ND	D and ND[a]	D and ND[b]	D and ND	D and ND	D and ND[a]
Sequencing technology	Base space	Base space	Base space	Base space	Base space	Base space
Sequencing mode	Single-end and paired-end	Single-end and paired-end	Single-end	Single-end and paired-end	Single-end and paired-end	Single-end
Best alignment criteria	Lowest number of non-BS mismatches	Lowest number of non-BS mismatches	Lowest number of mismatches	Lowest number of mismatches	Lowest number of non-BS mismatches	Lowest number of mismatches
Input file format	fasta/fastq	fasta/fastq	fasta, fastq, qseq, pure sequence	fasta/fastq/ SAM	fastq/fasta	fastq/fasta

Output file format	BAM/SAM	Brat/txt	BAM/SAM/ BS_Seeker	SAM/txt	BAM/SAM	BED
Output information	Mapping output, including methylation calls and extra tools	Mapping output and extra tools for methylation calls	Mapping output, including methylation calls	Mapping output	Mapping output and methylation call output	Mapping output
Advantages	Unbiased mapping; performance[c]	Unbiased mapping; performance	Unbiased mapping; performance	–	Unbiased mapping; performance	–
Drawbacks	–	Inflexible parameters	–	Performance[c]; biased mapping	–	Unbiased mapping only in C-G context; biased mapping in non-CG context

D, directional library; ND, nondirectional library; Non-BS, not bisulfite induced; –, not applicable.
[a]Requires two separate runs.
[b]Requires presence of a tag sequence.
[c]Performance here signifies run time on a reasonable timescale (i.e., a few hours, as compared with several days or even weeks for the same technique with a human dataset).
Source: Modified from table 1 in Ref. [40].

approaches make optimal use of the information present in the reads, they have high sensitivity (mapping efficiency). However, this type of aligner tends to overestimate methylation levels due to the higher mappability of methylated sequences (which contain A, G, T, and C) compared with their unmethylated counterparts (which only contain A, G, and T). Alternatively, a type of unbiased bisulfite sequencing alignment tools, which includes Bismark [7], BRAT [49,50], BS-Seeker [51] and MethylCoder [52], simplify the sequence mapping by converting all cytosines in both the bisulfite-treated reads and the reference genome to thymines before the alignment is performed. In this way, the alignment of each read is unaffected by its methylation status, and so the alignment can be carried out on a three-letter alphabet (A, G, and T) using standard aligners, such as Bowtie [41]. The drawback of this strategy is that it ignores the mappability information gained from the remaining cytosines in the bisulfite-converted reads and thereby further decreases the sequence complexity, resulting in overall reduced mapping efficiency.

The above-mentioned issues and trade-offs between mapping bias and low mapping efficiency associated with the current bisulfite sequencing alignment approaches are restricted to genomic regions with a high repetitive level, such as retrotransposons and segmental duplications, and can be alleviated by increasing the read length or using paired-end sequencing. The selection of the most suitable bisulfite aligner is also influenced by computational considerations, such as alignment speed, flexibility, memory consumption, ease of use, and information reported by the software.

3.4   TESTING FOR DIFFERENTIAL METHYLATION

After alignment of the bisulfite sequencing reads to the genome, the analysis often proceeds through several steps, including preprocessing, testing for differential methylation, visualization, annotation, and *post hoc* tests (Figure 3.2). Often, the main goal of genome-wide DNA methylation analysis is to identify differentially methylated CpG sites (DMCs) or regions. The bisulfite sequencing data can be summarized as counts of methylated (C) and unmethylated (T) reads at each CpG site, using special alignment pipelines, such as Bismark [7] or BS-Seeker2 [53], as described above. Typically, the methylation level is estimated as $\hat{p} = C/(C + T)$, and it is assumed that C given the coverage at the site, n, is distributed according to a binomial distribution with parameters n and p, that is treating $C|n \sim \mathrm{Bin}(n, p)$, where $n = C + T$. In humans, these values are typically obtained for a couple of million to tens of millions of CpG sites across the genome, depending on sequencing depth, mapability, and whether RRBS or WGBS was used.

One consideration of data preprocessing is whether to combine data from opposite strands. During DNA replication, under normal conditions the

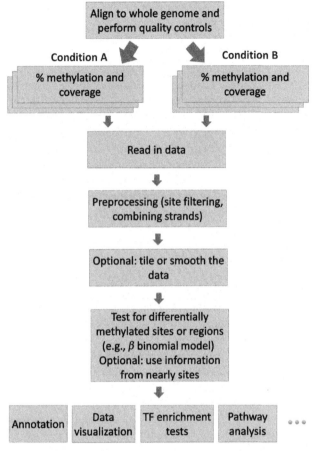

FIGURE 3.2 General workflow for the analysis of reduced representation or whole-genome bisulfite sequencing data.

methylation status is copied to the daughter strand by the enzyme Dnmt1 [54], such that the methylation status of a CpG site on opposite strands remains the same. Therefore, it is usually appropriate to aggregate the data from forward and reverse strands for cytosines one base-pair apart to form a single set of reads. This helps increase the estimation accuracy and statistical power by doubling the coverage on average and also increases the number of regions meeting the minimum coverage requirements. Another consideration is whether to conduct a differential methylation analysis based on information at each CpG site or after combining the information from nearby CpG sites. Due to the large number of CpG sites (~ 28 million in humans) and high cost of deep sequencing, the average coverage per CpG site is often low, particularly for WGBS. This makes combining information using local smoothing frameworks [31] or tiling data [30,54] from nearby CpG sites a desirable option to increase power.

Although not recommended, some experiments of bisulfite sequencing studies have been designed without replicates, or pooled samples from the same conditional group due to very limited numbers of cells from each replicate. In this type of one-versus-one sample comparison, a Fisher's exact test [4,30] or binomial test [30] has been proposed to identify DMCs. Another option is to use an empirical Bayes method [56] by applying a prior distribution for methylation levels, derived from the information in reads mapping across the genome. Although these methods are useful and may be sufficient for studies of highly homogeneous cell populations, such as cell lines, we suggest avoiding these methods when replicates exist in the comparison. Due to the heterogeneous nature of methylation levels among samples within a group with the same condition, ignoring biologic variations can result in highly inflated type-I error and unacceptably high false discovery rates (FDRs) [55].

Several methods have been published recently to identify DMC sites for RRBS or WGBS data. COHCAP [57], DMAP [58], and methylKit [30] either apply a Fisher's exact test, χ^2 test, or binomial-based test, which do not account for biologic variations, or they apply standard methods developed for continuous measures, such as analysis of variance or a t-test for each CpG site or region. BSmooth [31], BiSeq [59], and Bumphunter [60] use smoothing techniques, which are appropriate when focusing on differentially methylated regions (DMRs). The other methods were developed specifically for count data and use a beta-binomial approach from a Bayesian perspective [56], frequentist perspective [55,61], or a hybrid Bayesian and frequentist perspective [62]. The beta-binomial model is a commonly used model to introduce among-sample variation for count data using an overdispensation parameter. For $C|n \sim \mathrm{Bin}(n,p)$, if p is not fixed and $p \sim \mathrm{Beta}(\alpha,\beta)$, we say C has a beta-binomial distribution with probability mass function:

$$P(C = c) = \binom{n}{c} \frac{\Gamma(\alpha + c)\Gamma(\beta + n - c)\Gamma(\alpha + \beta)}{\Gamma(n + \alpha + \beta)\Gamma(\alpha)\Gamma(\beta)}$$

$$= \binom{n}{c} \frac{\Gamma(\mu\theta + c)\Gamma((1 - \mu)\theta + n - c)\Gamma(\theta)}{\Gamma(n + \theta)\Gamma(\mu\theta)\Gamma((1 - \mu)\theta)},$$

where the gamma function is $\Gamma(x) = \int_0^x t^{t-1}\mathrm{e}^{-t}\mathrm{d}t$, $\mu = \alpha/(\alpha + \beta)$, and $\theta = \alpha + \beta$. The mean and variance of C are $n\mu$ and $n\mu(1 - \mu)(1 + (n - 1)\phi)$. The overdispersion parameter, $\phi = 1/(1 + \theta)$, is used to account for biologic variations. This beta-binomial model can also be considered a Bayesian hierarchical model with fixed hyperparameters, α and β, as seen in Refs. [56,62].

Specifically, let $C_{igj} \sim \mathrm{Beta\text{-}Bin}(\mu_{ig}\theta_i,(1 - \mu_{ig})\theta_i, n_{igj})$, $g = 1,2; j = 1, 2,\ldots,J_g$, where C_{igj} and n_{igj} are the number of methylated cytosine reads and coverage at site i of the jth sample in group g, respectively. The goal is to test whether this CpG site, i, is differentially methylated in different groups. MOABS [56] introduced credible methylation difference, a new concept but considered a conservative estimation of methylation level difference [63], whereas DSS, RADmeth, and methylSig are using the hypothesis test, $H_{i0}:\mu_{i1} = \mu_{i2}$ against $H_{ia}:\mu_{i1} \neq \mu_{i2}$. DSS uses a Wald test based on the test statistic $t_i = \sum C_{i1j} / \sum n_{i1j} - \sum C_{i2j} / \sum n_{i2j}$, which is the same test statistic as for the binomial model. However, the variance estimation of t_i is based on a beta-binomial model, thus accounting for biologic variability while retaining the same estimated methylation difference from the binomial approach. Both RADmeth and methylSig apply a likelihood ratio test; RADmeth uses a traditional likelihood ratio test by fitting full and reduced beta-binomial regression models, whereas methylSig uses a testing procedure analogous to a t-test based on the beta-binomial likelihood directly. Due to the large number of tests performed, multiple comparisons become a critical issue. MethylSig explored type-I error rates using permutation with 21 acute myeloid leukemia (AML) samples and found that when at least three samples in both groups have data at the particular CpG site, the p-values from methylSig are close to the expected p-values at the tails of interest, that is, in the range of $10^{-6} \sim 0.05$.

To improve the power to detect significant differential methylation, local smoothing techniques are used to incorporate potential spatial correlation of methylation levels across the genome. BSmooth and BiSeq use smoothing estimation from each sample before applying it to a multiple-sample comparison, and methylSig provides a local information option by assuming a smoothness of group methylation levels within small windows across the genome. The focus may turn to the identification of DMRs after smoothing; however, regional consistent changes in methylation levels are generally more likely to reflect observable biologic effects and hence are preferred by many users. As a result, many methods have been developed to identify DMRs as opposed to DMCs. One approach is using predefined regions, such as sliding windows, CpG islands, or promoter regions (methylSig and methylKit). Another approach is defining DMRs based on the results from differential methylation analysis (Bsmooth, BiSeq, Bumphunder, and RADMeth). The latter approach is more difficult because there is no clear way to define and control the type-I error rates for DMRs. Some complicated methods have been proposed to control the FDR. For example, BiSeq uses two stage controls with a cluster-wise weighted FDR strategy and a test analogous to methods used for spatial signals [64]. Another approach is to combine p-values using either a weighted Z-test [61,65] or the Stouffer–Liptak test [66]. However, the combined p-values for a region using these methods is basically testing whether at least one CpG site is significant, and therefore does not necessarily imply multiple DMCs in the region. Therefore, it is still of interest to develop statistical methods that can clearly provide significance levels for the region to identify biologically meaningful DMRs.

3.5 IDENTIFYING ENRICHED OR DIFFERENTIALLY METHYLATED TRANSCRIPTION FACTOR–BINDING SITES

Hypermethylation is known to be able to prevent the binding of transcription factors (TFs), and therefore inhibit their regulatory role. For each TF in a given database, such as ENCODE uniform TF (http://genome.ucsc.edu/ENCODE/), it is particularly interesting to identify which TFs are different compared to others. There are multiple ways to consider what defines a difference, with two approaches described here. The first is to test which TFs are enriched, that is, which TFs' binding sites have a significantly larger proportion of DMCs than the overall proportion of DMCs. As an example, methylSig provides a simple test based on the binomial approach. Let N_T be the total number of CpG sites annotated into TF binding sites, and among these, let N_D be the total number of identified DMCs. For the TF_i, let n_i^T be the number of CpG sites and n_i^D be the number of DMCs within the binding sites of this TF_i. The test is based on the null hypothesis $H_{i0}:p_{Ti} = p_{Di}$ versus the alternative $H_{i1}:p_{Ti} \neq p_{Di}$, where $\hat{p}_{Ti} = N_D/N_T$ and $\hat{p}_{Ti} = n_i^D/n_i^T$. The results can be visualized as seen in Figure 3.3C. Alternatively, we can ask which TFs have a significant level of hypermethylation or hypomethylation across their binding sites, which could indicate whether the TF is having a weaker or stronger regulatory effect, respectively. To address this in methylSig, first for each sample, all reads are tiled (concatenated) from the regions to which a particular TF is predicted to bind. The same beta-binomial model as described above for tiled regions can then be applied to the data for each TF to identify TFs with significantly hyper or hypomethylated binding sites. This performs a self-contained hypothesis test, in that the level of differential methylation is compared with the null hypothesis of no differential methylation, as opposed to the level of differential methylation outside of the TF binding sites.

3.6 DATA VISUALIZATION AND ANNOTATION

DNA methylation levels in promoter regions often have a regulatory role in gene expression, with hypermethylation being generally associated with repressed expression. Genome-wide DNA methylation levels are also shown to be tissue specific and age specific. To access the potential biologic impact of differential methylation events, CpG sites can be annotated to the genomic context, such as CpG islands and shores, genic information (e.g., to the promoter, 5' and 3' untranslated regions, intron, or exon), noncoding RNA, intergenic regions, and TF binding sites or enhancers. This is important when interpreting DNA methylation differences because the presence of methylcytosine may have different effects on gene regulation, depending on the context of the genomic region [67]. MethylKit and methylSig provide multiple annotation and visulation functions, as seen in Figure 3.3.

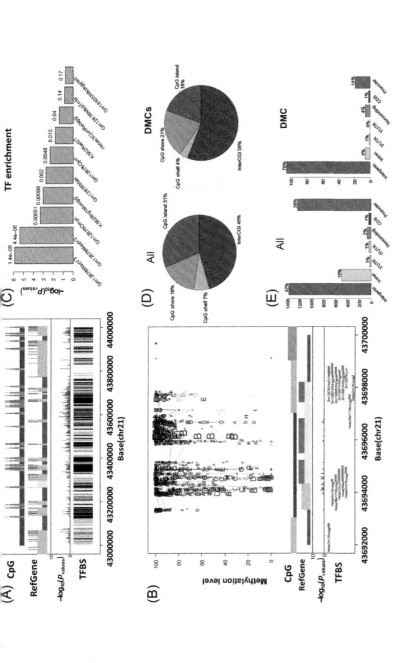

FIGURE 3.3 Visualization and annotation example of enhanced reduced representation bisulfite sequencing (ERRBS) data. (A) Visualization of data for a broad genomic range. When the selected chromosome range is large, location and significance levels are displayed, but raw data points are not shown. This view is helpful to identify interesting regions on which to zoom in. (B) When zoomed in to a small region, more information can be provided. In methylSig, the coverage levels, percent methylation levels, and group mean methylation estimates at each CpG site are shown in the same plot. (C) Identifying enriched transcription factors. (D) Annotating CpG sites to CpG islands. (E) Annotating CpG sites to RefGene models.

In addition, proper visualization of input data along with differential analysis results and annotation tracks may provide additional insight into the potential regulatory effects. MethylSig offers a unique two-tiered visualization of the methylation data, depending on the zoom level. For narrow regions (recommended for <100 kb) where at most 500 CpG sites have data reads, users can visualize sample-specific coverage levels and percent methylation at each site, together with group averages, significance levels, and a number of genomic annotations (Figure 3.3B). For broad regions, the visualization is simplified to the locations of multiple genomic features (e.g., CpG islands, enhancer regions, etc.) along with log 10-scale significance levels of the DMCs/DMRs (Figure 3.3A).

3.7   GENE SET ENRICHMENT STRATEGIES

Gene set, or functional, enrichment testing is a powerful approach that facilitates the interpretation of high-throughput results in terms of biologic pathways, cellular processes, and potential diseases. Beginning with methods and tools for microarray gene expression data, the field of functional enrichment testing has expanded to include methods and applications for GWAS [68], ChIP-seq [69,70], and DNA methylation data [71]. For bisulfite sequencing approaches, as for GWAS and ChIP-seq data, genomic sites need to first be associated with their expected target gene. This is most often done by associating sites, in this case CpGs or DMRs, with the nearest gene or the gene with the nearest transcription start site (TSS). Although this is an efficient method for proximal or greater promoter regions, more research is required to improve the regulatory associations that occur for sites in distal, intergenic regions. Currently, there is no gold standard for making these long-range association calls, and empirical validation is still required when a definitive answer is needed. Integration of epigenetic data sets, with data emerging from genomic studies that inform the three-dimensional organizational structure of the genome, can be useful in informing physical interactions between TSSs and distal enhancer regions. However, these data sets are still scarce and not yet available for most cell types.

Once the CpG sites or regions are associated with genes, a statistical test for enrichment of differential methylation can be performed against gene annotation databases, such as Gene Ontology or KEGG Pathways. Standard methods, such as Fisher's exact test, can be used when only considering differential methylation in promoter regions, but it is known to often lead to biased results when used with genome-wide DNA methylation data [72]. To avoid this bias, which often results in an overabundance of false positives, the test should take into account the relationship between the number of covered CpG sites in a gene's designated regulatory region, and either (i) the probability that the gene contains a differentially methylated site, or (ii) the strength or number of differentially methylated sites for the gene.

If testing larger-tiled windows for DMRs, then an enrichment testing program for ChIP-seq data, such as ChIP-Enrich (http://chip-enrich.med. umich.edu) [69], may be used. The epigenomics research community would benefit from tools developed specifically for performing gene set enrichment testing on ultra-deep bisulfite sequencing data.

3.8 PUBLISHED APPLICATIONS

3.8.1 Using ERRBS to Study Leukemia

RRBS and ERRBS have been successfully applied to numerous studies of stem cells and cancers (e.g., Refs [5,73–75]). Likely due to a combination of ease of sample collection and the high frequency of mutations that affect the DNA methylation machinery, acute leukemias are among the most frequently studied cancers since this technology was developed. The first study to apply this method to AMLs revealed the existence of divergent methylation landscapes among different subtypes of AML [27]. There the study reported the presence of profound hypermethylation, which prefentially targeted promoters and CpG island regions in leukemias carrying *IDH1/2* mutations, whereas widespread hypomethylation targeting distal intergenic regions and enhancers was the norm in leukemias with chromosomal rearrangements involving the *MLL* gene. Notably, despite these divergent methylomes, a small core of recurrently hypermethylated CpG sites common to both leukemias was identified. Thus, distinct forms of AML are characterized by the presence of unique DNA methylation profiles that clearly help distinguish molecular subtypes, whereas common epigenetic abnormalities may be involved in leukemia pathogenesis irrespective of disease subtype [27].

Schoofs et al. [76] used RRBS to study a special form of AML, acute promyelocytic leukemia (APL), which is characterized by the presence of the $t(15;17)$ translocation leading to the production of the fusion oncogene *PML-RARα* and compared DNA methylation in this tumor type to the epigenetic changes during promyelocyte differentiation. APL specimens had a very distinct DNA methylation profile, which tended to be more hypermethylated than those of controls, targeting promoters and gene body regions. Notably, despite previous reports that the PML-RARα fusion protein could recruit DNA methyltransferases to its target sites at gene promoters, no significant overlap was found between the DMRs in APL and the known binding sites for the oncogenic fusion protein.

Recently, mutations that target the DNA methylation machinery have been reported. Mutations in the *de novo* DNA methylatransferase DNMT3A occur in $\sim 22\%$ of AML cases and act as a dominant negative on the wild-type allele [77,78]. The Ten-Eleven translocation (TET) family of proteins lead to ultimate demethylation of cytosines through an Fe- and α-ketoglutarate dependent reaction. Loss-of-function mutations reported in

TET2 lead to hypermethylation and a block in myeloid differentiation [79]. In addition, IDH1 and IDH2 mutations occur in approximately 15% of AMLs and lead to the production of an oncometabolite, 2-hydroxyglutarate (2-HG), which acts as a competitive inhibitor of TET proteins, likewise leading to hypermethylation [79]. A recent study by Rampal et al. reports the existence of highly overlapping hypermethylation profiles between AMLs harboring IDH1, IDH2, and TET2 mutations. Hypermethylation was more profound in IDH-mutant AMLs than in cases with TET2 mutations, which is consistent with 2-HG leading to inhibition of all three TET proteins. Both mutation in IDH1/2 and TET2 also led to a corresponding decrease in 5hmC which correlated more strongly with changes in gene expression than 5mC, thus underscoring the importance that studying both types of cytosine modifications in these leukemias [16].

In the future, more studies looking at both 5mC and 5hmC will be required to fully understand the role that these distinct epigenetic marks play in gene regulation. The two methods, TAB-seq and OxBS [80,81], have been described to be capable of querying both 5mC and 5hmC at base-pair resolution, but these methods are cumbersome and expensive and are currently outside the reach for most labs. For now, the use of affinity-based methods, whether by selective chemical labeling [82] or by antibody enrichment [83], remain the most easily accessible methods for the determination of 5hmC.

3.8.2 WGBS in Cancers

Cancers have historically been viewed as diseases caused by genetic aberrations, with little attention paid to epigenetics. In recent years, the importance of epigenomics contributions to the development, progression, and recurrence of cancers has taken centerstage alongside genomics, with large national efforts, such as The Cancer Genome Atlas (TCGA), including epigenomics analysis, specifically DNA methylation, in their cancer-specific projects. Although TCGA has mainly used array-based approaches for DNA methylation analysis, several studies of cancer using WGBS have also been performed. One of the earliest WGBS publications assessed three colon cancer cases and three healthy colon samples at relatively low sequencing depth [84]. WGBS allowed Hansen et al. to examine the methylation levels and variability among samples in CpG islands, CpG island shores (the regions within 2 kb directly outside of CpG islands), and regions further out. From this, a clear picture of significantly higher variation among the cancer samples compared with normal samples emerged, indicating a lack of stability, and suggestive of the methylome's responsiveness to selective pressures in the cell's environment, which vary considerably among cancers. The study also illustrated a loss of control in the distinguishing signal in DNA methylation between the CpG island and non-CpG island regions. For example, boundaries between the mostly unmethylated CpG islands and methylated non-CpG islands sometimes shifted, or the CpG island became hypermethylated and the non-CpG island region was hypomethylated, resulting in a flat signal across an entire region in the cancer samples. Also observed were large genomic blocks of hypomethylation, whose methylation levels were also highly variable among individuals and which likely contributed to genomic instability.

More recently, in 2013, Bender et al. examined the epigenomics of pediatric high-grade gliomas (phGGs) by using WGBS and chromatin immunoprecipitation followed by deep sequencing [85]. About half of phGGs are known to harbor mutations in the histone *H3.3* gene, with the K27M mutation causing a genome-wide reduction of trimethylation of lysine 27 on Histone 3 (H3K27me3) [86]. Under normal conditions, the protein EZH2 in the PRC2 protein complex trimethylates H3K27 to repress a wide spectrum of mainly developmental genes, and a reduction in this histone mark, often concurrent with DNA hypermethylation, has been observed in several cancers [87,88]. In this study, gene expression analysis, ChIP-seq of H3K27me3 and WGBS were used to determine the global epigenomics effect of the K27M mutation on gene expression. They found substantial differences in genome-wide methylation among six H3.3 wildtype phGG and six K27M mutant phGG primary tumors, with significant overall hypomethylation in the K27M mutants compared with the wild type; this is in contrast to the hypermethylation often observed in cancers and suggestive of a different mechanism. Additional observations included pronounced hypomethylation immediately downstream of TSSs in many genes more highly expressed in K27M mutants, and higher methylation around TSSs for a smaller subset of the genes repressed in K27M mutants compared with the wild type. WGBS and ChIP-seq integrated with gene expression data enabled them to identify global and promoter region DNA hypomethylation and reduced H3K27me3 as the major determinants of gene upregulation in K27M mutant phGGs.

3.9 CONCLUSIONS AND FUTURE DIRECTIONS

WGBS and RRBS represent two powerful and complementary approaches to genome-wide DNA methylation assessment. The past year has seen a flurry of new methods introduced for analyzing RRBS or WGBS data, using some variant of a beta-binomial statistical model to test for differential CpG sites or regions; these software products will greatly enhance the ability of researchers to make informative conclusions based on these data. Currently, there is a lack of software for integrating RRBS or WGBS data with gene expression data and/ or other approaches for assessing 5hmC to distinguish these modifications to DNA and to study their functional consequences. To help fill this need, we are developing software based in Galaxy [89] that will analyze WGBS or RRBS data with methylSig [55], pull-down-based approaches, such as hmeDIP-seq and meDIP-seq using MACS2 [90] and PePr [91], and perform integrative analyses with gene expression data and publicly available resources. Along these same lines, Knijnenburg et al. recently developed software to identify genomic signals, such as from DNA methylation data, across a wide range of spatial resolutions, across the genome [92]. This and other future software will aid investigators in characterizing global epigenomic profiles and their transcriptional effects, identifying novel biomarkers for diseases, and discovering mechanistic links between environmental exposures and disease.

REFERENCES

[1] Frommer M, McDonald LE, Millar DS, Collis CM, Watt F, Grigg GW, et al. A genomic sequencing protocol that yields a positive display of 5-methylcytosine residues in individual DNA strands. Proc Natl Acad Sci USA 1992;89(5):1827−31.

[2] Mill J, Yazdanpanah S, Guckel E, Ziegler S, Kaminsky Z, Petronis A. Whole genome amplification of sodium bisulfite-treated DNA allows the accurate estimate of methylated cytosine density in limited DNA resources. BioTechniques 2006;41(5):603−7.

[3] Cokus SJ, Feng S, Zhang X, Chen Z, Merriman B, Haudenschild CD, et al. Shotgun bisulphite sequencing of the *Arabidopsis* genome reveals DNA methylation patterning. Nature 2008;452(7184):215−19.

[4] Lister R, Pelizzola M, Dowen RH, Hawkins RD, Hon G, Tonti-Filippini J, et al. Human DNA methylomes at base resolution show widespread epigenomic differences. Nature 2009;462(7271):315−22.

[5] Meissner A, Mikkelsen TS, Gu H, Wernig M, Hanna J, Sivachenko A, et al. Genome-scale DNA methylation maps of pluripotent and differentiated cells. Nature 2008;454(7205):766−70.

[6] Meissner A, Gnirke A, Bell GW, Ramsahoye B, Lander ES, Jaenisch R. Reduced representation bisulfite sequencing for comparative high-resolution DNA methylation analysis. Nucleic Acids Res 2005;33(18):5868−77.

[7] Krueger F, Andrews SR. Bismark: a flexible aligner and methylation caller for Bisulfite-Seq applications. Bioinformatics 2011;27(11):1571−2.

[8] Xi Y, Li W. BSMAP: whole genome bisulfite sequence MAPping program. BMC Bioinformatics 2009;10:232.

[9] Berman BP, Weisenberger DJ, Aman JF, Hinoue T, Ramjan Z, Liu Y, et al. Regions of focal DNA hypermethylation and long-range hypomethylation in colorectal cancer coincide with nuclear lamina-associated domains. Nat Genet 2012;44(1):40−6.

[10] Jeong M, Sun D, Luo M, Huang Y, Challen GA, Rodriguez B, et al. Large conserved domains of low DNA methylation maintained by Dnmt3a. Nat Genet 2014;46(1):17−23.

[11] Holmes EE, Jung M, Meller S, Leisse A, Sailer V, Zech J, et al. Performance evaluation of kits for bisulfite-conversion of DNA from tissues, cell lines, FFPE tissues, aspirates, lavages, effusions, plasma, serum, and urine. PLoS One 2014;9(4):e93933.

[12] Sims D, Sudbery I, Ilott NE, Heger A, Ponting CP. Sequencing depth and coverage: key considerations in genomic analyses. Nat Rev Genet 2014;15(2):121−32.

[13] Huang Y, Pastor WA, Shen Y, Tahiliani M, Liu DR, Rao A. The behaviour of 5-hydroxymethylcytosine in bisulfite sequencing. PLoS One 2010;5(1):e8888.

[14] Li W, Liu M. Distribution of 5-hydroxymethylcytosine in different human tissues. J Nucleic Acids 2011;2011:870726.

[15] Jin SG, Wu X, Li AX, Pfeifer GP. Genomic mapping of 5-hydroxymethylcytosine in the human brain. Nucleic Acids Res 2011;39(12):5015−24.

[16]   Rampal R, Alkalin A, Madzo J, Vasanthakumar A, Pronier E, Patel J, et al. DNA hydro-xymethylation profiling reveals that WT1 mutations result in loss of TET2 function in acute myeloid leukemia. Cell Rep 2014;9(5):1841−55.

[17]   Wu H, D'Alessio AC, Ito S, Wang Z, Cui K, Zhao K, et al. Genome-wide analysis of 5-hydroxymethylcytosine distribution reveals its dual function in transcriptional regulation in mouse embryonic stem cells. Genes Dev 2011;25(7):679−84.

[18]   Chadwick LH. The NIH Roadmap Epigenomics Program data resource. Epigenomics 2012;4(3):317−24.

[19]   Flusberg BA, Webster DR, Lee JH, Travers KJ, Olivares EC, Clark TA, et al. Direct detection of DNA methylation during single-molecule, real-time sequencing. Nat Methods 2010;7(6):461−5.

[20]   Smith ZD, Gu H, Bock C, Gnirke A, Meissner A. High-throughput bisulfite sequencing in mammalian genomes. Methods 2009;48(3):226−32.

[21]   Warnecke PM, Stirzaker C, Song J, Grunau C, Melki JR, Clark SJ. Identification and res-olution of artifacts in bisulfite sequencing. Methods 2002;27(2):101−7.

[22]   Gu H, Smith ZD, Bock C, Boyle P, Gnirke A, Meissner A. Preparation of reduced repre-sentation bisulfite sequencing libraries for genome-scale DNA methylation profiling. Nat Protoc 2011;6(4):468−81.

[23]   Lister R, O'Malley RC, Tonti-Filippini J, Gregory BD, Berry CC, Millar AH, et al. Highly integrated single-base resolution maps of the epigenome in *Arabidopsis*. Cell 2008;133(3):523−36.

[24]   Lin X, Sun D, Rodriguez B, Zhao Q, Sun H, Zhang Y, et al. BSeQC: quality control of bisulfite sequencing experiments. Bioinformatics 2013;29(24):3227−9.

[25]   Bock C, Tomazou EM, Brinkman AB, Muller F, Simmer F, Gu H, et al. Quantitative comparison of genome-wide DNA methylation mapping technologies. Nat Biotechnol 2010;28(10):1106−14.

[26]   Akalin A, Garrett-Bakelman FE, Kormaksson M, Busuttil J, Zhang L, Khrebtukova I, et al. Base-pair resolution DNA methylation sequencing reveals profoundly diver-gent epigenetic landscapes in acute myeloid leukemia. PLoS Genet 2012;8(6): e1002781.

[27]   Garrett-Bakelman, FE, Sheridan, CK, Kacmarczyk, TJ, Ishii, J, Betel, D, Alonso, A, et al. Enhanced reduced representation bisulfite sequencing for assessment of DNA methylation at base pair resolution. J Vis Exp 96, e52246.

[28]   Schatz MC, Delcher AL, Salzberg SL. Assembly of large genomes using second-generation sequencing. Genome Res 2010;20(9):1165−73.

[29]   Veal CD, Freeman PJ, Jacobs K, Lancaster O, Jamain S, Leboyer M, et al. A mechanistic basis for amplification differences between samples and between genome regions. BMC Genomics 2012;13:455.

[30]   Akalin A, Kormaksson M, Li S, Garrett-Bakelman FE, Figueroa ME, Melnick A, et al. methylKit: a comprehensive R package for the analysis of genome-wide DNA methylation profiles. Genome Biol 2012;13(10):R87.

[31]   Hansen KD, Langmead B, Irizarry RA. BSmooth: from whole genome bisulfite sequenc-ing reads to differentially methylated regions. Genome Biol 2012;13(10):R83.

[32]   Metzker ML. Sequencing technologies—the next generation. Nat Rev Genet 2010; 11(1):31−46.

[33]   Ni T, Corcoran DL, Rach EA, Song S, Spana EP, Gao Y, et al. A paired-end sequencing strategy to map the complex landscape of transcription initiation. Nat Methods 2010;7(7):521−7.

[34] Fullwood MJ, Wei CL, Liu ET, Ruan Y. Next-generation DNA sequencing of paired-end tags (PET) for transcriptome and genome analyses. Genome Res 2009;19(4):521−32.

[35] Bock C. Analysing and interpreting DNA methylation data. Nat Rev Genet 2012; 13(10):705−19.

[36] Genereux DP, Johnson WC, Burden AF, Stoger R, Laird CD. Errors in the bisulfite conversion of DNA: modulating inappropriate- and failed-conversion frequencies. Nucleic Acids Res 2008;36(22):e150.

[37] Schmieder R, Edwards R. Quality control and preprocessing of metagenomic datasets. Bioinformatics 2011;27(6):863−4.

[38] Cox MP, Peterson DA, Biggs PJ. SolexaQA: at-a-glance quality assessment of Illumina second-generation sequencing data. BMC Bioinformatics 2010;11:485.

[39] Bolger AM, Lohse M, Usadel B. Trimmomatic: a flexible trimmer for Illumina sequence data. Bioinformatics 2014;30(15):2114−20.

[40] Krueger F, Kreck B, Franke A, Andrews SR. DNA methylome analysis using short bisulfite sequencing data. Nat Methods 2012;9(2):145−51.

[41] Langmead B, Trapnell C, Pop M, Salzberg SL. Ultrafast and memory-efficient alignment of short DNA sequences to the human genome. Genome Biol 2009;10(3):R25.

[42] Li R, Li Y, Kristiansen K, Wang J. SOAP: short oligonucleotide alignment program. Bioinformatics 2008;24(5):713−14.

[43] Wu TD, Nacu S. Fast and SNP-tolerant detection of complex variants and splicing in short reads. Bioinformatics 2010;26(7):873−81.

[44] Frith MC, Mori R, Asai K. A mostly traditional approach improves alignment of bisulfite-converted DNA. Nucleic Acids Res 2012;40(13):e100.

[45] Coarfa C, Yu F, Miller CA, Chen Z, Harris RA, Milosavljevic A. Pash 3.0: a versatile software package for read mapping and integrative analysis of genomic and epigenomic variation using massively parallel DNA sequencing. BMC Bioinformatics 2010;11:572.

[46] Smith AD, Chung WY, Hodges E, Kendall J, Hannon G, Hicks J, et al. Updates to the RMAP short-read mapping software. Bioinformatics 2009;25(21):2841−2.

[47] Xi Y, Bock C, Muller F, Sun D, Meissner A, Li W. RRBSMAP: a fast, accurate and user-friendly alignment tool for reduced representation bisulfite sequencing. Bioinformatics 2012;28(3):430−2.

[48] Otto C, Stadler PF, Hoffmann S. Fast and sensitive mapping of bisulfite-treated sequencing data. Bioinformatics 2012;28(13):1698−704.

[49] Harris EY, Ponts N, Levchuk A, Roch KL, Lonardi S. BRAT: bisulfite-treated reads analysis tool. Bioinformatics 2010;26(4):572−3.

[50] Harris EY, Ponts N, Le Roch KG, Lonardi S. BRAT-BW: efficient and accurate mapping of bisulfite-treated reads. Bioinformatics 2012;28(13):1795−6.

[51] Chen PY, Cokus SJ, Pellegrini M. BS Seeker: precise mapping for bisulfite sequencing. BMC Bioinformatics 2010;11:203.

[52] Pedersen B, Hsieh TF, Ibarra C, Fischer RL. MethylCoder: software pipeline for bisulfite-treated sequences. Bioinformatics 2011;27(17):2435−6.

[53] Guo W, Fiziev P, Yan W, Cokus S, Sun X, Zhang MQ, et al. BS-Seeker2: a versatile aligning pipeline for bisulfite sequencing data. BMC Genomics 2013;14:774.

[54] Hermann A, Goyal R, Jeltsch A. The Dnmt1 DNA-(cytosine-C5)-methyltransferase methylates DNA processively with high preference for hemimethylated target sites. J Biol Chem 2004;279(46):48350−9.

[55] Park Y, Figueroa ME, Rozek LS, Sartor MA. MethylSig: a whole genome DNA methylation analysis pipeline. Bioinformatics 2014;30(17):2414−22.

[56] Sun D, Xi Y, Rodriguez B, Park HJ, Tong P, Meong M, et al. MOABS: model based analysis of bisulfite sequencing data. Genome Biol 2014;15(2):R38.

[57] Warden CD, Lee H, Tompkins JD, Li X, Wang C, Riggs AD, et al. COHCAP: an integrative genomic pipeline for single-nucleotide resolution DNA methylation analysis. Nucleic Acids Res 2013;41(11):e117.

[58] Stockwell PA, Chatterjee A, Rodger EJ, Morison IM. DMAP: differential methylation analysis package for RRBS and WGBS data. Bioinformatics 2014;30(13):1814−22.

[59] Hebestreit K, Dugas M, Klein HU. Detection of significantly differentially methylated regions in targeted bisulfite sequencing data. Bioinformatics 2013;29(13):1647−53.

[60] Jaffe AE, Murakami P, Lee H, Leek JT, Fallin MD, Feinberg AP, et al. Bump hunting to identify differentially methylated regions in epigenetic epidemiology studies. Int J Epidemiol 2012;41(1):200−9.

[61] Dolzhenko E, Smith AD. Using beta-binomial regression for high-precision differential methylation analysis in multifactor whole-genome bisulfite sequencing experiments. BMC Bioinformatics 2014;15:215.

[62] Feng H, Conneely KN, Wu H. A Bayesian hierarchical model to detect differentially methylated loci from single nucleotide resolution sequencing data. Nucleic Acids Res 2014;42(8):e69.

[63] Robinson MD, Kahraman A, Law CW, Lindsay H, Nowicka M, Weber LM, et al. Statistical methods for detecting differentially methylated loci and regions. Front Genet 2014;5:324.

[64] Benjamini Y, Heller R. False discovery rates for spatial signals. J Am Stat Assoc 2007;102(480):1272−81.

[65] Kechris KJ, Biehs B, Kornberg TB. Generalizing moving averages for tiling arrays using combined p-value statistics. Stat Appl Genet Mol Biol 2010;9:Article29.

[66] Li S, Garrett-Bakelman FE, Akalin A, Zumbo P, Levine R, To BL, et al. An optimized algorithm for detecting and annotating regional differential methylation. BMC Bioinformatics 2013;14(Suppl. 5):S10.

[67] van Vlodrop IJ, Niessen HE, Derks S, Baldewijns MM, van Criekinge W, Herman JG, et al. Analysis of promoter CpG island hypermethylation in cancer: location, location, location! Clin Cancer Res 2011;17(13):4225−31.

[68] Holmans P, Green EK, Pahwa JS, Ferreira MA, Purcell SM, Sklar P, et al. Gene ontology analysis of GWA study data sets provides insights into the biology of bipolar disorder. Am J Hum Genet 2009;85(1):13−24.

[69] Welch RP, Lee C, Imbriano PM, Patil S, Weymouth TE, Smith RA, et al. ChIP-Enrich: gene set enrichment testing for ChIP-seq data. Nucleic Acids Res 2014;42(13):e105.

[70] Cavalcante RG, Lee C, Patil S, Weymouth TE, Welch RP, Scott LJ, et al. Broad-Enrich: functional interpretation of large sets of broad genomic regions. Bioinformatics 2014;30(17):i393−400.

[71] Kim JH, Karnovsky A, Mahavisno V, Weymouth T, Pande M, Dolinoy DC, et al. LRpath analysis reveals common pathways dysregulated via DNA methylation across cancer types. BMC Genomics 2012;13:526.

[72] Geeleher P, Hartnett L, Egan LJ, Golden A, Raja Ali RA, Seoighe C. Gene-set analysis is severely biased when applied to genome-wide methylation data. Bioinformatics 2013;29(15):1851−7.

[73] Beerman I, Bock C, Garrison BS, Smith ZD, Gu H, Meissner A, et al. Proliferation-dependent alterations of the DNA methylation landscape underlie hematopoietic stem cell aging. Cell Stem Cell 2013;12(4):413−25.

[74] Bock C, Beerman I, Lien WH, Smith ZD, Gu H, Boyle P, et al. DNA methylation dynamics during in vivo differentiation of blood and skin stem cells. Mol Cell 2012;47(4):633−47.

[75]   Shearstone JR, Pop R, Bock C, Boyle P, Meissner A, Socolovsky M. Global DNA demethylation during mouse erythropoiesis *in vivo*. Science 2011;334(6057):799−802.

[76]   Schoofs T, Rohde C, Hebestreit K, Klein HU, Gollner S, Schulze I, et al. DNA methylation changes are a late event in acute promyelocytic leukemia and coincide with loss of transcription factor binding. Blood 2013;121(1):178−87.

[77]   Ley TJ, Ding L, Walter MJ, McLellan MD, Lamprecht T, Larson DE, et al. DNMT3A mutations in acute myeloid leukemia. N Engl J Med 2010;363(25):2424−33.

[78]   Russler-Germain DA, Spencer DH, Young MA, Lamprecht TL, Miller CA, Fulton R, et al. The R882H DNMT3A mutation associated with AML dominantly inhibits wild-type DNMT3A by blocking its ability to form active tetramers. Cancer Cell 2014;25(4):442−54.

[79]   Figueroa ME, Abdel-Wahab O, Lu C, Ward PS, Patel J, Shih A, et al. Leukemic IDH1 and IDH2 mutations result in a hypermethylation phenotype, disrupt TET2 function, and impair hematopoietic differentiation. Cancer Cell 2010;18(6):553−67.

[80]   Booth MJ, Branco MR, Ficz G, Oxley D, Krueger F, Reik W, et al. Quantitative sequencing of 5-methylcytosine and 5-hydroxymethylcytosine at single-base resolution. Science 2012;336(6083):934−7.

[81]   Yu M, Hon GC, Szulwach KE, Song CX, Jin P, Ren B, et al. Tet-assisted bisulfite sequencing of 5-hydroxymethylcytosine. Nat Protoc 2012;7(12):2159−70.

[82]   Song CX, Szulwach KE, Fu Y, Dai Q, Yi C, Li X, et al. Selective chemical labeling reveals the genome-wide distribution of 5-hydroxymethylcytosine. Nat Biotechnol 2011;29(1):68−72.

[83]   Tan L, Xiong L, Xu W, Wu F, Huang N, Xu Y, et al. Genome-wide comparison of DNA hydroxymethylation in mouse embryonic stem cells and neural progenitor cells by a new comparative hMeDIP-seq method. Nucleic Acids Res 2013;41(7):e84.

[84]   Hansen KD, Timp W, Bravo HC, Sabunciyan S, Langmead B, McDonald OG, et al. Increased methylation variation in epigenetic domains across cancer types. Nat Genet 2011;43(8):768−75.

[85]   Bender S, Tang Y, Lindroth AM, Hovestadt V, Jones DT, Kool M, et al. Reduced H3K27me3 and DNA hypomethylation are major drivers of gene expression in K27M mutant pediatric high-grade gliomas. Cancer Cell 2013;24(5):660−72.

[86]   Lewis PW, Muller MM, Koletsky MS, Cordero F, Lin S, Banaszynski LA, et al. Inhibition of PRC2 activity by a gain-of-function H3 mutation found in pediatric glioblastoma. Science 2013;340(6134):857−61.

[87]   Widschwendter M, Fiegl H, Egle D, Mueller-Holzner E, Spizzo G, Marth C, et al. Epigenetic stem cell signature in cancer. Nat Genet 2007;39(2):157−8.

[88]   Avissar-Whiting M, Koestler DC, Houseman EA, Christensen BC, Kelsey KT, Marsit CJ. Polycomb group genes are targets of aberrant DNA methylation in renal cell carcinoma. Epigenetics 2011;6(6):703−9.

[89]   Blankenberg D, Coraor N, Von Kuster G, Taylor J, Nekrutenko A, Galaxy T. Integrating diverse databases into an unified analysis framework: a Galaxy approach. Database 2011;2011:bar011.

[90]   Feng J, Liu T, Qin B, Zhang Y, Liu XS. Identifying ChIP-seq enrichment using MACS. Nat Protoc 2012;7(9):1728−40.

[91]   Zhang Y, Lin YH, Johnson TD, Rozek LS, Sartor MA. PePr: a peak-calling prioritization pipeline to identify consistent or differential peaks from replicated ChIP-Seq data. Bioinformatics 2014;30(18):2568−75.

[92]   Knijnenburg TA, Ramsey SA, Berman BP, Kennedy KA, Smit AF, Wessels LF, et al. Multiscale representation of genomic signals. Nat Methods 2014;11(6):689−94.

Chapter 4

Bioinformatics Tools in Epigenomics Studies

Gustavo F. Bayón[1], Agustín F. Fernández[1] and Mario F. Fraga[2]
[1]*Cancer Epigenetics Laboratory, Institute of Oncology of Asturias, HUCA, Universidad de Oviedo, Oviedo, Spain,* [2]*Nanomaterials and Nanotechnology Research Center, Oviedo, Spain*

Chapter Outline

M. Fraga & A.F. Fernandez (Eds): Epigenomics in Health and Disease.
DOI: http://dx.doi.org/10.1016/B978-0-12-800140-0.00004-2
© 2016 Elsevier Inc. All rights reserved.

73

4.1 INTRODUCTION

Epigenomics, the study of all the epigenetic modifications for an entire genome [1], is similar to other "−omics" approximations in, among other factors, the amount of information produced by the average experiment. Next-generation sequencing (NGS) platforms, along with microarray-based technologies are, in part, responsible for this phenomenon. Faced with these huge amounts of data, researchers must take advantage of the latest computational tools in order to reveal any biologic meaning present in their experimental results.

Bioinformatics tools are crucial these days for the analysis of epigenomics data. The high dimensionality of the data sets makes them too complex to be handled from a classical point of view. It is, then, necessary to address the problem from a number of different perspectives at once. Without forgetting the biologic and global question the researcher wants to answer, new requirements arise in meeting the technologic challenges this kind of data presents.

In this chapter, we describe some of the aspects that define a bioinformatics approach to the analysis of epigenomics data. First, we introduce the major epigenetic modifications that are often studied in this kind of analysis and the possible implications and limitations they impose on the toolchain selected for the analysis. Next, we divide the tools according to goal, level of complexity, and purpose.

It is almost impossible to describe all of the methods available to the researcher, especially in this field of knowledge, where new applications and technologies seem to appear daily. Therefore, we focus on the most common applications a researcher is likely to find when faced with data from an epigenomics experiment. A researcher should be able to choose a coherent approach for the analysis and thus try to discover the biologic meaning underlying the experiment in hand.

4.2 TYPES OF EXPERIMENTS AND DATA CHARACTERISTICS

Despite the vast amount of tools and software designed for data analysis in other fields of Biology, we have to be careful to study the various applications and libraries in order to check if their assumptions are still valid when the data to be analyzed come from epigenetic studies. For example, some of the normalization methods used in the more common gene expression microarrays rely on the assumption that only a tiny fraction of genes are going to be expressed. This, however, is not necessarily true for a general DNA Methylation microarray. This implies that, despite the number of tools, or however promising they might appear, the most important part of the data analyst's job is to fully understand the problem at hand and translate its description to the most appropriate set of questions, tasks, and assumptions.

4.2.1 DNA Methylation

As already described in previous chapters, DNA Methylation is an epigenetic modification, whereby a methyl group is added to a DNA nucleotide. In vertebrates, this modification mainly occurs in the cytosines that precede guanines and are usually referred to as CpGs. Although most CpGs in the genome are methylated, there are CpG-dense regions, called CpG islands, that overlap the promoters of 60−70% of all genes, and it has been suggested that most are unmethylated [2−4].

DNA methylation plays a crucial role in cell differentiation and genomic imprinting, among other processes [5−8]. However, DNA methylation alterations can also occur in various pathologies, developmental processes, and aging [9−11], and it has been widely studied in human cancer, where the hypermethylation of CpG islands in promoter regions is the typical hallmark of a cancer cell [10,12,13].

It is, in fact, probably the most studied epigenetic modification, and so it is easy to find an overwhelming set of tools and libraries aimed at analyzing this kind of data. We, as data analysts, are especially interested in the characteristics of the data generated by the different experimental methods.

4.2.1.1 Whole-Genome Bisulfite Sequencing

A high-throughput, genome-wide analysis of DNA methylation, whole-genome bisulfite sequencing (WGBS) is a technique made up of two steps. First, the input DNA is treated with sodium bisulfite. This converts unmethylated cytosines to uracil, whereas the methylated ones remain as cytosines. The resulting DNA is sequenced on an NGS platform, which usually produces a huge number of short reads [14].

The main characteristic of the data generated from a WGBS experiment is that the sequences generated do not correspond to a real reference genome. In fact, they represent small slices from a virtual, bisulfite-converted reference genome that has a great number of variants. This needs to be taken into account in analyzing the reads, either by using an automatically converted reference genome and manually analyzing the different pathways a cytosine can take, depending on its methylation status, or by employing a specialized alignment tool, such as Bismark or BSMAP.

4.2.1.2 Reduced-Representation Bisulfite Sequencing

WGBS experiments are still very expensive, due, in part, to the number of reads necessary to achieve a given sequencing depth. Reduced-representation bisulfite sequencing (RRBS) is an alternative technique devised to minimize the drawbacks of WGBS while allowing the experimenters to retain the benefits of a sequencing-based approach [15].

RRBS adds an initial stage in which the input DNA is digested using a methylation-insensitive restriction enzyme specially targeted for CpG-enriched fragments. The reduced fragments, which represent around 1% of the genome, still include most promoter regions as well as repeated sequences. This is one of the main advantages of this technique, as repeated sequences are very difficult to analyze in a typical sequencing experiment.

After digestion and generation of fragments, DNA is treated with sodium bisulfite and sequenced on an NGS platform. As in WGBS, tools specially designed to work with bisulfite-treated genomes must be used in order to estimate the amount of DNA methylation at the targeted sites.

4.2.1.3 Methylated DNA Immunoprecipitation

In methylated DNA immunoprecipitation (MeDIP), the input DNA is first sheared into fragments of varying lengths, usually by sonication. After the fragments are generated, an additional immunoprecipitation step is conducted, where methylated fragments are isolated by using a specific antibody. The technique usually follows one of two pathways from this point. The resulting DNA fragments can be sequenced on an NGS platform (MeDIP-seq) or hybridized to a microarray (MeDIP-chip) [16,17]. Another closely related technique is MethylCap [18].

In contrast to WGBS and RRBS, MeDIP-seq generated reads can be aligned to a reference genome without the drawbacks associated with the bisulfite conversion—based experiments. There are several aligners suitable for the task, some of which are reviewed later in the chapter. Downstream analysis of the generated signal and coverage must take into account the possible biases due to the variable density of CpG throughout the whole genome.

In a MeDIP-chip experiment, fractions of DNA before and after the immunoprecipitation step are labeled with Cy5 and Cy3, respectively, and then co-hybridized on a two-channel genomic microarray in order to estimate the relative abundance of methylation at the gene or transcript scale. Tiling arrays can also be used in order to attain more precision or access to previously unexplored regions.

4.2.1.4 DNA Methylation Microarrays

DNA Methylation microarrays constitute a cheaper alternative to the methods mentioned so far while providing good resolution, even at a genome-wide scale in some models. The arrays are designed to work with bisulfite-treated DNA and contain specific oligonucleotides intended to hybridize to the input DNA, depending on its methylation status. Additionally, those oligonucleotides are labeled with fluorescent dyes. This allows us to get a relative measure of methylation for a given CpG [19].

When analyzing data from DNA methylation microarrays, it is important to keep some facts in mind. First, the probes have to be designed in a specific way, preventing cross-hybridizations that would distort the final data. This is especially difficult when trying to study repetitive regions. Additionally, for some probe types, it is assumed that the closest CpGs in distance to the one being studied, the ones that are included in the oligonucleotides designed for that particular probe, have the same methylation status. The data analyst must be aware of these assumptions when trying to extract conclusions from this kind of data.

4.2.2   Histone Modifications

Histones are proteins controlling the structure of the DNA in the nuclei of eukaryotic cells. They package the DNA in structures called "nucleosomes" and play a fundamental role in chromatin regulation. The way in which the DNA is structured by this mechanism, along with other modifications, such as DNA methylation, regulates the expression of genes. This structural regulation also helps package the DNA in a very efficient way [20].

Histones are subject to post-translational modifications, which happen to alter their relationship with other molecules. The set of all possible combinations of these modifications constitutes the histone code, a regulation control code of great complexity [21].

4.2.2.1   Chromatin Immunoprecipitation

Similarly to the MeDIP method, chromatin immunoprecipitation (ChIP) starts by shearing the DNA, either by sonication or enzyme digestion. Then, and depending on the protein interaction with DNA we want to study, the interesting fragments are immunoprecipitated using a specific antibody. With this technique, it is possible to study post-translational modifications on the histone tails, transcription factor binding sites, and methyl-binding domains [22].

After purification of the immunoprecipitated DNA, we can either hybridize the resulting DNA with a tiling microarray (ChIP-chip) or sequence it on a NGS platform (ChIP-seq). The advantages and drawbacks of the two pathways are equivalent to those of MeDIP and are related to the resolution, depth of coverage, and cost of the methods [23]. It is very important in this kind of experiment to generate an input sample, extracted from the DNA before immunoprecipitation, to account for the different biases associated with an ultra-deep sequencing project. Technical replicates are also needed in order to produce high-quality results [24].

4.3 BIOINFORMATICS TOOLS

An incredible amount of work is currently being done in epigenomics data analysis, and a complete review of all the tools, databases, and libraries available to the bioinformatics community, along with examples of use and extensive comparatives, is well beyond the scope of this chapter. Instead, we will attempt to provide an up-to-date description of the most relevant tools being used in epigenomics studies (Figure 4.1).

This section is divided on the basis of the nature of the tools and the way they are used. Online tools are conveniently separated from the tools the user needs to properly install, configure, and use (most of the time as a command line statement, although there are tools with more intuitive interfaces). R/Bioconductor has its own section, as it is becoming the lingua franca for biostatistics and bioinformatics due to its huge amount of packages, broad coverage of topics, many facilities for data visualization, and great expressivity/efficiency ratio.

Moreover, there is an additional section on general concepts that are crucial to the future development of the field. Orthogonal validation is key in the progressive acceptance of agnostic statistical results from next-generation methods. Provenance helps researchers ask themselves questions about their own experiments superseding the concept of a lab notebook in the era of big data and parallel computing. And Open Source publishing of scientific software, something that is currently more common in such fields as Machine Learning or Physics, will add an additional layer of confidence on the bulk results of big scale bioinformatics projects.

4.3.1 R/Bioconductor

The R language [25] was designed for and by statisticians. It is a free software implementation of the awarded S statistical language [26]. It is especially suited for statistical analysis and data mining tasks and has great visualization capabilities. It is a flexible computer language, and embraces several programming paradigms, such as object-oriented design or functional programming. Although it is sometimes considered less efficient than other languages, due to its interpreted nature, this can be easily rectified with the help of its foreign language interface and its combination with more efficient languages, such as C++ (Figure 4.2).

The Comprehensive R Archive Network (CRAN)[1] is a central repository where R-packages are regularly added, providing more functionality to the user. At the time of writing, CRAN holds more than 6800 packages, covering practically all areas of statistical analysis and data visualization.

---

1. The CRAN master site at WU (Wirtschaftsuniversität Wien) in Austria can be found at http://cran.r-project.org/.

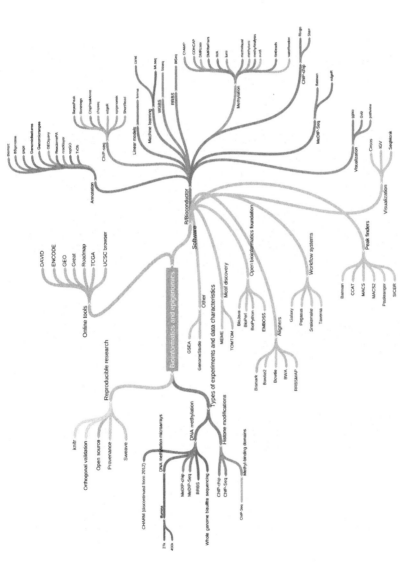

FIGURE 4.1 A nonexhaustive set of bioinformatics workbench tools displayed as leafs in a tree structure. Branches are shown in different colors, resembling the main sections in which the chapter is divided. The heterogeneous nature of these tools represents a challenge for the epigenomics data analyst, especially when facing a new technology or experimental technique.

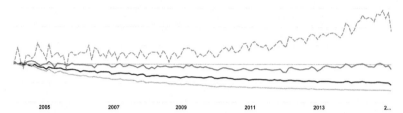

| 2005 | 2007 | 2009 | 2011 | 2013 | 2... |

FIGURE 4.2   Popularity trends of the R (dashed blue), Python (red), and Perl (yellow) programming languages against the general programming category (blue). R is becoming the lingua franca for data analysis and, as such, it is an invaluable tool for the epigenomics researcher.

The package installation process is simple and fast, helping the data analyst to interactively explore his experimental data using the latest algorithms and techniques.

Bioconductor [27] is another repository, containing at the time of writing over 1000 packages related to Bioinformatics and Biostatistics. There are packages for data analysis, annotation, normalization, information related to whole-genome transcripts, and auxiliary definitions for specific technologies. Implementations of some of the latest Bioinformatics algorithms are often released in Bioconductor at the time of their publication. This allows the researcher to try relatively new techniques and algorithms on his data even before the official software from the manufacturers is able to implement them. Bioconductor has packages for nearly all types of techniques, and in the following section, we will review some packages that are especially suited for the work with epigenomics data.

4.3.1.1   Whole-Genome Bisulfite Sequencing

The *bsseq* package allows the researcher to implement the BSmooth pipeline [28] for the processing of WGBS data. It provides functions for reading data from BSmooth and Bismark aligners, and for its posterior downstream analyses. The main advantage of the BSmooth pipeline is that it achieves great accuracy even in low-coverage (4×) situations. It also accounts for biologic replicates, something that is crucial to avoid sequencing biases. Moreover, it is able to detect differentially methylated regions (DMRs), that is, regions of the genome that present homogeneous and different levels of methylation between two conditions. *bsseq* comes with a companion package, named *bsseqData*, which contains an example colon cancer WGBS data set for practice purposes.

4.3.1.2   MeDIP-seq

Data from a MeDIP-seq experiment can be seen as a continuous signal over the whole genome. The coverage of the sequencing experiment is indicative of the methylation status of the loci being interrogated. This is a typical context in several sequencing projects, such as ChIP-seq or even RNA-seq. In fact, some of the tools used for the latter can be used in this context. The package *edgeR* [29] implements a powerful algorithm for the detection of differentially expressed (for RNA-seq) regions, but it can also be used for MeDIP-seq. It uses a negative binomial distribution to model the counts throughout the genome and an Empirical Bayes framework in order to detect the DMRs.

4.3.1.3   Reduced-Representation Bisulfite Sequencing

The *BiSeq* package in Bioconductor is specially designed for working with targeted bisulfite methylation data. As such, it provides classes and functions that facilitate working with this kind of experiments. It allows the user to read data directly produced by the Bismark alignment tool. The package includes functionality to check the quality and coverage of the studied samples, as well as an algorithm [30] for detecting DMR between two groups of samples.

4.3.1.4   ChIP-chip

We have selected two packages that deal with the analysis of ChIP-chip data. Immunoprecipitated DNA is usually analyzed with tiling arrays, where the probes are laid in contiguous regions, allowing the experimenter to work on regions that are sequenced but have unknown functionality. The first package, *Ringo* [31], is specialized on the analysis of data coming from Nimblegen two-color microarrays, although the authors state that it can also work with other manufacturers' products (for Nimblegen one-color arrays, they point the researcher to the *oligo* package [32]), or even custom data structures prepared by the data analyst. It includes functions for data importing, quality assessment, visualization, and identification of genomic regions showing enrichment for the protein under research.

The other package, *Starr* [33], is similar to *Ringo* but is specially designed for the work with Affymetrix arrays. Besides the common importing, quality assessment, and visualization functionalities, it also provides functions for the association of ChIP signals with annotated features or gene expression results, and peak finding with the CMARRT [34] algorithm. Functions in *Starr* can be used to explore data preprocessed by the *Ringo* package, and vice versa.

4.3.1.5   ChIP-seq

As ChIP-seq experiments are quite common nowadays for the study of histone modifications, it is not difficult to find many tools for analysis of this type of data. The alignment of the short reads is usually accomplished by means of tools external to Bioconductor, such as the Bowtie aligner, but there are other packages, such as *ShortRead* [35], that offer several functions for importing, assessing quality, and filtering short reads. Additionally, the *Biostrings* package can be used to align the reads to a reference genome. Although this is a crucial step in terms of performance and external tools are usually preferred, sometimes it is advisable to use the flexible methods implemented in the Bioconductor packages.

The *chipseq* package provides some basic functionality built on top of the *ShortRead* and *IRanges* [36] Bioconductor packages for analyzing coverages, managing reads, and finding peaks with a simple model. If the researcher wants to apply more elaborate models in order to detect enrichment peaks, he or she can use the *edgeR* package, which was designed for RNA-seq experiments but can be used for any data in the form of genome-wide counts.

Another peak finding algorithm is implemented in the *BayesPeak* [37,38] package, where a Markov model is fitted to the counts data across several disjointed regions of the genome called "jobs" and "bins" (each job, about 6 Mb length is divided into bins, 100 bases each in the default configuration). The model is fitted by using Markov Chain Monte Carlo methods, and it produces, for each bin, a posterior probability of being enriched. This probability is not only useful for peak calling but also as a quantitative measure in other downstream analyses.

The *ChIPpeakAnno* [39] package provides functionality for annotating peaks obtained from ChIP-seq experiments. Built on the *BiomaRt*, *BSgenome*, *GO.db*, and other annotation packages, it can help the researcher relate the found peaks with several types of different features, including gene ontologies, motifs, bidirectional promoters or gene bodies. It is a helpful front end for some of the typical annotations performed in most studies. More flexible annotation of the peaks can be achieved by using the functionality implemented in the core Bioconductor packages.

An interesting approach to the study of ChIP-seq results is that of the package *epigenomix* [40]. This package implements an integrative analysis of the results of a ChIP-seq experiment and gene expression data, obtained from RNA-seq or microarray technology. By using either a maximum-likelihood-estimate or a Bayesian-mixture model, it is able to find genes showing significant differences in both types of experiments. This novel approach might prove very useful in uncovering new mechanisms of coordinated control.

4.3.1.6   DNA Methylation

There are numerous packages available to the researcher for DNA methylation analysis, due, in part, to its being the most studied epigenetic mechanism to date. An interesting approach that offers the researcher a complete pipeline for the analysis of Illumina 450K microarrays (although it can also be used for analyzing WGBS and RRBS data) is that of the *RnBeads* package [41]. The functions and classes in this package implement several types of data formats, perform quality control, normalize and filter the resulting methylation data sets, identify and correct for batch effects present in the data, and find differentially methylated probes and regions. This functionality is integrated in a single package with a full-featured output as comprehensive and highly interpretable HTML reports, which makes *RnBeads* a useful solution for an initial overview and quality assessment of a DNA methylation data set.

In the Bioconductor suite, there are several related packages implementing functions and classes for the analysis of DNA methylation data. Such packages as *lumi* [42], *methylumi*, and *minfi* [43] provide functionality for data importing, quality control, pre-processing, and normalization, as well as analysis of differential methylation. The differences among all of these packages lie in the underlying algorithms, and the final decision on which one to use depends on the special needs for the experiment at hand. For example, if the researcher wants to take advantage of the latest implementation of the Houseman method for the estimation of subtype populations in heterogeneous blood samples [44], the *minfi* package could be a good choice. The *IMA* package [45], not in the Bioconductor suite, is very similar to the packages described above. It can also import data directly from *minfi* classes.

The *ChAMP* package [46] also provides useful functionality for the analysis of Infinium 450k DNA methylation microarrays. Apart from the SWAN normalization method [47], it implements peak-based correction [48], and β-mixture quantile normalization [49]. It supports both SVD and ComBat [50] for the study of batch effects in methylation data. Additionally, a Probe Lasso algorithm is included for the identification of DMRs. *ChAMP* was one of the first R-packages to estimate copy number aberrations from signal intensities of Infinium 450k probes.

Another example for an integrated pipeline is the *COHCAP* [51] package. It provides a unified interface for the analysis of single-nucleotide resolution methylation data. This includes not only methylation microarrays but also targeted BS-seq data and similar experimental techniques. The main feature of this package is that it is quite self-contained, including common annotations and the possibility of performing an integrative analysis of methylation and expression data.

The *wateRmelon* package [52] implements 15 data quality metrics and normalization methods that can be used on Infinium 450K microarrays. It is a wrapper package, implementing the methods described on the package authors' paper, based on several well-known characteristics of methylation data, such as their role in X-chromosome inactivation or methylation alterations in genomic imprinted regions. It also provides the necessary functionality for importing methylation data from other packages, such as *minfi*, *methylumi*, or *IMA*.

A simple package, oriented to the discovery of DMRs, is *methyAnalysis*. It is focused toward a chromosome location−based DNA methylation analysis. It also provides functionality for slide window smoothing of DNA methylation levels and the detection and visualization of DMRs. *methVisual*, on the other hand, is a package specially designed for the visualization and interactive, downstream exploration of binary methylation data (the kind of data generated from bisulfite sequencing methods). It provides functions for data importing, methylation calls, co-occurrence and lollipop diagrams, clustering, and correspondence analysis.

The *DMRcate* package is oriented toward the discovery of DMRs from Illumina Infinium 450k samples via kernel density modeling. It relies on other packages for data importing and filtering of bad probes and samples. It provides functionality for filtering probes possibly confounded by SNPs.

The *DMRforPairs* [53] package takes a different approach for the detection of DMRs. It is designed for studies in which only a limited number of unique samples are available. In most traditional methods, the differential methylation status of a given probe is determined by the comparison of a fixed number of observation groups of individuals or biologic samples, whereas *DMRforPairs* builds the comparison groups from genomic ranges with sufficient probes located in close proximity to each other. Optionally, the researcher can also annotate these sets of probes to the same functional class. This type of analysis is useful when measurement density is low. The package also provides functionality for the analysis of the DMRs identified, along with annotation and plotting functions able to integrate the results with data from Ensembl.

4.3.1.7 Linear Models

Data from a DNA methylation microarray experiment is usually presented (after the importing and pre-processing steps) in the form of a numerical matrix. The number of elements in this matrix tends to be huge, especially if we are working with the more dense Infinium 450K microarrays. Moreover, data from other type of experiments, such as WGBS or MeDIP-seq, can be further preprocessed and aggregated with respect to genomic location-dependent units in order to produce a similar matrix.

In this context, it is common to try to fit a linear model [54] for each of the sites or aggregated units to study if there is a significant difference among a fixed number of sample groups or conditions. Linear models, while remaining simple in concept, are very powerful tools and allow the researcher to model the data in very flexible ways.

Usually, the package of choice for the definition and fitting of linear models on microarray data is *limma* [55]. It allows the researcher to fit a linear model to all the probes in a simple and efficient manner by defining a design matrix representing the model to be fit. The design matrix approach allows for a lot of flexibility in the representation of terms, interactions, and nonlinear functions of the original predictors. Most common parametric hypothesis tests, such as t-tests or analysis of variance can be described as linear models and analyzed accordingly.

Limma employs a hierarchical, or empirical Bayes [56,57], framework in order to fit the linear models. Thus, it assumes a prior distribution over the standard deviation of all the variables and estimates the prior parameters from the available data. As a result, the standard deviation estimates for all variables are shrunk toward the population standard deviation, which often leads to better results.

Both β-value and M-value metrics are commonly used to measure methylation levels in microarrays. In order to fit a statistical model that assumes homoscedasticity (as the one used in *limma*), M-value methods are better suited for the task due to their better performance on both highly methylated and unmethylated CpG sites [58].

The *limma* package also provides functionality for contrasts definition, M-estimators for robust fitting of the models [59], and the possibility of defining a trend-based standard deviation prior to distribution, which is very useful when working with methylation data, where the relationship between the mean and standard deviation is usually strong. Moreover, the *limma* documentation is a great resource for studying linear models and experimental design, thanks to its thorough and broad view of the package capabilities.

4.3.1.8 Annotation

We have presented an overview of packages designed for the pre-processing and analysis of epigenomics data, a stage that is usually called "upstream analysis." However, when a set of interesting probes, genes, or functional units are found to be interesting, it is very important to study the relationship between that set and whatever information the researcher might think of as relevant for the condition being studied. For example, once a group of probes has been characterized as being differentially hypermethylated in a condition with respect to a control group, it might be interesting to study if those genes are enriched for a given biologic pathway, or if they are all located on the same chromosomal region. This later stages of the analysis are usually called "downstream analysis," and there are many packages in Bioconductor that can be used for the task.

The package *GenomicRanges* in Bioconductor can be thought of as a cornerstone for nearly all annotation procedures, as it implements a lot of useful and necessary functions for working with genome ranges. It not only provides the functionality for representing and analyzing ranges or alignments, but it also does it in a clear, flexible, and efficient way. This is very useful, especially when used together with the *GenomicFeatures* package, which allows the researcher to easily manage transcript-centric annotations.

The researcher can download the annotations from the UCSC browser or a BiomaRt database. The information in these annotations keeps track of the relationships among genes, transcripts, exons, and coding regions and is easily queried using the methods implemented in the package.

The *FDb.InfiniumMethylation.hg19* package is based on *GenomicFeatures*, and it contains a precompiled version of the Infinium 27k and 450k annotations for the hg19 reference genome. Similarly, the *TxDb.Hsapiens.UCSC.hg19. knownGene* contains the corresponding transcript-oriented annotation information for the hg19 reference genome. Similar packages for different organisms and genomes do exist. The *BSgenome* family of packages is based on the *Biostrings* functionality in order to provide researchers with full genome sequences for different genomes. For example, the *BSgenome.Hsapiens.UCSC. hg19* package contains the full human genome sequence corresponding to the hg19 version.

It is also possible to use Bioconductor packages to access online databases and query data of different types in order to annotate the researcher results. For example, the *GEOquery* [60] package can be used to download a data set stored at the Gene Expression Omnibus (GEO) just by providing its GEO identification number. This allows the researcher to easily download results from other projects in order to compare or try to replicate results.

BiomaRt [61] is a similar package. It provides a common interface for accessing several collections of databases in a uniform and intuitive way. Some examples of *BiomaRt* databases are Ensembl, COSMIC, Uniprot, or dbSNP. Similarly, the *rtracklayer* [62] package gives the researcher access to the information stored in the UCSC browser database. It is important to document precisely where the researcher is getting his information from, as the formats can differ for different databases. For example, genomic locations in the UCSC database are 0-based, whereas the corresponding information in Ensembl is 1-based. Although the Bioconductor packages can deal with these inconsistencies and fix data beforehand, it is convenient to take into account all the characteristics of the data being analyzed.

A common task in downstream analysis is that of testing gene sets or pathways. Once the researcher has selected a subset of relevant units, it is possible to test if, for a group of predefined sets, there is a significant relationship (enrichment, impoverishment or both) with respect to the selected subset. If the predefined sets correspond to manually curated collections of genes or elements with a clear biologic function, the researcher can infer that the selected subset is related to the function.

A quite common collection of curated gene sets is the Gene Ontology (GO) Project [63]. Genes are grouped according to three different biologic domains: molecular function, biologic process, and cellular component. The researcher can use the *topGO* [64] and *GOstats* [65] packages in Bioconductor to perform this kind of gene set testing. *GOstats* provides functions to test for significant gene ontologies conditioned on the GO structure by using a hypergeometric test. On the other hand, *topGO* allows for more testing algorithms and the use of all of the scores of the genes in order to get more accurate results.

KEGG [66,67] and Reactome [68,69] are both biologic pathway databases. The former, although still very popular, has recently changed its licensing model to a more restrictive one. The package *KEGG.db* contains a set of annotation maps with the latest information available before the change of license in 2011. Reactome, in contrast, is an open-source, open-access, manually curated, and peer-reviewed database. The package *ReactomePA* provides functionality for researchers to analyze whether their data sets of interest are associated with any biologic pathway from Reactome. It also includes functions for Gene Set Enrichment Analysis [70,71] association tests, and pathway visualization.

An additional package for gene set enrichment and pathway analysis is *gage* [72]. The package provides functionality to the researcher that is independent of microarray or RNA-seq data attributes, such as sample sizes or experimental designs. It includes functions for multiple analyses in a batch and the combination of results from different sources.

4.3.1.9 Visualization

An intuitive and effective presentation of downstream analysis results is a fundamental part in every research project. In R/Bioconductor, there are a lot of packages for data and results visualization. Some of them are quite general in scope, and others are specialized in the presentation of data obtained from specific methodologies.

The *Gviz* package is based on the idea underlying most genome browsers visual layout of information. Results are arranged on tracks showing data of a specific type and aligned to a reference genome. *Gviz* graphs are usually focused on one chromosome at a time. There are tracks for annotation information, usually defined as simple genomic regions, tracks representing ideograms, tracks able to represent genes and transcripts relationships, and several other types allowing for different kinds of data. The modular design helps the researcher build complex genomic graphs in a flexible way.

The experienced R-developer is usually aware of the *ggplot2* package [73] and its incredible graphical capabilities. It employs a powerful paradigm, the Grammar of Graphics, which allows for succinct and clear ways of defining graphs of great logical complexity. The package *ggbio* [74] is built upon *ggplot2* and extends its functionality for creating genomic graphs. It provides functionality for leveraging Bioconductor core objects, and flexible layout transformations to create not only track-based genomic graphs but also circular or Manhattan plots.

Another interesting visualization package is *pathview* [75]. It offers functions for the integration of data and biologic pathway diagrams. Thus, researchers can provide the results from their experiments, specify a target pathway, and obtain a visual representation of the global alterations across its functional units. The package seamlessly integrates with existing pathway and gene set enrichment tools.

4.3.1.10   Machine Learning

Machine Learning is the field of Artificial Intelligence focused on the study of systems that are able to learn from data. The information extracted from an epigenomics experiment can be used either to build a predictor with good generalization power or to understand the hidden complexities ruling the relationships among the different variables (epigenetic measures and phenotype information) under study. There are several families of algorithms and techniques in the field, classified according to the nature of their inputs, the type of used models and the methodology employed for input data evaluation. A good introduction to Machine Learning can be found in Bishop [76].

The R-package *caret* [77] implements several Machine Learning algorithms, as well as the most common ways of measuring their performance. It also provides functions for fine-tuning the algorithms, data preparation, preprocessing, and feature selection. It is focused on building predictive models, that is, the kind of models that try to build a prediction over a phenotype or clinical outcome from the epigenomics information under study.

In Bioconductor, the *MLSeq* package extends *caret* in order to provide a set of Machine Learning algorithms for the analysis of RNA-seq data. It requires an input data set in the shape of a count matrix, where the number of reads are summarized at a transcript level for all the samples. This setup makes it possible to use the *MLSeq* package also in the context of ChIP-seq data and other types of sequencing experimental results.

4.3.2   Open Bioinformatics Foundation

Although this chapter is highly focused on the R/Bioconductor suite, there are several other alternatives for different programming languages and environments. An interesting set of them is grouped under the Open Bioinformatics Foundation (OBF),[2] a global nonprofit initiative dedicated to the promotion of the Open Source and Open Science philosophies within the biologic science research community. We have selected some of the most popular projects inside the OBF as a representative sample.

4.3.2.1   BioPerl

Perl is an interpreted language, originally designed for language processing tasks. It is used in such contexts as finances and bioinformatics, due to its

---

2. http://www.open-bio.org.

expressive power, fast development cycle, and capabilities of handling big data. It has traditionally been one of the most used programming languages in the bioinformatics research community. The BioPerl project [78] is a set of tools written in Perl, developed and maintained by a community of users, for working with biologic data. The Sanger Center and the University of Nebraska are examples of users of BioPerl. It consists of a central core of packages and several additional sets that extend the functionality of the suite.

4.3.2.2 BioJava

The Java language is omnipresent in the business world of big projects and intensive database queries, where its cross-platform nature and efficiency is very much appreciated, along with the different frameworks that extend its core functionality beyond the initial scope of the language. BioJava [79] is, similar to BioPerl, an open-source project dedicated to building a Java framework for processing biologic data. It provides functionality to the researcher in Bioinformatics through its extensive set of modules. The Java language might not be as widely used as Perl for epigenetic data analysis, but it has a huge developer base worldwide.

4.3.2.3 Biopython

Besides statistical programming languages, such as R, the Perl language was initially the most popular programming language in the Bioinformatics community.[3] Nowadays, despite the unprecedented amount of bioinformatics packages and tools based on Perl, it seems that the language is losing its edge over modern languages, such as Python.[4,5] Python is an interpreted, multiplatform and multiparadigm language that allows the developer to code complex functions in an efficient way while keeping the code clean and legible. There are several implementations of Python, allowing it to work on the .NET platform or use JIT compiling techniques in order to improve efficiency.

The Biopython project [80] is a community-based initiative, similar to the BioPerl and BioJava projects, focused on developing tools in Python for working in bioinformatics research projects. It provides functionality for parsing specific bioinformatics file formats into Python data structures, accessing online databases, performing common operations on sequences, and even interactive tools for performing some of these tasks.

---

3. http://www.bioinformatics.org/poll/.
4. http://www.tiobe.com/index.php/content/paperinfo/tpci/Perl.html.
5. http://www.tiobe.com/index.php/content/paperinfo/tpci/Python.html.

4.3.2.4 *The European Molecular Biology Open Software Suite*

The European Molecular Biology Open Software Suite (EMBOSS) [81] is a free Open Source software suite implementing several tools for the analysis of data from molecular biology experiments. It is a very popular solution, being currently used a lot in production environments. A major new version of EMBOSS is released every year.

EMBOSS includes tools for the alignment of sequences, phylogenetic studies, protein structures, and PCR primer design, among others. The applications are organized into logical groups to help the researcher find the more convenient tool for the task at hand. Although EMBOSS could be more oriented toward the molecular biology researcher, such tools as Cpgplot for the identification and visualization of CpG islands can be used in epigenomics studies.

4.3.3   Online Tools

The low-level approach for the analysis of epigenomics data that is achieved by using a programming language or command line tool provides the most flexible tool set for the researcher. However, the researcher might at some time want a fast and direct answer, especially at the first stages of the downstream part of the processing pipeline, when doing a lot of exploratory data analysis. In this scenario, an online, ready-to-use tool is very valuable for the task at hand. There are many online applications that can be used for epigenomics analyses, sometimes implementing functionality that might seem too complex for a researcher with limited programming skills.

4.3.3.1 *Databases*

Online databases are a great source of data for those researchers who cannot afford to analyze a set of samples big enough to discover any relevant relationship. Additionally, they can be used as validation data sets to check the results obtained from in-house experiments. By sharing results from different types of experiments, integrative analyses can also be easily conducted in order to find more discoveries from already existing data.

The GEO [82,83] is an online public repository where researchers store experimental results of various kinds. It provides the necessary functionality for uploading, storing, and querying result data sets from high-throughput experiments. Data in GEO are organized in terms of platforms, samples, and series. Platforms define technologies used to measure a certain biologic characteristic. Samples are collections of measures as obtained from a given platform. Series represent an experiment, a set of samples that have been measured using the same platform. Each entity in GEO is referenced by a unique identifier, which can be used to download or provide a reference to it.

Another important online database is The Cancer Genome Atlas (TCGA) [84,85]. The main focus of TCGA is to increase our understanding of cancer, by aggregation of different experimental technology sources into a world-wide accessible database. TCGA was started by the National Cancer Institute[6] and the National Human Genome Research Institute[7] in 2006. TCGA is organized into components, such as Genome Characterization Centers and Data Coordinating Centers, which coordinate the experiments on a fixed set of different cancer types over a common and shared clinical information framework. TCGA data types include DNA methylation experimental results for WGBS and Infinium 27k and 450k platforms. TCGA is a very useful source of information to perform integrative studies comprising methylation, CNV (Copy Number Variation), expression, and other kind of experimental data.

The Encyclopedia of DNA Elements (ENCODE),[8] launched in 2003, is a public research consortium focused on the identification of all functional elements in the human genome sequence. It featured an original pilot phase, where only a fraction of the human genome was studied. Results from this initial phase [86,87] showed that the pilot project had been successful and allowed for the project to carry on, this time with the aim of deciphering the whole-genome sequence. Additional results from this second phase can be found in the publication by the ENCODE Project Consortium [88].

From an epigenomics researcher point of view, ENCODE is a valuable resource of information for validation and annotation of experimental results. It contains data from ChIP-seq experiments, DNase-seq, RRBS, and DNA methylation microarrays platforms, among others. Researchers can use these data to annotate their results with respect to the functional elements provided by the project.

The NIH Roadmap Epigenomics Program (Roadmap),[9] initiated in 2008, is another project focused on the study of functional elements of the human genome, this time centered on the in-depth analysis of epigenomics experimental results over a fixed list of prioritized human cell types. The aim of the consortium is to produce genome-wide histone modifications and DNA methylation analysis complete epigenomics mappings. In fact, two complete DNA methylomes (for the H1 hESC and IMR90 cell lines) have already been published [89].

These projects are important not only for the information they contain but also for their capacity to generate new methods and standards. They help state-of-the-art methodologies to move forward in order to improve our knowledge about the elements that rule gene regulation, developing new

6. http://www.cancer.gov.
7. http://www.genome.gov.
8. http://www.genome.gov/encode.
9. http://www.roadmapepigenomics.org.

technologies for epigenomics analyses. Results from the Roadmap project are usually released immediately and stored in the GEO database.

4.3.3.2 Annotation

We have already seen how the packages in R/Bioconductor can help epigenomics researchers to annotate their results in a flexible way, allowing them to extract conclusions about possible findings. However, sometimes it is preferable to use an online tool, providing at least the same functionality but with a more friendly interface for the researcher who does not want to learn advanced coding skills. Moreover, online services eliminate the need to download and process large data sets locally. We have selected a pair of tools for the annotation of results that can be of much interest to the epigenomics researcher.

The Database for Annotation, Visualization, and Integrated Discovery (DAVID) [90,91] comprises a set of tools for the functional annotation of results. The Gene Functional Classification tool is able to group an input set of genes into functional categories, helping the researcher to find relevant biologic meaning in the results he or she has produced.

Probably the most used feature in DAVID is the Functional Annotation tool. From an initial set of genes, it is able to test its enrichment against a big list of functional categories. The researcher can employ DAVID to check for enriched GO ontologies or biologic pathways, along with curated sets related to disease classification or simple genomic regions.

A list of genes is usually supplied by Gene Symbol or Entrez ID, but DAVID is also capable of translating from several different types of manufacturer-dependent probe definitions. It is also possible to specify a background list of genes in order to check for differential enrichment on a smaller and more restricted universe. Results are usually presented to the researcher in the form of a table sorted by the significance level of the enrichment.

A limitation of DAVID when working on epigenomics projects is that the epigenetic mechanisms more often studied, such as DNA methylation and histone modifications, are associated to individual loci and can present an heterogeneous behavior across the same gene. Sometimes, lists of genes are built just by selecting those that contain a probe or given loci. Criteria can vary, but in any step that translates from regions to genes, we are certainly losing information.

GREAT (Genomic Regions Enrichment of Annotations Tool) [92] is a tool similar to DAVID but oriented to the annotation and analysis of genomic regions instead of genes. In this way, no information about the real location of the epigenomics alterations is lost. GREAT assigns first a regulatory domain to each gene, this is, a genomic range where the regulatory elements are directly related to the given gene. Then, it associates the input genomic regions to the regulatory regions they overlap. By using this model, GREAT is able to model accurately those situations where other tools would assign an input region to a very distant gene because it was the nearest one, although it might not be the correct option.

4.3.3.3 Visualization

Good data visualization is crucial for the analysis of experimental results in epigenomics. It is useful not only for the analysis, annotation, and filtering of the genomic regions that the researcher has extracted from his experiment but also in the initial phases of exploratory data analysis, when the visualization of data in the correct context can lead to the formulation of better models and questions. Online visualization tools can help the researcher in performing these tasks. Given that they are usually deployed on a corporation server basis, one of the greatest advantages of online tools is that they can store a lot of information relevant to the questions being asked.

In this sense, the UCSC Genome Browser [93] is a valuable tool. Not only does it provide the user the functionality necessary for the visualization of his results, but it also makes available an impressive list of data resources (including data from the ENCODE project, for example) that can be used throughout the exploration in order to uncover some biologic meaning.

The UCSC Genome Browser can display user data as long as it can be associated with a genomic location. Thus, data extracted from microarray experiments needs to be defined on a set of probes that can be located in the genome. DNA methylation arrays from Illumina fulfill this requisite, since every probe is associated to a unique genome locus. Information tracks, depending on their sizes, may have to be indexed and stored on an external site.

There are several other online visualization tools, such as the Washington University Human Epigenome Browser [94] and EpiExplorer [95], which can be used to visually compare and extract knowledge from several sources of information. The former, visually similar to the UCSC browser, is able to perform statistical tests on selected data, and the latter focuses on the interactive exploration of large-scale genomic data sets and functional region-based association analysis in a way similar to the UCSC Table Browser.

4.3.4   Software Tools

Online tools are ready-to-use, powerful resources for the researcher in epigenomics. However, the functionality available through external libraries, such as Bioconductor or Biopython, gives the developer much more power in terms of flexibility and customization. However, there are still some tasks for which efficiency is the key, and specialized tools written in low-level and mid-level programming languages can excel in this aspect. It is not only a matter of efficiency but also of their configuration possibilities and ability to integrate into existing pipelines of data processing.

This type of tool is usually available in the form of command line applications. Of course, that can be a big drawback for those researchers not used to working in scientific environments, where the UNIX family of operating systems is very common and command line tools are customary. However, they are usually highly customizable and prevent the user from having to learn to code in a lower level programming language.

4.3.4.1 Aligners

For historical reasons, there is a lot of alignment software in the form of external, independent applications. The goal of alignment software is to take a large collection of small fragments (called "reads" or "tags") produced by a sequencer and map them to a reference genome. The mapping needs not to be exact, and different algorithms and strategies can be implemented.

The Burrows−Wheeler Alignment tool [14,96] is an alignment application made up of three algorithms. One of the algorithms is especially suited for Illumina sequences under 100 bp, whereas the others deal better with longer reads. It is based on the Burrows−Wheeler Transform (BWT), a mathematical indexing operation that allows for better performance. In this sense, it is superior to first generation aligners, such as MAQ,[10] which was much slower and did not support gapped alignment of single-end reads.

Bowtie [97] is another short read aligner. It is very fast and uses a BWT in order to use as low memory as possible. The authors estimate a rate of 25 million of 35 bp reads per hour. Output from Bowtie is usually in the form of standard SAM (Sequence Alignment/Map) files, containing the alignments to the reference genome. Bowtie is also the cornerstone of several popular tools, such as TopHat [98] and Cufflinks [99].

Bowtie 2 [100] is the new version of the Bowtie aligner. It is not just a version change, but a major modification of the tool. This makes both aligners still useful, each on the field where it excels. Bowtie 2 implements an FM Index (based on the BWT), and it can cope with gapped, local, and paired alignment modes. It outputs the alignments in the standard SAM file format, as its predecessor. Bowtie 2 is generally faster than Bowtie for longer reads, supports local alignment, and has no upper limit on the size of reads.

For bisulfite-treated genome methylation experiments, specific alignment software is needed. These tools need to take into account the different ways a cytosine can change according to its methylation status. BSMAP [101] is an application, based on a modified version of the SOAP aligner, able to

---

10. http://maq.sourceforge.net.

work with bisulfite-treated data. Its output is a list of reads, along with their matching status and suggested methylation positions for unique matches.

Another aligner able to cope with bisulfite-treated data is Bismark.[11] It is written in Perl, as a wrapper for calling the Bowtie aligner in parallel over the set of possible combinations of cytosines with respect to their methylation status. The output contains all the matched reads and needs to be further processed in order to obtain the methylated CpGs.

RRBSMAP is a special version of the BSMAP tool for RRBS experiments. As it only indexes the genome on the enzyme digestion sites, it reduces the runtime of the analysis. It also avoids the need for any special pre-processing or postprocessing steps. In Ref. [102], the authors compared RRBSMAP to a MAQ-base pipeline and found similar accuracy but a great difference in runtime performance.

4.3.4.2   Peak Finders

The different alignment software tools differ in the algorithms implemented, or their runtime efficiency, more than in their final result. In such sense, they are all somewhat similar and tend to produce a set of alignments with low variability. However, peak finding algorithm and tools are highly heterogeneous and can produce very different results. It is vital for researchers to take this into account when trying to find peaks in their sequencing experimental results. It is usually recommended to run at least two peak finding algorithms over the aligned reads and to filter spurious results that cannot be replicated.

One of the most popular peak finding tools for ChIP-Seq experiments is MACS (Model-based Analysis for Chip-Seq) [103], a model-based approach able to estimate empirically the length of sequenced fragments. MACS uses a Poisson distribution to model the counts for ChIP-Seq reads. It also offers the possibility of running it with or without control samples.

MACS2 is a major reimplementation of MACS. It features a different peak finding algorithm, and it can produce different results. It has not been tested as much as its predecessor, but some of its new options are very useful for specific scenarios of ChIP-Seq experiments. For example, MACS2 has an option to force the algorithm to search for broad peaks, which is something crucial when studying histone modifications that are known to behave that way.

PeakRanger [104] is a suite comprising several utilities for the analysis of sequencing data. It can estimate the signal-to-noise ratio and the library complexity of the alignments and generate coverage reports. With respect to peak finding, the suite provides two different algorithms. The PeakRanger ChIP-Seq peak caller is able to identify enriched genomic regions and

---

11. http://www.bioinformatics.babraham.ac.uk/projects/bismark/.

summits inside those regions. PeakRanger uses a binomial distribution to model the genomic counts. Additionally, the suite implements another peak finding tool, an implementation of the CCAT [105] algorithm, especially tuned to identify broad peaks.

SICER [106] is a clustering approach to peak finding in ChIP-Seq experiments developed in Python. According to its authors, the method is able to identify spatial clusters of signals by pooling the enrichment information from neighboring nucleosomes. This helps improve the signal-to-noise ratio and identify very broad peaks, which otherwise would not be significant on their own.

Batman [16] is a tool specially devised to work with MeDIP-chip and MeDIP-seq data. It uses a Bayesian framework to develop a model of the varying densities of methylated CpGs on MeDIP enrichment signals. The tool is not especially user-friendly, and its results need to be postprocessed in order to account for CNV alterations. It also has strong assumptions about the location of DNA methylation.

4.3.4.3 Motif Discovery

Once the researcher has identified a coherent set of enriched regions, he or she can annotate it and search for functional annotations. The MEME (Multiple EM for Motif Elicitation) suite [107] provides functionality to search for motifs that are very common throughout a set of sequences. It includes several tools, not only related to motif finding but also for the direct assignment of motifs to annotation databases such as The Gene Ontology or JASPAR (transcription factor binding sites motifs) [108].

The main tool in the suite, MEME, tries to identify new motifs on a set of sequences by performing Expectation Maximization on a motif model of a fixed width and sorting the results according to their probability. There is an alternative motif finding tool in the suite, called DREME (Discriminative Regular Expression Motif Elicitation), which follows a discriminative approach and tries to identify motifs that are common on a set of sequences with respect to a fixed control set.

There are also other tools in the suite that can use the motifs discovered by MEME or DREME in order to extract biologic meaning, such as GOMO (Gene Ontology for Motifs), for identifying gene ontology terms enriched for the input motifs, or TomTom, which tries to find similar patterns of enrichment in databases, such as JASPAR.

4.3.4.4 Workflow Definition

A bioinformatics analysis of epigenomics data in the NGS era can easily be made up of several tools, written in different languages, using heterogeneous input and output formats. Moreover, some of the steps in the processing pipeline can have a long runtime and should be executed only if necessary.

Ideally, the whole process should take advantage of the parallel capabilities of the platform the pipeline is running on.

Fortunately, there are tools that can help the researcher define the flow of data, the relationships among the different steps in the pipeline, and even the amount of resources that can be used by each step. Some of these workflow definition tools are institution-based initiatives, aimed at improving the definition of processes inside a big organization. However, a single researcher can find these tools useful in order to save resources and improve the productivity of his experiments.

A typical example of a workflow definition tool is Galaxy [109−111], a web-oriented tool offering the researcher a set of the most common operations on bioinformatics data. It can be used interactively, but it also allows definition of workflows, which can be parameterized, stored, and reused. Galaxy is highly configurable, and it is not uncommon to find customized versions of it deployed in different research groups.

Taverna [112] and Pegasus [113] are two powerful tools for the definition of pipelines. The first one is highly visual and relies a lot on the definition of functionalities via web services. It can also define tasks that interface with a local R/Bioconductor installation. Pegasus, on the other hand, is very powerful, but it is more challenging to configure. It is not visually oriented, and its goal is to let scientific workflows run on different scenarios, including clusters, desktops, and clouds.

The RUbioSeq [114] software is another tool for the definition of pipelines which includes out-of-the-box, state-of-the-art workflows for the analysis of SNV, CNV, and Bisulfite-seq data analysis. The methylation pipeline is based on the Bismark aligner. RUbioSeq can also be obtained on a live DVD, which makes its functionalities ready to use with minimal effort by any researcher.

In simple scenarios, it is also common to use the standard Make tool,[12] originally intended to help build applications from source code in order to define and implement processing pipelines. Despite its flexibility, it has some drawbacks that can make it difficult to express some complex tasks. The Snakemake tool [115] builds upon the Makefile philosophy and is especially suited for bioinformatics workflows. Developed in Python, it offers task definition, resource limitation, parallel execution, R-integration, the possibility of running Python code directly on the rules, and many more advantages, which make Snakemake a powerful tool for small research groups.

4.3.4.5 Visualization

Local, non-online, tools for the visualization of information are useful for the researcher who does want a fast visual analysis of the experimental

---

12. http://www.gnu.org/software/make/.

results and does not have any special needs on accessing large data, such as the ENCODE information. In this sense, IGV (Integrative Genomics Viewer) [116,117] is a very powerful visualization program. Developed in the Broad Institute, it could be resembled as a local version of the UCSC Genome Browser. Although it can import data from servers, its main function is to rapidly visualize the results from experiments.

SeqMonk[13] is another visualization tool that is focused on mapped data from sequencing experiments. It can import data in the standard SAM/BAM formats, as well as the direct output from Bowtie or other aligners. It also allows for creation of reports containing data and genome annotations, as well as statistical analyses on the data in order to find regions of interest.

Circos [118] is a famous package for the visualization of information in a circular layout. It is not only limited to genomic information, but it can also display any data relationship that can be layered in a table format. Circos is focused on the creation of great quality figures, and it is controlled by plain text configuration files, which makes it possible to include Circos diagram generation in any pipeline definition system (Figure 4.3).

4.4 REPRODUCIBLE RESEARCH

Nowadays, data from bioinformatics experiments is being generated at an increasingly higher rate. The algorithms implemented for the analysis of this huge amount of data are increasing in complexity, and a new generation of *in silico* researchers are taking on ever more important roles in projects. This introduces more complexity when trying to replicate experiments or use existing data. We think that some basic steps should be taken to guarantee the reproducibility of computational analysis tasks and prevent false discoveries on the basis of hidden information or incorrectly specified algorithms.

4.4.1 Literate Programming

Literate programming [119] is an approach focused on the creation of self-contained documents made up of code and a natural language explanation regarding the code intention. In this way, there is no separation between the data processing and the conclusions extracted by the researcher, who can also be sure that the results in the document are up to date, portable, and reproducible. Literate programs flow is similar to that of human reasoning. There are tools dependent on a specific programming language and others that are language agnostic.

A typical tool found in bioinformatics projects is Sweave [120], which is built upon the *noweb*-literate programming system [121]. It allows the researcher to include R-code inside LaTeX documents. The data processing

---

13. http://www.bioinformatics.babraham.ac.uk/projects/seqmonk/.

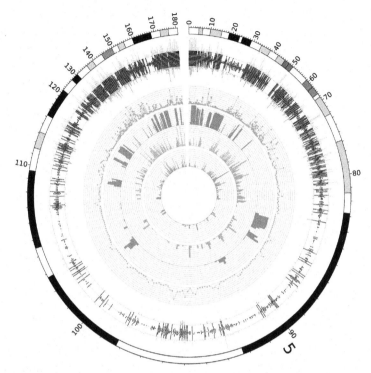

FIGURE 4.3   An example figure produced by the Circos visualization tool in the context of an epigenetics data analysis project. Outer tracks, based on tile layouts, are used in this figure to describe the relative positions of CpG sites. The innermost tracks, on the other hand, use line and histogram types of charts to represent the averaged peak scores for ENCODE Broad Histone data.

is performed when compiling the document with Sweave, and the results are then mixed with the LaTeX source code in order to produce a self-contained document. knitr [122–124] is a revision of Sweave, more modular and extensible and able to work with more formats, such as Markdown.[14]

4.4.2   Orthogonal Validation

The term *orthogonal* applied to the validation of results can take different meanings, depending on the field of knowledge we are working in. Here, Orthogonal Validation is defined as the practice of validating biologic findings from bioinformatics data analysis projects by using different and independent (hence orthogonal) techniques and methods. Different computational biology algorithms and tools can produce different results, and it is very important to double-check our conclusions to prevent methodology biases.

---

14. http://daringfireball.net/projects/markdown/.

Bioinformatics results should, whenever possible, be validated and backed up by wet-lab procedures. This ensures that the biologic findings are reproducible. But, when looking from a bioinformatics point of view, it is not only important to validate the results using the former techniques but also crucial to use different tools for each task in order to eliminate possible biases. A great example of this situation is that of peak finding in ChIP-Seq experimental projects. Peak finders often produce different sets of peaks, and researchers often focus only on those results found to be coherent.

4.4.3 Provenance

"Provenance" refers to the complete history of a data object. In the case of bioinformatics data analysis pipelines, it is important to gather all possible information about a data set produced by the pipeline or an intermediate task. For example, if we have a set of interesting DNA methylation probe IDs, we would like to be able to know not only the data set and the models that have been employed to produce them but also the normalization method used to generate the data set, the creation dates, or the people in charge of the different task definitions.

Data sets are reliable and hence are of good provenance when they are documented in such a level of detail that it allows another researcher to be able to reproduce the complete experiment. Provenance is an active field of research in data science. Some systems, such as Taverna, already implement provenance capabilities in their workflows. There is even a standard model [125] focused on allowing the exchange of provenance information among different systems.

4.4.4 Open Source

In several other fields of knowledge, such as Machine Learning or Physics, it is now customary to include the source code of every application or script that has been used in order to generate results. We think this is a very good practice to enable experimental reproducibility. It is hard for a researcher to discuss or even reimplement the results from an experiment if he or she has no exact knowledge of the computational methods employed. Even when a statistical method or analysis algorithm is well described in the results, there is always the possibility of a hidden bug or mistake driving the conclusions.

The philosophies of Open Source[15] and Free Software[16] can help the research community in bioinformatics to achieve experimental reproducibility while keeping control of the produced code and ensuring that any benefit or technology advance is returned to the community. Some journals, such as

---

15. http://opensource.org.
16. http://www.fsf.org.

Bioinformatics or *Genome Biology*, encourage authors to publish a full working implementation of the published algorithm under an open source or free license.

4.5 CONCLUSION

In this chapter, we gave a brief description of the enormous landscape of bioinformatics tools available to the epigenomics researcher. The tools and applications presented are only a small part of the techniques available to analyze experimental results. Moreover, we did not go into detail about their customization parameters, nor did we make extensive comparisons of the different alternatives, since such information could fill a book on its own.

However, we believe we have introduced a handy set of tools and methods available to researchers, whether they are computational data analysts trying to uncover significant patterns in NGS data, or biologists who want to try a complex model on their microarray experiment without learning to code complicated algorithms in a programming language. After being introduced to this huge toolbox, researchers are given a starting point to select the approach they think will fit better to their experimental setup.

The amount of information produced in epigenomics experiments is increasing continuously, and the research community is fully aware of that. Computational tools, languages, applications, and paradigms are already part of the Biology experimental framework. The future of the field is clearly multidisciplinary, involving teams of heterogeneous profiles in order to uncover new biologic meaning by means of orthogonal methods and technologies.

REFERENCES

[1] Beck S, Olek, Walter J. From genomics to epigenomics: a loftier view of life. Nat Biotechnol 1999;17(12):1144.

[2] Bird A. DNA methylation patterns and epigenetic memory. Genes Dev 2002;6−21.

[3] Bernstein BE, Meissner A, Lander ES. The mammalian epigenome. Cell 2007;669−81.

[4] Illingworth RS, Bird AP. CpG islands—"a rough guide". FEBS Lett 2009;583:1713−20.

[5] Hirabayashi Y, Gotoh Y. Epigenetic control of neural precursor cell fate during development. Nat Rev Neurosci 2010;11:377−88.

[6] Calvanese V, Fernández AF, Urdinguio RG, Suárez-Alvarez B, Mangas C, Pérez-García V, et al. A promoter DNA demethylation landscape of human hematopoietic differentiation. Nucleic Acids Res 2012;40:116−31.

[7] Feinberg AP, Cui H, Ohlsson R. DNA methylation and genomic imprinting: insights from cancer into epigenetic mechanisms. Semin Cancer Biol 2002;12:389−98.

[8] Payer B, Lee JT. X chromosome dosage compensation: how mammals keep the balance. Annu Rev Genet 2008;42:733−72.

[9] Urdinguio RG, Sanchez-Mut JV, Esteller M. Epigenetic mechanisms in neurological diseases: genes, syndromes, and therapies. Lancet Neurol 2009;1056−72.

[10] Fernandez AF, Assenov Y, Martin-Subero JI, Balint B, Siebert R, Taniguchi H, et al. A DNA methylation fingerprint of 1628 human samples. Genome Res 2012;22:407−19.

[11] Fernández AF, Bayón GF, Urdinguio RG, Toraño EG, García MG, Carella A, et al. H3K4me1 marks DNA regions hypomethylated during aging in human stem and differentiated cells. Genome Res 2015;25(1):27−40.

[12] Esteller M. Epigenetics in cancer. N Engl J Med 2008;358(11):1148−59.

[13] Jones PA, Baylin SB. The fundamental role of epigenetic events in cancer. Nat Rev Genet 2002;3:415−28.

[14] Li H, Durbin R. Fast and accurate long-read alignment with Burrows-Wheeler transform. Bioinformatics 2010;26:589−95.

[15] Gu H, Bock C, Mikkelsen TS, Jäger N, Smith ZD, Tomazou E, et al. Genome-scale DNA methylation mapping of clinical samples at single-nucleotide resolution. Nat Methods 2010;7:133−6.

[16] Down TA, Rakyan VK, Turner DJ, Flicek P, Li H, Kulesha E, et al. A Bayesian deconvolution strategy for immunoprecipitation-based DNA methylome analysis. Nat Biotechnol 2008;26:779−85.

[17] Weber M, Hellmann I, Stadler MB, Ramos L, Pääbo S, Rebhan M, et al. Distribution, silencing potential and evolutionary impact of promoter DNA methylation in the human genome. Nat Genet 2007;39:457−66.

[18] Brinkman AB, Simmer F, Ma K, Kaan A, Zhu J, Stunnenberg HG. Whole-genome {DNA} methylation profiling using MethylCap-seq. Methods 2010;52(3):232−6.

[19] Bibikova M, Barnes B, Tsan C, Ho V, Klotzle B, Le JM, et al. High density DNA methylation array with single CpG site resolution. Genomics 2011;98:288−95.

[20] Fraga MF, Esteller M. Towards the human cancer epigenome: a first draft of histone modifications. Cell Cycle 2005;1377−81.

[21] Jenuwein T, Allis CD. Translating the histone code. Science 2001;293:1074−80.

[22] Massie CE, Mills IG. ChIPping away at gene regulation. EMBO Rep 2008;9:337−43.

[23] Ho JWK, Bishop E, Karchenko PV, Nègre N, White KP, Park PJ. ChIP-chip versus ChIP-seq: lessons for experimental design and data analysis. BMC Genomics 2011;12:134.

[24] Park PJ. ChIP-seq: advantages and challenges of a maturing technology. Nat Rev Genet 2009;10(10):669−80.

[25] R Core Team. R: a language and environment for statistical computing [internet]. Vienna, Austria: R Foundation for Statistical Computing; 2014. Available from: <http://www.R-project.org/>.

[26] Chambers JM. Programming with data: a guide to the S language. New York: Springer-Verlag; 2004.

[27] Gentleman RC, Carey VJ, Bates DM, et al. Bioconductor: open software development for computational biology and bioinformatics. Genome Biol 2004;5:R80. Available from: <http://genomebiology.com/2004/5/10/R80>.

[28] Hansen KD, Langmead B, Irizarry RA. BSmooth: from whole genome bisulfite sequencing reads to differentially methylated regions. Genome Biol 2012;13:R83.

[29] Robinson MD, McCarthy DJ, Smyth GK. edgeR: a bioconductor package for differential expression analysis of digital gene expression data. Bioinformatics 2010;26:139−40.

[30] Hebestreit K, Dugas M, Klein HU. Detection of significantly differentially methylated regions in targeted bisulfite sequencing data. Bioinformatics 2013;1647−53.

[31] Toedling J, Skylar O, Krueger T, Fischer JJ, Sperling S, Huber W. Ringo−an R/Bioconductor package for analyzing ChIP-chip readouts. BMC Bioinformatics 2007;8:221.

[32] Carvalho BS, Irizarry RA. A framework for oligonucleotide microarray preprocessing. Bioinformatics 2010;26:2363−7.

[33] Zacher B, Kuan PF, Tresch A. Starr: Simple Tiling ARRay analysis of Affymetrix ChIP-chip data. BMC Bioinformatics 2010;11:194.

[34] Kuan PF, Chun H, Keleş S. CMARRT: a tool for the analysis of ChIP-chip data from tiling arrays by incorporating the correlation structure. Pacific symposium on biocomputing pacific symposium on biocomputing; 2008. pp. 515−26.

[35] Morgan M, Anders S, Lawrence M, Aboyoun P, Pagès H. Gentleman R. ShortRead: a bioconductor package for input, quality assessment and exploration of high-throughput sequence data. Bioinformatics 2009;25:2607−8.

[36] Lawrence M, Huber W, Pagès H, Aboyoun P, Carlson M, Gentleman R, et al. Software for computing and annotating genomic ranges. PLoS Comput Biol 2013;9.

[37] Spyrou C, Stark R, Lynch AG, Tavaré S. BayesPeak: Bayesian analysis of ChIP-seq data. BMC Bioinformatics 2009;10:299.

[38] Cairns J, Spyrou C, Stark R, Smith ML, Lynch AG, Tavaré S. BayesPeak—an R package for analysing ChIP-seq data. Bioinformatics 2011;27:713−14.

[39] Zhu LJ, Gazin C, Lawson ND, Pagès H, Lin SM, Lapointe DS, et al. ChIPpeakAnno: a Bioconductor package to annotate ChIP-seq and ChIP-chip data. BMC Bioinformatics 2010;11:237.

[40] Klein H-U, Schäfer M, Porse BT, Hasemann MS, Ickstadt K, Dugas M. Integrative analysis of histone ChIP-seq and transcription data using Bayesian mixture models. Bioinformatics 2014;1−9.

[41] Assenov Y, Muller F, Lutsik P, Walter J, Lengauer T, Bock C. Comprehensive analysis of DNA methylation data with RnBeads. Nat Methods 2014;11(11):1138−40.

[42] Du P, Kibbe WA, Lin SM. lumi: a pipeline for processing Illumina microarray. Bioinformatics 2008;24:1547−8.

[43] Aryee MJ, Jaffe AE, Corrada-Bravo H, Ladd-Acosta C, Feinberg AP, Hansen KD, et al. Minfi: a flexible and comprehensive bioconductor package for the analysis of Infinium DNA methylation microarrays. Bioinformatics 2014;1−8.

[44] Houseman EA, Accomando WP, Koestler DC, Christensen BC, Marsit CJ, Nelson HH, et al. DNA methylation arrays as surrogate measures of cell mixture distribution. BMC Bioinformatics 2012;13:86.

[45] Wang D, Yan L, Hu Q, Sucheston LE, Higgins MJ, Ambrosone CB, et al. IMA: an R package for high-throughput analysis of Illumina's 450K Infinium methylation data. Bioinformatics 2012;28:729−30.

[46] Morris TJ, Butcher LM, Feber A, Teschendorff AE, Chakravarthy AR, Wojdacz TK, et al. ChAMP: 450k chip analysis methylation pipeline. Bioinformatics 2014;30:428−30.

[47] Makismovic J, Gordon L, Oshlack A. SWAN: subset-quantile within array normalization for illumina infinium HumanMethylation450 BeadChips. Genome Biol 2012;R44.

[48] Dedeurwaerder S, Defrance M, Calonne E, Denis H, Sotiriou C, Fuks F. Evaluation of the infinium methylation 450K technology. Epigenomics 2011;771−84.

[49] Teschendorff AE, Marabita F, Lechner M, Bartlett T, Tegner J, Gomez-Cabrero D, et al. A Beta-mixture quantile normalisation method for correcting probe design bias in illumina infinium 450k DNA methylation data. Bioinformatics 2012;1−8.

[50] Johnson WE, Li C, Rabinovic A. Adjusting batch effects in microarray expression data using empirical Bayes methods. Biostatistics 2007;8:118−27.

[51] Warden CD, Lee H, Tompkins JD, Li X, Wang C, Riggs AD, et al. COHCAP: an integrative genomic pipeline for single-nucleotide resolution DNA methylation analysis. Nucleic Acids Res 2013;41.

[52] Pidsley R, Y Wong CC, Volta M, Lunnon K, Mill J, Schalkwyk LC. A data-driven approach to preprocessing Illumina 450K methylation array data. BMC Genomics 2013;14:293.

[53] Rijlaarsdam M, Zwan Y, van der, Dorssers L, Looijenga L. DMR2 + : identifying differentially methylated regions between unique samples using array based methylation profiles. BMC Bioinformatics 2014;15(1):141. Available from: <http://www.biomedcentral.com/1471-2105/15/141>

[54] Agresti A. Categorical data analysis [internet]; 2002. Available from: <http://www.loc.gov/catdir/toc/wiley024/2002068982.html>.

[55] Smyth GK. Limma: linear models for microarray data. In: Gentleman R, Carey VJ, Huber W, Irizarry RA, Dudoit S, editors. Bioinformatics and computational biology solutions using R and bioconductor [internet]. New York: Springer; 2005. p. 397−420. Available from: < http://link.springer.com/10.1007/0-387-29362-0 > , 978-0-387-25146-2.

[56] Casella G. An introduction to empirical bayes data analysis. Am Stat 1985;39:83−7.

[57] Efron B, Tibshirani R, Storey JD, Tusher V. Empirical bayes analysis of a microarray experiment. J Am Stat Assoc 2001;1151−60.

[58] Du P, Zhang X, Huang C-C, Jafari N, Kibbe WA, Hou L, et al. Comparison of beta-value and M-value methods for quantifying methylation levels by microarray analysis. BMC Bioinformatics 2010;11.

[59] Huber PJ. Robust statistics. Statistics 2004;60:1−11.

[60] Sean D, Meltzer PS. GEOquery: a bridge between the Gene Expression Omnibus (GEO) and BioConductor. Bioinformatics 2007;23:1846−7.

[61] Smedley D, Haider S, Ballester B, Holland R, London D, Thorisson G, et al. BioMart−biological queries made easy. BMC Genomics 2009;10:22.

[62] Lawrence M, Gentleman R, Carey V. rtracklayer: an R package for interfacing with genome browsers. Bioinformatics 2009;25:1841−2.

[63] Ashburner M, Ball CA, Blake JA, Botstein D, Butler H, Cherry JM, et al. Gene ontology: tool for the unification of biology. The Gene Ontology Consortium. Nat Genet 2000;25:25−9.

[64] Alexa A, Rahnenführer J, Lengauer T. Improved scoring of functional groups from gene expression data by decorrelating GO graph structure. Bioinformatics 2006;22:1600−7.

[65] Falcon S, Gentleman R. Using GOstats to test gene lists for GO term association. Bioinformatics 2007;23:257−8.

[66] Ogata H, Goto S, Sato K, Fujibuchi W, Bono H, Kanehisa M. KEGG: Kyoto encyclopedia of genes and genomes. Nucleic Acids Res 1999;29−34.

[67] Kanehisa M, Goto S, Sato Y, Kawashima M, Furumichi M, Tanabe M. Data, information, knowledge and principle: back to metabolism in KEGG. Nucleic Acids Res 2014;42.

[68] Croft D, O'Kelly G, Wu G, Haw R, Gillespie M, Matthews L, et al. Reactome: a database of reactions, pathways and biological processes. Nucleic Acids Res 2011;39.

[69] Milacic M, Haw R, Rothfels K, Wu G, Croft D, Hermjakob H, et al. Annotating cancer variants and anti-cancer therapeutics in reactome. Cancers 2012;4:1180−211.

[70] Mootha VK, Lindgren CM, Eriksson K-F, Subramanian A, Sihag S, Lehar J, et al. PGC-1alpha-responsive genes involved in oxidative phosphorylation are coordinately downregulated in human diabetes. Nat Genet 2003;34:267−73.

[71] Subramanian A, Subramanian A, Tamayo P, Tamayo P, Mootha VK, Mootha VK, et al. Gene set enrichment analysis: a knowledge-based approach for interpreting genome-wide expression profiles. Proc Natl Acad Sci USA 2005;102:15545−50.

[72] Luo W, Friedman MS, Shedden K, Hankenson KD, Woolf PJ. GAGE: generally applicable gene set enrichment for pathway analysis. BMC Bioinformatics 2009;10:161.

[73] Wickham H. ggplot2: elegant graphics for data analysis [internet]. New York, NY: Springer; 2009. Available from: <http://had.co.nz/ggplot2/book>.

[74] Yin T, Cook D, Lawrence M. ggbio: an R package for extending the grammar of graphics for genomic data. Genome Biol 2012;R77.

[75] Luo W, Brouwer C. Pathview: an R/Bioconductor package for pathway-based data integration and visualization. Bioinformatics 2013;29:1830−1.

[76] Bishop CM. In: Jordan M, Kleinberg J, Schölkopf B, editors. Pattern recognition and machine learning. New York: Springer-Verlag; 2006.

[77] Kuhn M. Building predictive models in R using the caret Package. J Stat Softw 2008;28:1−26.

[78] Stajich JE, Block D, Boulez K, Brenner SE, Chervitz SA, Dagdigian C, et al. The Bioperl toolkit: Perl modules for the life sciences. Genome Res 2002;12(10):1611−18.

[79] Prlić A, Yates A, Bliven SE, Rose PW, Jacobsen J, Troshin PV, et al. BioJava: an open-source framework for bioinformatics in 2012. Bioinformatics 2012;28:2693−5.

[80] Cock PJA, Antao T, Chang JT, Chapman BA, Cox CJ, Dalke A, et al. Biopython: freely available Python tools for computational molecular biology and bioinformatics. Bioinformatics 2009;25:1422−3.

[81] Rice P, Longden I, Bleasby A. EMBOSS: the European Molecular Biology Open Software Suite. Trends Genet 2000;16:276−7.

[82] Edgar R, Domrachev M, Lash AE. Gene Expression Omnibus: NCBI gene expression and hybridization array data repository. Nucleic Acids Res 2002;30:207−10.

[83] Barrett T, Wilhite SE, Ledoux P, Evangelista C, Kim IF, Tomashevsky M, et al. NCBI GEO: archive for functional genomics data sets—update. Nucleic Acids Res 2013;41.

[84] Stratton MR, Campbell PJ, Futreal PA. The Cancer Genome Atlas. Nature 2009; 719−24.

[85] Weinstein JN, Collisson EA, Mills GB, Shaw KRM, Ozenberger BA, Ellrott K, et al. The Cancer Genome Atlas Pan-Cancer analysis project. Nat Genet 2013;45:1113−20.

[86] The ENCODE Project Consortium. Identification and analysis of functional elements in 1% of the human genome by the ENCODE pilot project. Nature 2007;447:799−816.

[87] Weinstock GM. ENCODE: more genomic empowerment. Genome Res 2007;17:667−8.

[88] The ENCODE Project Consortium. An integrated encyclopedia of DNA elements in the human genome. Nature 2012;489:57−74.

[89] Lister R, Pelizzola M, Dowen RH, Hawkins RD, Hon G, Tonti-Filippini J, et al. Human DNA methylomes at base resolution show widespread epigenomic differences. Nature 2009;462:315−22.

[90] Huang DW, Sherman BT, Lempicki RA. Systematic and integrative analysis of large gene lists using DAVID bioinformatics resources. Nat Protoc 2009;4:44−57.

[91] Huang DW, Sherman BT, Lempicki RA. Bioinformatics enrichment tools: paths toward the comprehensive functional analysis of large gene lists. Nucleic Acids Res 2009;37:1−13.

[92] McLean CY, Bristor D, Hiller M, Clarke SL, Schaar BT, Lowe CB, et al. GREAT improves functional interpretation of *cis*-regulatory regions. Nat Biotechnol 2010;28:495−501.

[93] Kent WJ, Sugnet CW, Furey TS, Roskin KM, Pringle TH, Zahler AM, et al. The Human Genome Browser at UCSC. Genome Res 2002;996−1006.

[94] Zhou X, Maricque B, Xie M, Li D, Sundaram V, Martin EA, et al. The Human Epigenome Browser at Washington University. Nat Methods 2011;8(12):989−90.

[95] Halachev K, Bast H, Albrecht F, Lengauer T, Bock C. EpiExplorer: live exploration and global analysis of large epigenomic datasets. Genome Biol 2012;13(10):R96.

[96] Li H, Durbin R. Fast and accurate short read alignment with Burrows-Wheeler transform. Bioinformatics 2009;25:1754−60.

[97] Langmead B, Trapnell C, Pop M, Salzberg S. Ultrafast and memory-efficient alignment of short DNA sequences to the human genome. Genome Biol 2009;10(3):R25.

[98] Kim D, Pertea G, Trapnell C, Pimentel H, Kelley R, Salzberg SL. TopHat2: accurate alignment of transcriptomes in the presence of insertions, deletions and gene fusions. Genome Biol 2013;14:R36.

[99] Trapnell C, Williams BA, Pertea G, Mortazavi A, Kwan G, van Baren MJ, et al. Transcript assembly and quantification by RNA-Seq reveals unannotated transcripts and isoform switching during cell differentiation. Nat Biotechnol 2010;28:511−15.

[100] Langmead B, Salzberg SL. Fast gapped-read alignment with Bowtie 2. Nat Methods 2012;357−9.

[101] Xi Y, Li W. BSMAP: whole-genome bisulfite sequence MAPping program. BMC Bioinformatics 2009;10:232.

[102] Xi Y, Bock C, Müller F, Sun D, Meissner A, Li W. RRBSMAP: a fast, accurate and user-friendly alignment tool for reduced representation bisulfite sequencing. Bioinformatics 2012;28:430−2.

[103] Zhang Y, Liu T, Meyer CA, Eeckhoute J, Johnson DS, Bernstein BE, et al. Model-based analysis of ChIP-Seq (MACS). Genome Biol 2008;9:R137.

[104] Feng X, Grossman R, Stein L. PeakRanger: a cloud-enabled peak caller for ChIP-seq data. BMC Bioinformatics 2011;12:139.

[105] Xu H, Handoko L, Wei X, Ye C, Sheng J, Wei CL, et al. A signal-noise model for significance analysis of ChIP-seq with negative control. Bioinformatics 2010;26: 1199−204.

[106] Zang C, Schones DE, Zeng C, Cui K, Zhao K, Peng W. A clustering approach for identification of enriched domains from histone modification ChIP-Seq data. Bioinformatics 2009;25:1952−8.

[107] Bailey TL, Boden M, Buske FA, Frith M, Grant CE, Clementi L, et al. MEME SUITE: tools for motif discovery and searching. Nucleic Acids Res 2009;37.

[108] Mathelier A, Zhao X, Zhang AW, Parcy F, Worsley-Hunt R, Arenillas DJ, et al. JASPAR 2014: an extensively expanded and updated open-access database of transcription factor binding profiles. Nucleic Acids Res 2014;42.

[109] Giardine B, Riemer C, Hardison RC, Burhans R, Elnitski L, Shah P, et al. Galaxy: a platform for interactive large-scale genome analysis. Genome research. Cold Spring Harbor Lab 2005;15(10):1451−5.

[110] Goecks J, Nekrutenko A, Taylor J, Team TG. Galaxy: a comprehensive approach for supporting accessible, reproducible, and transparent computational research in the life sciences. Genome Biol 2010;11(8):R86.

[111] Blankenberg D, Kuster GV, Coraor N, Ananda G, Lazarus R, Mangan M, et al. Galaxy: a web-based genome analysis tool for experimentalists. Curr Protoc Mol Biol 2010; John Wiley & Sons, Inc; Chapter 19: Unit 19.10.1-21.

[112] Wolstencroft K, Haines R, Fellows D, Williams A, Withers D, Owen S, et al. The Taverna workflow suite: designing and executing workflows of Web Services on the desktop, web or in the cloud. Nucleic Acids Res 2013;41.

[113] Deelman E, Singh G, Su M-h, Blythe J, Gil Y, Kesselman C, et al. Pegasus: a framework for mapping complex scientific workflows onto distributed systems. Sci Program J 2005;13:219−37.

[114] Rubio-Camarillo M, Gómez-López G, Fernández JM, Valencia A, Pisano DG. RUbioSeq: a suite of parallelized pipelines to automate exome variation and bisulfite-seq analyses. Bioinformatics 2013;1687—9.

[115] Köster J, Rahmann S. Snakemake—a scalable bioinformatics workflow engine. Bioinformatics 2012;28:2520—2.

[116] Robinson JT, Thorvaldsdóttir H, Winckler W, Guttman M, Lander ES, Getz G, et al. Integrative genomics viewer. Nat Biotechnol 2011;24—6.

[117] Thorvaldsdóttir H, Robinson JT, Mesirov JP. Integrative Genomics Viewer (IGV): high-performance genomics data visualization and exploration. Brief Bioinform 2013;14:178—92.

[118] Krzywinski M, Schein J, Birol I, Connors J, Gascoyne R, Horsman D, et al. Circos: an information aesthetic for comparative genomics. Genome Res 2009;19:1639—45.

[119] Knuth DE. Literate programming. Comput J 1984;27(2):97—111.

[120] Leisch F. Sweave: dynamic generation of statistical reports using literate data analysis. In: COMPSTAT 2002 proceedings in computational statistics [internet]; 2002. pp. 575—80. Available from: <http://www.stat.uni-muenchen.de/~leisch/Sweave>.

[121] Ramsey N. Literate programming simplified. IEEE Softw 1994;11:97—105.

[122] Xie Y. Dynamic documents with R and knitr [internet]. Boca Raton, FL: Chapman and Hall/CRC; 2013. Available from: <http://yihui.name/knitr/>.

[123] Xie Y. knitr: a general-purpose package for dynamic report generation in R [internet]. 2014. Available from: <http://yihui.name/knitr/>.

[124] Xie Y. knitr: a comprehensive tool for reproducible research in R. In: Stodden V, Leisch F, Peng RD, editors. Implementing reproducible computational research [internet]. Chapman and Hall/CRC; 2014.

[125] Moreau L, Clifford B, Freire J, Futrelle J, Gil Y, Groth P, et al. The Open Provenance Model core specification (v1.1). Future Gener Comput Syst 2011;743—56.

Chapter 5

Noncoding RNA Regulation of Health and Disease

Nicolas Léveillé[1], Carlos A. Melo[1,2] and Sonia A. Melo[3,4]

[1]*Division of Gene Regulation, The Netherlands Cancer Institute, Amsterdam, The Netherlands,*
[2]*Doctoral Programme in Biomedicine and Experimental Biology, Centre for Neuroscience and Cell Biology, Coimbra, Portugal,* [3]*Department of Cancer Biology, Metastasis Research Center, University of Texas MD Anderson Cancer Center, Houston, TX, USA,* [4]*Instituto de Investigação e Inovação em Saúde and Institute of Pathology and Molecular Immunology of the University of Porto, Porto, Portugal*

Chapter Outline

5.1 INTRODUCTION

Gene expression is a complex and highly regulated process that allows the cell to maintain its homeostasis and to respond to various environmental cues. The control of transcription is a critical step to modulate and coordinate gene expression. Among transcriptional regulators, noncoding RNAs (ncRNAs) have recently emerged as important factors involved in numerous key cellular processes, such as metabolism, cell cycle control, apoptosis, and differentiation [1−4]. Mechanistically, ncRNAs can influence transcription by recruiting or blocking factors to or from specific genomic loci as well as by altering the

M. Fraga & A.F. Fernandez (Eds): Epigenomics in Health and Disease.
DOI: http://dx.doi.org/10.1016/B978-0-12-800140-0.00005-4
© 2016 Elsevier Inc. All rights reserved.

epigenetic state of the DNA. Interestingly, deregulation of ncRNA expression has been observed and associated associated with various human pathologies, such as Alzheimer disease, heart disease, and cancer [5–8]. In this chapter, we review the features and mechanisms of action of a subset of ncRNAs as well their deregulation in human disease, and present their potential as future predictive biomarkers.

5.2 NONCODING RNAs

The genome is a rapidly evolving entity that constantly reshapes its structure by acquiring, losing or rearranging its DNA. This great plasticity certainly contributes to the evolution and increased complexity of the human genome. With the recent explosion of refined genomic and large-scale sequencing technologies, we are now facing the task to redefine certain concepts and definitions no longer in consonance with collected experimental data. The central dogma of molecular biology reflects the idea that the genetic information flows from DNA to RNA to protein [9]. Although this concept is largely accurate in prokaryotes, where genomes are mainly composed of protein-coding genes, it does not reflect the reality of higher eukaryotes, where protein-coding genes occupy less than 3% of the genome. This discrepancy suggests a correlation between the emerging complexities of the genome with the expansion of its noncoding portion. Initially believed to be transcriptionally inactive, the predominant fraction of the genome is in reality pervasively transcribed into thousands of different ncRNAs. Although this finding challenges the established view of RNA as a modest intermediary in protein synthesis, it also brings a paradigm shift toward the recognition of RNAs as functional molecules that shape the evolution of complex organisms.

ncRNAs are transcripts that contain little or no open reading frame and are, in general, divided into two main categories: small ncRNAs (sncRNAs; <200 bp) and long ncRNAs (lncRNAs; >200 bp). The initial observation by Andrew Fire and Craig C. Mello in the 1990s that double-stranded RNAs (dsRNAs) were involved in gene silencing when exogenously introduced into *Caenorhabditis elegans*, generated a great interest for small ncRNAs [10–12]. The family of sncRNA has since expanded into various classes, including guide RNAs (small nucleolar RNAs), splicing RNAs (small nuclear RNAs), housekeeping RNAs (ribosomal RNAs and transfer RNAs), regulatory RNAs (microRNAs (miRNAs), PIWI-RNAs), and several transcription-associated RNAs (promoter upstream transcripts, promoter-associated RNAs, transcription initiation RNAs, and termini-associated small RNAs).

The curiosity for lncRNAs also spurted in the 1990s with the initial characterization of two important genes: the imprinted H19 [13] and the X-inactive specific transcript (*Xist*) [14]. In the first case, Tilghman et al. identified H19 (clone 19 in row H) while screening for genes regulated by α-fetoprotein in the liver. This discovery led to the description of H19 as a 2.3-kb capped, spliced, and polyadenylated transcript that has no conserved open reading frames and further established the locus as an archetype of imprinted genes [13]. In the second case, Xist was identified as an ncRNA produced at the X chromosome inactivation center (XIC). Only expressed from the inactive X chromosome, it was then demonstrated that the RNA recruits the polycomb-group proteins to catalyze the deposition of histone 3 lysine 27 trimethylation (H3K27me3) in *cis* to epigenetically repress the X chromosome [14,15]. During the following decades, lncRNAs increased in number and evolved in different classes, such as long intergenic ncRNAs (lincRNAs), enhancer RNAs (eRNAs), and activating ncRNAs (ncRNAs-a).

These recent developments led to the discovery of a plethora of functional ncRNAs with surprising and varied functions, ranging from the simple structural module to potent transcriptional, translational, and epigenetic regulators. Here, we overview the function of the most extensively studied sncRNAs and lncRNAs, with special attention to their implication in diverse cellular processes and deregulation in diseases.

5.2.1   Long Noncoding RNAs

5.2.1.1   Origin and Epigenetic Signatures

Recent technologic advances in high-throughput functional genomics have resulted in the identification of new regulatory RNA molecules termed "long non-coding RNAs." Although intense investigation is being done to identify and understand their functions, only few reports have addressed their origin. Initial analysis of their sequences suggested that lncRNAs are less conserved than protein-coding genes but slightly more than neutrally evolving genomic regions [16,17]. However, efforts to find distant homologs to human lncRNAs yielded modest successes compared with protein-coding genes, suggesting a rapid and species-specific evolution [18−21]. Interestingly, different hypotheses have been proposed to explain the formation of lncRNAs, including transformation of protein-coding genes, duplication of another lncRNA, and emergence from transposable element (TE) sequences [18,19,22−26]. Although examples have been found for each option, TEs seem to be an interesting option, since they occupy at least half of the human genome [27]. TEs are mobile genetic elements divided in two classes: class I (retrotransposons) and class II (DNA transposons). Although the class I TEs are copied in two steps involving transcription (DNA → RNA) and reverse transcription (RNA → DNA), the class II TEs use a cut-and-paste transposition (no RNA intermediate) that is performed by various transposases. Importantly, several reports have already highlighted TEs as the source of different genomic regulatory domains, such as promoters, enhancers, and insulators [28−30]. It was also demonstrated that the lncRNA Xist is derived from a mixture between a decayed protein-coding gene and the progressive accumulation of various TEs [31]. This isolated case has recently been expanded to the genomic scale, where it was found that 83% of lincRNAs (a class of lncRNAs) contain TEs, whereas TEs constitute 42% of lincRNAs [32]. In addition, Kelley and Rinn showed that TEs preferentially accumulate at the transcription start sites (TSS) of lincRNAs. This feature is biologically relevant as lincRNAs devoid of TEs are generally expressed at greater levels. A growing body of evidence now suggests that TEs are omnipresent in lncRNAs, suggesting that they might have shaped their evolution and function.

Besides their sequence composition, their epigenetic and RNA features characterize lncRNAs. LncRNAs are classically capped and polyadenylated transcripts produced by RNA polymerase II. Their genomic location usually harbor epigenetic features of protein-coding genes, such as histone 3 lysine 4 trimethylation (H3K4me3) at the transcription start site and histone 3 lysine 36 trimethylation (H3K36me3) throughout the gene body [26]. These features regroup several classes of lncRNAs, such as lincRNAs, transcribed ultra-conserved regions (T-UCRs) as well as numerous antisense RNAs. However, eRNAs have different epigenetic and RNA features. Enhancer RNAs are unidirectional (1D-eRNA) or bidirectional (2D-eRNA) capped and nonpolyadenylated transcripts produced by RNA polymerase II. The genomic features of enhancers are often characterized by a combination of a higher ratio between monomethylation of histone H3 at lysine 4 (H3K4me1) over trimethylation of histone H3 at lysine 4 (H3K4me3). This ratio, in combination with the presence of cyclic adenosine monophosphate (cAMP) response element (CRE)-binding protein (CBP) and p300, is commonly used to identify putative enhancer domains [33]. Furthermore, the presence of other epigenetic marks, such as acetylation of histone H3 at lysine 27 and lysine 9 (H3K27ac, H3K9ac), is characteristic of active enhancers [34,35].

5.2.2 Mechanisms of Action

5.2.2.1 LncRNAs as Chromatin Modifiers

Regulation of gene expression is a highly dynamic process that primarily depends on the chromatin accessibility, which, in turn, is dictated by the intricate interactions existing between transcription factors (TFs), histone modifications, chromatin modifiers, and chromatin remodelers. Thus, the epigenetic features of a genomic region will largely influence the balance between a more open state and a closed chromatin state. Active genes are known to have promoter regions decorated with H3K4me3 and H3K27 acetylation (H3K27ac), whereas repressed or inactive gene promoters are classically marked by H3K9me3 and H3K27me3 [35]. The exact mechanisms that govern the dynamic variation of epigenetic states remain mostly unknown. However, it is becoming clear that lncRNAs have an essential role in the process. An interesting example is the lncRNA HOTTIP (HOXA transcript at the distal tip) and its impact on the HOXA locus [36]. The gene encoding HOTTIP is located upstream of HOXA13 at the 5′ tip of the HOXA locus. The transcription of HOTTIP generates a 3.7-kb transcript that influences the activation of several 5′ HOXA genes. Importantly, it was shown that the *HOTTIP* gene is physically brought in close proximity with the 5′ HOXA gene promoters, by DNA looping. Once produced, HOTTIP RNA can interact with the WD repeat-containing protein 5 (WDR5) protein, which then recruits the histone methyltransferase mixed-lineage leukemia (MLL). The tethered WDR5-MLL complex finally catalyzes the methylation of local H3K4 and activates the transcription of the HOXA locus, thereby influencing development and cell fate. In contrast, lncRNAs have also been found to negatively regulate gene expression. Gupta et al. recently reported that the lncRNA HOTAIR was able to repress the HOXD cluster genes by a *trans*-acting mechanism (Figure 5.1A) [6]. HOTAIR operates by the help of two functional domains required to interact with polycomb repressive complex 2 (PRC2) and lysine-specific demethylase 1 (LSD1/CoREST1) complexes. By tethering these two repressive chromatin modifiers, HOTAIR organizes their action and facilitates their recruitment to specific promoter regions, where they deposit repressive epigenetic marks, such as H3K27me3 [37]. Notably, the epigenetic remodeling mediated by HOTAIR has an important role in skin differentiation [38,39]. A similar repressive mechanism was also employed

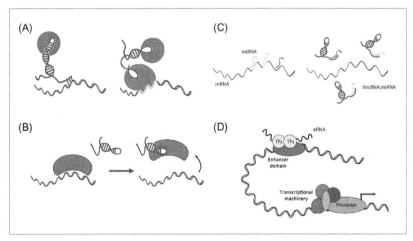

FIGURE 5.1   Mechanisms of long noncoding RNA (lncRNA) function (A) lncRNAs can act as molecular guides, tethering specific chromatin-modifying complexes to DNA; this can occur through direct RNA−DNA interaction or mediated by a DNA-binding protein adapter (green); (B) lncRNAs can function as decoys, binding and sequestering transcription factors from their target DNA sequence; (C) lncRNAs can interact with miRNAs, serving as molecular sponges and preventing them from binding to mRNA; (D) eRNAs, transcribed from enhancer domains, directly affect the expression of neighboring genes, through chromatin-looping. TFs, transcription factors.

by the lncRNA-p21. It was found that the tumor suppressor p53 controls the expression of lincRNA-p21 in order to repress a subset of genes related to apoptosis [4]. Although the mechanism was not entirely clarified, it was suggested that the repressive function of lincRNA-p21 was mediated at promoters and by its interaction with the heterogeneous nuclear ribonucleoprotein K (hnRNPK). An elusive aspect of lncRNAs function resides in their capacity to reach or interact with specific chromatin regions. Although several hypotheses, such as formation of specific RNA−DNA hybrids, interaction with an adapter protein, or direct association to a TF, have been described, only few reports have so far properly addressed this question. One well-characterized case involves the lncRNA Mistral. Transcribed from a region located between the *Hoxa6* and *Hoxa7* genes, the lncRNA Mistral was shown to locally recruit and anchor the MLL complex by the formation of an RNA−DNA hybrid (Figure 5.1A). Interestingly, the $3'$ region of Mistral can form a stem-loop structure that is specifically recognized by the SET domain of MLL1, which then gain the ability to bind single-strand DNA (ssDNA). The local accumulation of the MLL complex favors the deposition of active H3K4me3, which ultimately promotes the transcriptional regulation of *Hoxa6* and *Hoxa7* genes and influences stem cell differentiation [40].

5.2.2.2 LncRNAs as Decoy

The transcriptional activity of a gene depends on the balance between silencing and activating cues. Because of their ability to associate with RNAs and proteins, lncRNAs are interesting candidates to fine-tune or impact such balance. Beside their direct and local influence on promoters, lncRNAs can also act as decoys by keeping activators or repressors away from their targets. One example of lncRNA decoy was discovered with the lncRNA braveheart (Bvht). It was observed that Bvht is necessary for the activation of a gene network involved in cardiac lineage commitment [41]. Because Bvht interacts with SUZ12 (a component of PRC2) during cardiomyocyte differentiation, the authors suggested that Bvht mediates epigenetic changes toward cardiac commitment, by sequestering PRC2 from its target genes. In addition to protein decoy, lncRNAs can cross-talk with other RNAs by competing for shared miRNAs (Figure 5.1B). The muscle-specific lncRNA MD1 (linc-MD1) participates in the timing of muscle differentiation by acting as a decoy for miRNAs (Figure 5.1C) [42]. Cesana et al. demonstrated that linc-MD1 functions as a competing endogenous RNA (ceRNA) for the miR-133 and miR-135, which are negatively regulating mastermind-like-1 (MAML1) and myocyte-specific factor 2C (MEF2C), two prominent factors involved in myogenesis. In support of the crucial role of linc-MD1 in the control of the myogenic program, it was found that linc-MD1 expression is reduced in myoblasts derived from patients with Duchenne muscular dystrophy (DMD), which corroborate with their limited ability to undergo terminal differentiation. Importantly, reintroduction of the linc-MD1 in DMD myoblast could partially rescue the timing and the expression of myogenic factors required for the differentiation program. These observations highlight the critical function of lncRNAs in many cellular processes and stress their potential influence on the delicate balance that often distinguished between health and disease states.

5.2.2.3 LncRNAs in Disease

Accumulating evidences suggest that lncRNAs are critical regulators of several key biologic processes, such as transcription, splicing, RNA stability/decay, and translation. With regard to their potential and wide range of action, it is not surprising to find an increasing number of deregulated lncRNAs in diverse diseases, including Beckwith—Wiedemann syndrome (h19), diabetes (cdkn2b-as), Huntington disease (har1a), Alzheimer disease (51a), and various type of cancers (uca1, hotair, malat1, anril), among others.

In the context of cancer, several lines of evidence suggest that inactivation of tumor-suppressive lncRNAs is a driving force either in the initiation or the progression of the disease. For instance, the lncRNA growth arrest specific transcript 5 (GAS5) can interact with the glucocorticoid receptor (GR) and suppress its ability to reach and regulate its target genes [43]. Glucocorticoids are known to influence important cellular processes, such as cell growth, metabolism, and survival. Importantly, although GAS5 overexpression induces growth arrest and apoptosis in breast cancer cell lines, its expression levels are significantly reduced in breast tumors [44]. Moreover, GAS5 can interact with other steroid hormone receptors known to be involved in breast cancer, such as the androgen receptor (AR) and the progesterone receptor (PR), further supporting the implication of GAS5 in tumorigenesis [45,46]. The imprinted lncRNA MEG3 is known to inhibit cell growth by regulating the levels of the tumor suppressor p53 protein [47]. Similar to that of GAS5, MEG3 expression was undetectable in a panel of human tumors [48]. In addition, MEG3-null mice had an increased vascularization of the brain, suggesting that MEG3 may prevent tumor growth by inhibiting angiogenesis [49].

Alternatively, lncRNAs with oncogenic function have been found to be upregulated in cancer. The lncRNA urothelial carcinoma associated 1 (lncRNA-UCA1) was initially identified as being upregulated in bladder cancer and more recently observed to be overexpressed in breast cancer specimens as well [50,51]. lncRNA-UCA1 promotes the progression of tumor growth by reducing the translation of the tumor suppressor p27 (kip1). It was demonstrated that UCA1 can interact with the heterogeneous nuclear ribonucleoprotein I (hnRNPI) protein and prevent the latter (by decoy) from binding to and enhancing the translation of p27 messenger RNA (mRNA). In support of these data, reducing the expression of UCA1 in MCF-7 cells, resulted in a decreased tumor growth when transplanted into mice. Use of large panel of tissue samples in combination with high-throughput sequencing led to the identification of 121 lncRNAs associated with prostate cancer. Among these prostate cancer-associated transcripts (PCAT), the functional characterization of PCAT-1 pinpointed its implication in cancer cell proliferation. Mechanistically, PCAT-1 is a transcriptional repressor that preferentially regulates several tumor suppressor genes, such as *BRCA2* [52]. ANRIL is another upregulated lncRNA in prostate cancer tissues that influences the expression of potent tumor suppressors [53,54]. It was shown that ANRIL interacts with CBX7 (a subunit of PRC1) and SUZ12 (a subunit of PRC2), to mediate in *cis* the transcriptional repression of INK4a-ARF-INK4b locus. Importantly, genome-wide association studies (GWAS) have also shown that the ANRIL locus is associated with increased susceptibility to such diseases as intracranial aneurysm, type 2 diabetes, and different types of cancers [55].

Mutations or sequence variations within lncRNA gene loci have also been associated with heritable diseases. For example, mutations in the *TERC* gene have been associated with dyskeratosis congenita (DKC) [56]. Some of these mutations destabilize TERC or change its interaction with the telomerase reverse transcriptase (hTERT), which results in a poor telomere maintenance. Affected individuals have an increased risk of developing life-threatening conditions, such as cancer and pulmonary fibrosis [57]. In the autosomal dominant facio-scapulo-humeral dystrophy (FSHD) syndrome, a reduction in the copy number of the repressive D4Z4 repeat causes an epigenetic switch that leads to the activation of a lncRNA named DBE-T. The de-repressed DBE-T recruits the Trithorax group protein Ash1L and coordinates long-range gene regulation that contributes to the progression of the disease [58].

5.2.3 Enhancer RNAs

Enhancers were originally defined as DNA elements capable of modulating gene expression over long distances, independently of their orientation [59]. Interestingly, recent findings have revealed the existence and the production of ncRNAs at regulatory enhancer regions. These genomic elements are actively engaged in the regulation of transcription, where they contribute to fine-tune tissue-specific gene expression. Moreover, several reports have suggested that enhancers can be subject to coordinated regulation. This idea is in agreement with the detection of many TFs binding sites at enhancers [60,61]. Recent advances in high-throughput sequencing technologies have allowed us to expand our understanding of genomic and epigenetic features at enhancers. For instance, it was observed that many of these regulatory elements possess an open chromatin state, marked by the presence of DNase hypersensitivity regions, which favors the binding of TFs [62].

Transcription at enhancers was first reported in the locus control region (LCR) of the *β-globin* cluster [63]. In the following years, genome-wide studies help demonstrate that transcription at enhancers is a widespread phenomenon. Two landmark studies specifically demonstrated that RNA polymerase II (RNA Pol II) is recruited to active enhancers in order to mediate transcription [64,65]. Kim et al. further demonstrated that stimulated neurons (potassium chloride) were producing bidirectional transcription at enhancers. These RNAs extended ~1.2 kb away from the TF binding sites and their expression correlated with neighboring gene expression [64]. Enhancer RNAs are most commonly bidirectional (2D-eRNAs) and nonpolyadenylated; however, it was observed that eRNAs can also be unidirectional (1D-eRNAs) and polyadenylated [65]. Follow-up studies revealed that several TFs were involved in the coordination and regulation of these RNA molecules. For example, chromatin immuno-precipitation followed by deep sequencing (ChIP-seq) revealed that the tumor suppressor p53 binds with high affinity to enhancer domains harboring p53 response elements (p53REs). Interestingly, p53 activation (Nutlin-3a treatment) produced the transcriptional induction of eRNAs at p53-bound enhancer regions (p53BERs). Moreover, p53BERs depletion, through small interfering RNA (siRNA), was able to abrogate the induction of nearby protein-coding genes, which had an important influence on p53-dependent cell cycle arrest [3]. The authors further demonstrated that eRNAs are tethered to promoters through the formation of DNA loops, by using circularized chromosome conformation capture combined with deep sequencing (4C technology). Intriguingly, the DNA loops were predetermined, as depletion of p53 did not disturb the established chromatin structure. Collectively, the discovery and the examination of p53BERs provided the first insights into the functional role of eRNAs (Figure 5.1D). Additionally, the fact that a major tumor suppressor, such as p53, regulates several of its target genes through enhancers highlights the therapeutic potential of these eRNA producing domains as novel targets for cancer treatments.

Global run-on sequencing (Gro-seq) [66] has recently emerged as a powerful tool to detect newly synthesized transcripts. Numerous studies involving Gro-Seq technology validated the widespread importance of eRNA regulation in macrophages, diabetes, and cancer [67−71]. For instance, the nuclear receptors Rev-Erbs, which function as transcriptional repressors that regulate the expression of genes involved in many processes, such as metabolism and inflammatory responses [71]. In mouse macrophages, these transcriptional repressors bind predominantly to extragenic and intragenic enhancer domains and repress their neighboring protein-coding genes. Approximately half of these enhancers were enriched with RNA Pol II, and Gro-seq showed that a majority of Rev-Erb-bound enhancers produce eRNAs. Depletion of eRNA at these regions directly impaired the expression of neighboring genes, supporting previous findings [3]. It was also observed that eRNA transcription was de-repressed in Rev-Erb deficient macrophages, whereas constitutive expression of Rev-Erb reduced eRNA transcription. Notably, binding of Rev-Erb to these sites, recruited co-repressor histone deacetylase (HDAC) complexes, reducing the levels of H3K9ac and, consequently, the eRNA and protein-coding target gene levels. Moreover, the characterization of an eRNA, by using luciferase reporter assays, demonstrated that only one strand was required for the activating function. Altogether, Rev-Erbs regulates macrophage gene expression by negatively influencing the transcriptional activity of enhancers. This example clearly demonstrates the relevance of eRNAs in physiologic and pathologic inflammatory responses, such as wound healing and atherosclerosis. Numerous other reports further demonstrate the importance of eRNA production influencing target gene transcription, in part through the stabilization of enhancer–promoter looping, as suggested by Li et al. [70]. Surprisingly, the authors observed that depletion of eRNA could reduce the looping between enhancers and promoters, supporting a model where the eRNA can facilitate and stabilize long-range interactions, thereby influencing target gene expression [70]. Understanding the eRNA dynamics during several cellular events, such as cancer, is a matter of immediate importance. Hence, while key findings started to shed light on the mechanisms and biologic relevance of these RNA molecules, a long walk lies ahead shadowed by what we still do not know. Future study will certainly clarify these events and potentially create new therapeutic approaches to disease.

5.3 CIRCULATING NONCODING RNAs

Circulating noncoding RNAs, mostly miRNAs, have been isolated from most human cells and body fluids, including serum, saliva, urine, breast milk, and cerebrospinal fluid, among others [72,73]. Placental-specific miRNAs were found in the serum of pregnant women that are downregulated upon childbirth [74]. Cancer-associated miRNAs are higher in serum from cancer patients than healthy individuals, indicating that they could be used as biomarkers [75].

In 2007, Valadi et al. have reported for the first time miRNAs to be contained inside exosomes [76]. Our group has recently described for the first time miRNA biogenesis in exosomes derived from cancer cells as a way to obtain stoichiometric amounts of mature miRNAs that will efficiently alter gene expression in recipient cells [77]. Exosomes are small membrane vesicles secreted by all cell types, with a diameter ranging from 50 to 150 nm [78] (Figure 5.2). After the first report that demonstrated the existence of miRNAs in exosomes, several reports confirmed the existence of miRNAs in apoptotic bodies [79], high-density/low-density lipoproteins (HDL/LDL) [80], and RNA-binding proteins [81].

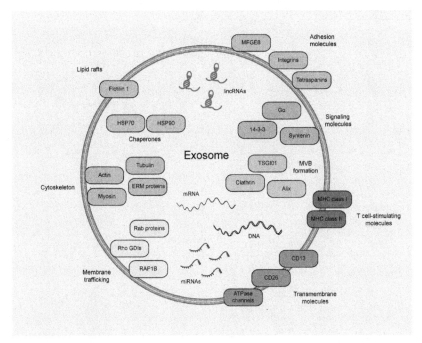

FIGURE 5.2   Schematic representation of an exosome and the most predominant proteins and nucleic acids described so far.

5.4   EXOSOMES: BIOGENESIS AND CANCER BIOMARKERS

Exosomes are formed as intraluminal vesicles (ILVs) by inward budding from the limiting membrane into early endosomes and multivesicular bodies (MVBs) (Figure 5.3). Several molecules are involved in the biogenesis of ILVs, such as the Endosomal Sorting Complex Required for Transport (ESCRT) machinery, and lipids, such as ceramide and tetraspanins. ESCRT consists of four complexes plus associated proteins: ESCRT-0 is responsible for cargo clustering in a ubiquitin-dependent manner, ESCRT-I and ESCRT-II induce bud formation, ESCRT-III drives vesicle scission, and the accessory proteins allow dissociation and recycling of the ESCRT machinery. Multivesicular bodies can either fuse with lysosomes or with the plasma membrane, which allows the release of their content into the extracellular environment. RAB proteins (RAB11, RAB27, and RAB35) have been shown to be involved in the transport of MVBs to the plasma membrane and in exosomes secretion. This RAB family of small GTPase proteins controls different steps of intracellular vesicular trafficking, such as vesicle budding, vesicle and organelle mobility through cytoskeleton interaction, and docking of vesicles to their target compartment, leading to membrane fusion [82]. In addition, SNAREs (soluble *N*-ethylmaleimide-sensitive factor attachment protein receptors) are probably involved in the fusion of these MVBs with the plasma membrane.

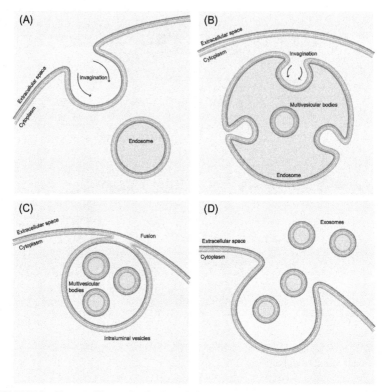

FIGURE 5.3 Exosome biogenesis: (A) the cell, throughout its life cycle periodically engulfs small amounts of intracellular fluid by invagination of the plasma membrane, forming a small intercellular body called endosome; (B) as the endosome matures, it becomes characterized by the formation of intraluminal vesicles (ILVs) ranging from 30 to 100 nm in diameter. The ILVs are formed by inward budding of the endosomal membrane, engulfing portions of the cytosolic contents and incorporating transmembrane and peripheral proteins into the invaginating membrane. The late endosome containing ILVs is called multivesicular body (MVB); (C) the MVBs may fuse with the plasma membrane in a proposed ion-dependent manner, releasing the ILVs in an exocytotic fashion to the extracellular environment; (D) these vesicles are then referred to as "exosomes."

Exosomes, as carriers of genetic information (DNA, mRNA, and miRNAs), have long attracted diagnostic interests. The heterogeneity of circulating exosomes may provide clues as to the differential representation of specific cell types within the tumor microenvironment. Additionally, exosomes provide diagnostic opportunities for the profiling of cancer subtypes (virtually a *liquid biopsy*), unraveling new therapeutic targets and predicting therapeutic responses when used as biomarkers. Therefore, a novel concept of cancer biomarker called "liquid biopsy" has been proposed [83]. A liquid biopsy would be useful for numerous diagnostic applications and avoid the need for invasive tumor biopsies. Current studies have shown that genomic alterations in solid cancers can be characterized through the massively parallel sequencing of circulating cell-free tumor DNA released from cancer cells [83]. Additionally, it has been shown that circulating serum exosomes are positive for a mutant/variant of epidermal growth factor receptor vIII (EGFRvIII) when the parental glioblastoma cells also expressed this mutant/variant [84]. Therefore, determining EGFR status from a small serum sample-derived exosomes, instead of a need for primary tumor biopsy that involved invasive brain surgery, is of immense benefit for patients. This suggests that also circulating ncRNAs, as miRNAs, are also good candidates for liquid biopsy, as the quantities and sequences of miRNAs convey information for diagnosis. Particularly, circulating miRNAs, which have been previously shown to function in cell−cell communication, might be good candidates for this application. Therefore it is important to investigate the function of secretory miRNAs in cell−cell communication, and in parallel explore the usefulness of these molecules as biomarkers using animal models.

5.5 CONCLUSION

Epigenomics emerging role in health brings to the forefront unpreceded opportunities for the development of new therapies. More specifically, ncRNAs have recently attracted interest because of their important roles in various cellular processes. Pervasively transcribed across the human genome, ncRNAs are abundant, and present a variety of features and epigenetic mechanisms. Interestingly, emerging evidences seem to indicate a direct connection between the deregulation of ncRNA expression and multiple diseases in humans. Because of their key cellular functions and extensive diversity, ncRNAs represent interesting target molecules for the development of new therapeutic strategies. Alternatively, ncRNAs have a great potential for the identification of novel diagnostic biomarkers. For instance, ncRNAs transported by cell-derived vesicles (e.g., exosomes) were recently found in biologic fluids, such as blood and urine. This important discovery opens the possibility of ncRNAs as biomarkers and certainly urges further investigation toward the development of "liquid biopsies" [83]. Collectively, shedding light on the "dark matter" of the genome will certainly help to deepen our molecular understanding of normal and pathologic conditions, and may bring innovative therapeutic possibilities.

REFERENCES

[1] Kornfeld JW, Bruning JC. Regulation of metabolism by long, non-coding RNAs. Front Genet 2014;5:57.

[2] Leveille N, Melo CA, Rooijers K, Diaz-Lagares A, Melo SA, Korkmaz G, et al. Genome-wide profiling of p53-regulated enhancer RNAs uncovers a subset of enhancers controlled by a lncRNA. Nat Commun 2015;6:6520.

[3] Melo CA, Drost J, Wijchers PJ, van de Werken H, de Wit E, Oude Vrielink JA, et al. eRNAs are required for p53-dependent enhancer activity and gene transcription. Mol Cell 2013;49(3):524−35.

[4] Huarte M, Guttman M, Feldser D, Garber M, Koziol MJ, Kenzelmann-Broz D, et al. A large intergenic noncoding RNA induced by p53 mediates global gene repression in the p53 response. Cell 2010;142(3):409−19.

[5] Kaneko S, Li G, Son J, Xu CF, Margueron R, Neubert TA, et al. Phosphorylation of the PRC2 component Ezh2 is cell cycle-regulated and up-regulates its binding to ncRNA. Genes Dev 2010;24(23):2615−20.

[6] Gupta RA, Shah N, Wang KC, Kim J, Horlings HM, Wong DJ, et al. Long non-coding RNA HOTAIR reprograms chromatin state to promote cancer metastasis. Nature 2010; 464(7291):1071−6.

[7] Sang X, Zhao H, Lu X, Mao Y, Miao R, Yang H, et al. Prediction and identification of tumor-specific noncoding RNAs from human UniGene. Med Oncol 2010;27 (3):894−8.

[8] Taft RJ, Pang KC, Mercer TR, Dinger M, Mattick JS. Non-coding RNAs: regulators of disease. J Pathol 2010;220(2):126−39.

[9] Crick FH, Barnett L, Brenner S, Watts-Tobin RJ. General nature of the genetic code for proteins. Nature 1961;192:1227−32.

[10] Fire A, Xu S, Montgomery MK, Kostas SA, Driver SE, Mello CC. Potent and specific genetic interference by double-stranded RNA in Caenorhabditis elegans. Nature 1998;391 (6669):806−11.

[11] Hammond SM, Bernstein E, Beach D, Hannon GJ. An RNA-directed nuclease mediates post-transcriptional gene silencing in Drosophila cells. Nature 2000;404(6775):293−6.

[12] Zamore PD, Tuschl T, Sharp PA, Bartel DP. RNAi: double-stranded RNA directs the ATP-dependent cleavage of mRNA at 21 to 23 nucleotide intervals. Cell 2000; 101(1):25−33.

[13] Bartolomei MS, Zemel S, Tilghman SM. Parental imprinting of the mouse H19 gene. Nature 1991;351(6322):153−5.

[14] Brown CJ, Ballabio A, Rupert JL, Lafreniere RG, Grompe M, Tonlorenzi R, et al. A gene from the region of the human X inactivation centre is expressed exclusively from the inactive X chromosome. Nature 1991;349(6304):38−44.

[15] Simon MD, Pinter SF, Fang R, Sarma K, Rutenberg-Schoenberg M, Bowman SK, et al. High-resolution Xist binding maps reveal two-step spreading during X-chromosome inactivation. Nature 2013;504(7480):465−9.

[16] Marques AC, Ponting CP. Catalogues of mammalian long noncoding RNAs: modest conservation and incompleteness. Genome Biol 2009;10(11):R124.

[17] Guttman M, Garber M, Levin JZ, Donaghey J, Robinson J, Adiconis X, et al. Ab initio reconstruction of cell type-specific transcriptomes in mouse reveals the conserved multi-exonic structure of lincRNAs. Nat Biotechnol 2010;28(5):503−10.

[18] Cabili MN, Trapnell C, Goff L, Koziol M, Tazon-Vega B, Regev A, et al. Integrative annotation of human large intergenic noncoding RNAs reveals global properties and specific subclasses. Genes Dev 2011;25(18):1915−27.

[19] Ulitsky I, Shkumatava A, Jan CH, Sive H, Bartel DP. Conserved function of lincRNAs in vertebrate embryonic development despite rapid sequence evolution. Cell 2011;147 (7):1537−50.

[20]   Nam JW, Bartel DP. Long noncoding RNAs in *C. elegans*. Genome Res 2012;22 (12):2529–40.

[21]   Kutter C, Watt S, Stefflova K, Wilson MD, Goncalves A, Ponting CP, et al. Rapid turnover of long noncoding RNAs and the evolution of gene expression. PLoS Genet 2012; 8(7):e1002841.

[22]   Rinn JL, Chang HY. Genome regulation by long noncoding RNAs. Annu Rev Biochem 2012;81:145–66.

[23]   Jia H, Osak M, Bogu GK, Stanton LW, Johnson R, Lipovich L. Genome-wide computational identification and manual annotation of human long noncoding RNA genes. RNA 2010; 16(8):1478–87.

[24]   Derrien T, Johnson R, Bussotti G, Tanzer A, Djebali S, Tilgner H, et al. The GENCODE v7 catalog of human long noncoding RNAs: analysis of their gene structure, evolution, and expression. Genome Res 2012;22(9):1775–89.

[25]   Dinger ME, Amaral PP, Mercer TR, Pang KC, Bruce SJ, Gardiner BB, et al. Long noncoding RNAs in mouse embryonic stem cell pluripotency and differentiation. Genome Res 2008;18(9):1433–45.

[26]   Guttman M, Amit I, Garber M, French C, Lin MF, Feldser D, et al. Chromatin signature reveals over a thousand highly conserved large non-coding RNAs in mammals. Nature 2009;458(7235):223–7.

[27]   Lander ES, Linton LM, Birren B, Nusbaum C, Zody MC, Baldwin J, et al. Initial sequencing and analysis of the human genome. Nature 2001;409(6822):860–921.

[28]   Faulkner GJ, Kimura Y, Daub CO, Wani S, Plessy C, Irvine KM, et al. The regulated retrotransposon transcriptome of mammalian cells. Nat Genet 2009;41(5):563–71.

[29]   Bejerano G, Lowe CB, Ahituv N, King B, Siepel A, Salama SR, et al. A distal enhancer and an ultraconserved exon are derived from a novel retroposon. Nature 2006;441 (7089):87–90.

[30]   Schmidt D, Schwalie PC, Wilson MD, Ballester B, Goncalves A, Kutter C, et al. Waves of retrotransposon expansion remodel genome organization and CTCF binding in multiple mammalian lineages. Cell 2012;148(1-2):335–48.

[31]   Duret L, Chureau C, Samain S, Weissenbach J, Avner P. The Xist RNA gene evolved in eutherians by pseudogenization of a protein-coding gene. Science 2006;312(5780): 1653–5.

[32]   Kelley D, Rinn J. Transposable elements reveal a stem cell-specific class of long noncoding RNAs. Genome Biol 2012;13(11):R107.

[33]   Heintzman ND, Stuart RK, Hon G, Fu Y, Ching CW, Hawkins RD, et al. Distinct and predictive chromatin signatures of transcriptional promoters and enhancers in the human genome. Nat Genet 2007;39(3):311–18.

[34]   Creyghton MP, Cheng AW, Welstead GG, Kooistra T, Carey BW, Steine EJ, et al. Histone H3K27ac separates active from poised enhancers and predicts developmental state. Proc Natl Acad Sci USA 2010;107(50):21931–6.

[35]   Ernst J, Kheradpour P, Mikkelsen TS, Shoresh N, Ward LD, Epstein CB, et al. Mapping and analysis of chromatin state dynamics in nine human cell types. Nature 2011;473 (7345):43–9.

[36]   Wang KC, Yang YW, Liu B, Sanyal A, Corces-Zimmerman R, Chen Y, et al. A long noncoding RNA maintains active chromatin to coordinate homeotic gene expression. Nature 2011;472(7341):120–4.

[37] Tsai MC, Manor O, Wan Y, Mosammaparast N, Wang JK, Lan F, et al. Long noncoding RNA as modular scaffold of histone modification complexes. Science 2010;329 (5992):689–93.

[38] Rinn JL, Kertesz M, Wang JK, Squazzo SL, Xu X, Brugmann SA, et al. Functional demarcation of active and silent chromatin domains in human HOX loci by noncoding RNAs. Cell 2007;129(7):1311–23.

[39] Chuong CM. Homeobox genes, fetal wound healing, and skin regional specificity. J Invest Dermatol 2003;120(1):9–11.

[40] Bertani S, Sauer S, Bolotin E, Sauer F. The noncoding RNA Mistral activates *Hoxa6* and *Hoxa7* expression and stem cell differentiation by recruiting MLL1 to chromatin. Mol Cell 2011;43(6):1040–6.

[41] Klattenhoff CA, Scheuermann JC, Surface LE, Bradley RK, Fields PA, Steinhauser ML, et al. Braveheart, a long noncoding RNA required for cardiovascular lineage commitment. Cell 2013;152(3):570–83.

[42] Cesana M, Cacchiarelli D, Legnini I, Santini T, Sthandier O, Chinappi M, et al. A long noncoding RNA controls muscle differentiation by functioning as a competing endogenous RNA. Cell 2011;147(2):358–69.

[43] Kino T, Hurt DE, Ichijo T, Nader N, Chrousos GP. Noncoding RNA gas5 is a growth arrest- and starvation-associated repressor of the glucocorticoid receptor. Sci Signal 2010;3(107):ra8.

[44] Mourtada-Maarabouni M, Pickard MR, Hedge VL, Farzaneh F, Williams GT. GAS5, a non-protein-coding RNA, controls apoptosis and is downregulated in breast cancer. Oncogene 2009;28(2):195–208.

[45] Brisken C. Progesterone signalling in breast cancer: a neglected hormone coming into the limelight. Nat Rev Cancer 2013;13(6):385–96.

[46] Chang C, Lee SO, Yeh S, Chang TM. Androgen receptor (AR) differential roles in hormone-related tumors including prostate, bladder, kidney, lung, breast and liver. Oncogene 2014;33(25):3225–34.

[47] Braconi C, Kogure T, Valeri N, Huang N, Nuovo G, Costinean S, et al. microRNA-29 can regulate expression of the long non-coding RNA gene MEG3 in hepatocellular cancer. Oncogene 2011;30(47):4750–6.

[48] Zhang X, Zhou Y, Mehta KR, Danila DC, Scolavino S, Johnson SR, et al. A pituitary-derived MEG3 isoform functions as a growth suppressor in tumor cells. J Clin Endocrinol Metab 2003;88(11):5119–26.

[49] Gordon FE, Nutt CL, Cheunsuchon P, Nakayama Y, Provencher KA, Rice KA, et al. Increased expression of angiogenic genes in the brains of mouse meg3-null embryos. Endocrinology 2010;151(6):2443–52.

[50] Huang J, Zhou N, Watabe K, Lu Z, Wu F, Xu M, et al. Long non-coding RNA UCA1 promotes breast tumor growth by suppression of p27 (Kip1). Cell Death Dis 2014;5:e1008.

[51] Wang XS, Zhang Z, Wang HC, Cai JL, Xu QW, Li MQ, et al. Rapid identification of UCA1 as a very sensitive and specific unique marker for human bladder carcinoma. Clin Cancer Res 2006;12(16):4851–8.

[52] Prensner JR, Iyer MK, Balbin OA, Dhanasekaran SM, Cao Q, Brenner JC, et al. Transcriptome sequencing across a prostate cancer cohort identifies PCAT-1, an unannotated lincRNA implicated in disease progression. Nat Biotechnol 2011;29(8):742–9.

[53]   Yap KL, Li S, Munoz-Cabello AM, Raguz S, Zeng L, Mujtaba S, et al. Molecular inter-
       play of the noncoding RNA ANRIL and methylated histone H3 lysine 27 by polycomb
       CBX7 in transcriptional silencing of INK4a. Mol Cell 2010;38(5):662–74.

[54]   Kotake Y, Nakagawa T, Kitagawa K, Suzuki S, Liu N, Kitagawa M, et al. Long non-
       coding RNA ANRIL is required for the PRC2 recruitment to and silencing of p15
       (INK4B) tumor suppressor gene. Oncogene 2011;30(16):1956–62.

[55]   Pasmant E, Sabbagh A, Vidaud M, Bieche I. ANRIL, a long, noncoding RNA, is an unex-
       pected major hotspot in GWAS. Faseb J 2011;25(2):444–8.

[56]   Vulliamy TJ, Marrone A, Knight SW, Walne A, Mason PJ, Dokal I. Mutations in dyskera-
       tosis congenita: their impact on telomere length and the diversity of clinical presentation.
       Blood 2006;107(7):2680–5.

[57]   Marrone A, Sokhal P, Walne A, Beswick R, Kirwan M, Killick S, et al. Functional char-
       acterization of novel telomerase RNA (TERC) mutations in patients with diverse clinical
       and pathological presentations. Haematologica 2007;92(8):1013–20.

[58]   Cabianca DS, Casa V, Bodega B, Xynos A, Ginelli E, Tanaka Y, et al. A long ncRNA
       links copy number variation to a polycomb/trithorax epigenetic switch in FSHD muscular
       dystrophy. Cell 2012;149(4):819–31.

[59]   Bulger M, Groudine M. Functional and mechanistic diversity of distal transcription enhan-
       cers. Cell 2011;144(3):327–39.

[60]   Heinz S, Benner C, Spann N, Bertolino E, Lin YC, Laslo P, et al. Simple combinations of
       lineage-determining transcription factors prime cis-regulatory elements required for mac-
       rophage and B cell identities. Mol Cell 2010;38(4):576–89.

[61]   Lupien M, Eeckhoute J, Meyer CA, Wang Q, Zhang Y, Li W, et al. FoxA1 translates epige-
       netic signatures into enhancer-driven lineage-specific transcription. Cell 2008;132(6):958–70.

[62]   He HH, Meyer CA, Shin H, Bailey ST, Wei G, Wang Q, et al. Nucleosome dynamics
       define transcriptional enhancers. Nat Genet 2010;42(4):343–7.

[63]   Tuan D, Kong S, Hu K. Transcription of the hypersensitive site HS2 enhancer in erythroid
       cells. Proc Natl Acad Sci USA 1992;89(23):11219–23.

[64]   Kim TK, Hemberg M, Gray JM, Costa AM, Bear DM, Wu J, et al. Widespread transcrip-
       tion at neuronal activity-regulated enhancers. Nature 2010;465(7295):182–7.

[65]   De Santa F, Barozzi I, Mietton F, Ghisletti S, Polletti S, Tusi BK, et al. A large fraction of
       extragenic RNA pol II transcription sites overlap enhancers. PLoS Biol 2010;8(5):e1000384.

[66]   Core LJ, Waterfall JJ, Lis JT. Nascent RNA sequencing reveals widespread pausing and
       divergent initiation at human promoters. Science 2008;322(5909):1845–8.

[67]   Hah N, Danko CG, Core L, Waterfall JJ, Siepel A, Lis JT, et al. A rapid, extensive, and
       transient transcriptional response to estrogen signaling in breast cancer cells. Cell
       2011;145(4):622–34.

[68]   Kaikkonen MU, Spann NJ, Heinz S, Romanoski CE, Allison KA, Stender JD, et al.
       Remodeling of the enhancer landscape during macrophage activation is coupled to
       enhancer transcription. Mol Cell 2013;51(3):310–25.

[69]   Step SE, Lim HW, Marinis JM, Prokesch A, Steger DJ, You SH, et al. Anti-diabetic rosi-
       glitazone remodels the adipocyte transcriptome by redistributing transcription to
       PPARgamma-driven enhancers. Genes Dev 2014;28(9):1018–28.

[70]   Li W, Notani D, Ma Q, Tanasa B, Nunez E, Chen AY, et al. Functional roles of enhancer
       RNAs for oestrogen-dependent transcriptional activation. Nature 2013;498(7455):516–20.

[71] Lam MT, Cho H, Lesch HP, Gosselin D, Heinz S, Tanaka-Oishi Y, et al. Rev-Erbs repress macrophage gene expression by inhibiting enhancer-directed transcription. Nature 2013;498(7455):511−15.

[72] Kosaka N, Iguchi H, Ochiya T. Circulating microRNA in body fluid: a new potential biomarker for cancer diagnosis and prognosis. Cancer Sci 2010;101(10):2087−92.

[73] Baglio SR, Pegtel DM, Baldini N. Mesenchymal stem cell secreted vesicles provide novel opportunities in (stem) cell-free therapy. Front Physiol 2012;3:359.

[74] Lawrie CH, Gal S, Dunlop HM, Pushkaran B, Liggins AP, Pulford K, et al. Detection of elevated levels of tumour-associated microRNAs in serum of patients with diffuse large B-cell lymphoma. Br J Haematol 2008;141(5):672−5.

[75] Chim SS, Shing TK, Hung EC, Leung TY, Lau TK, Chiu RW, et al. Detection and characterization of placental microRNAs in maternal plasma. Clin Chem 2008;54(3):482−90.

[76] Valadi H, Ekstrom K, Bossios A, Sjostrand M, Lee JJ, Lotvall JO. Exosome-mediated transfer of mRNAs and microRNAs is a novel mechanism of genetic exchange between cells. Nat Cell Biol 2007;9(6):654−9.

[77] Melo SA, Sugimoto H, O'Connell JT, Kato N, Villanueva A, Vidal A, et al. Cancer exosomes perform cell-independent microRNA biogenesis and promote tumorigenesis. Cancer cell 2014;26(5):707−21.

[78] Raposo G, Stoorvogel W. Extracellular vesicles: exosomes, microvesicles, and friends. J Cell Biol 2013;200(4):373−83.

[79] Zernecke A, Bidzhekov K, Noels H, Shagdarsuren E, Gan L, Denecke B, et al. Delivery of microRNA-126 by apoptotic bodies induces CXCL12-dependent vascular protection. Sci Signal 2009;2(100):ra81.

[80] Vickers KC, Palmisano BT, Shoucri BM, Shamburek RD, Remaley AT. MicroRNAs are transported in plasma and delivered to recipient cells by high-density lipoproteins. Nat Cell Biol 2011;13(4):423−33.

[81] Arroyo JD, Chevillet JR, Kroh EM, Ruf IK, Pritchard CC, Gibson DF, et al. Argonaute2 complexes carry a population of circulating microRNAs independent of vesicles in human plasma. Proc Natl Acad Sci USA 2011;108(12):5003−8.

[82] Stenmark H. Rab GTPases as coordinators of vesicle traffic. Nat Rev Mol Cell Biol 2009;10(8):513−25.

[83] Forshew T, Murtaza M, Parkinson C, Gale D, Tsui DW, Kaper F, et al. Noninvasive identification and monitoring of cancer mutations by targeted deep sequencing of plasma DNA. Sci Transl Med 2012;4(136):136ra68.

[84] Skog J, Wurdinger T, van Rijn S, Meijer DH, Gainche L, Sena-Esteves M, et al. Glioblastoma microvesicles transport RNA and proteins that promote tumour growth and provide diagnostic biomarkers. Nat Cell Biol 2008;10(12):1470−6.

Chapter 6

Genome-Wide DNA Methylation Changes During Aging

Kevin C. Johnson and Brock C. Christensen
Department of Epidemiology, Geisel School of Medicine at Dartmouth, Hanover, NH, USA; Department of Pharmacology and Toxicology, Geisel School of Medicine at Dartmouth, Hanover, NH, USA; Department of Community and Family Medicine, Geisel School of Medicine at Dartmouth, Hanover, NH, USA

Chapter Outline

6.1 INTRODUCTION

The aging process is characterized by a progressive decline in a cell's ability to properly respond to damage, which, in turn, results in impaired cellular function [1]. Deteriorative aging represents a substantial increased risk factor for disease and, ultimately, death. Recent technologic advancements have been applied to aging research in humans and the results have enhanced an appreciation for the complexity surrounding the molecular aspects of aging. The common features of aging that have been defined across tissues are just beginning to be described and include telomere shortening, genomic instability, deregulated nutrient sensing, mitochondrial dysfunction, altered intracellular signaling,

M. Fraga & A.F. Fernandez (Eds): Epigenomics in Health and Disease.
DOI: http://dx.doi.org/10.1016/B978-0-12-800140-0.00006-6
© 2016 Elsevier Inc. All rights reserved.
127

cellular senescence, loss of proteostasis, stem cell exhaustion, and epigenetic alterations [2]. Despite the identification of aging hallmarks, each feature in isolation fails to fully explain the aging phenotype, and there remains a lack of consensus regarding the relative contributions of these processes to aging. Importantly, epigenetic alterations, including DNA methylation, may play critical roles in controlling the changes in cellular gene expression, stem cell exhaustion, and genomic instability that constitute the aging phenotype [3]. This chapter highlights current evidence that DNA methylation changes are associated with the aging process and occur in a genomic context-specific manner. The focus of this chapter is centered on human studies that use genome-wide approaches to measure DNA methylation and its relationship to age.

6.2 OBSERVED DIFFERENCES IN DNA METHYLATION PATTERNS WITH AGING

DNA methylation is an essential epigenetic mark that controls gene expression or gene expression potential and is almost exclusively confined to cytosines in the context of cytosine followed by guanine (CpG) nucleotide [4]. Patterns of DNA methylation are generally stable over short periods because of the heritable nature of DNA methylation from the mother cell to the daughter cell, but marked changes of methylation in eukaryotes ranging from plants to humans have been observed in association with aging [5–7]. Here, we define methylation changes as those observed differences in DNA methylation between subjects along a continuum of infancy to old age. In histologically normal cells, early studies reported age-associated DNA methylation changes in normal tissues for select cancer-related genes, such as the insulin-like growth factor 2 (*IGF2*) and estrogen receptor 1 (*ESR1*) genes [8,9]. More broadly, Fraga et al. brought attention to the impact that both age and environmental factors can have on the epigenome through study of monozygotic twins who exhibited divergence in epigenetic profiles with increasing age and with differences in lifestyles [10]. Evidence of both gene-related promoter hypermethylation and repeat element (global) hypomethylation indicates a relationship between DNA methylation and aging that is dependent on genomic context [11,12]. Roughly 70−90% of all CpG sites in the human genome are located outside CpG islands, and these sites are typically methylated, whereas sites in CpG islands are typically unmethylated in normal cells [13,14]. Of course, patterns of aberrant loss or gain of DNA methylation is dependent on the basal methylation state, which is strongly related to genomic context, such as distance and direction from transcription start site and CpG density. In aging, repeat element and nongenic CpG sites lose methylation (global methylation levels decrease), and CpG island CpGs sites may become methylated [15,16]. Early work has made it clear that both losses and gains of DNA methylation are observed with aging and that these changes are context dependent.

Preliminary studies in mammalian species noted that global (genome-wide) 5-methylcytosine content decreases in aging tissues [17,18]. The term "global methylation" has been used to describe both overall 5-methylcytosine content (i.e., aggregate methylation measurement methods that are not sequence dependent) as well as sequence-dependent measures of repeat elements, such as long interspersed nuclear elements (*LINE-1*), and ALU. Due to the distribution of CpG sites in the genome, sequence-independent methods of assessing global methylation predominantly comprise CpG sites that are not in CpG islands and that are typically methylated, similar to measures of *LINE-1* and ALU. Indeed, significant correlations between HPLC-based measures of total 5-methylcytosine content and *LINE-1* and ALU repeat element methylation have been observed in peripheral blood DNA [19]. Yet, unambiguous communication of a global methylation measurement method and clear presentation of the advantages and limitations of various approaches that measure "global methylation" are important considerations for comparisons across studies [20].

Consistent with patterns observed in age-related diseases, an age-related decline in repeat element DNA methylation has been observed in disease-free individuals. The Normative Aging Study based in the Boston area of Massachusetts in the United States is a longitudinal study of aging that measured over 1000 DNA samples from peripheral blood for *LINE-1* and ALU repeat element methylation with polymerase chain reaction (PCR) pyrosequencing. In a cross-sectional analysis, significant declines in *LINE-1* and ALU repeat element methylation were observed with aging, whereas a longitudinal analysis of samples collected 8 years apart showed significant intraindividual declines in ALU methylation but not *LINE-1* [21]. Another longitudinal study of intraindividual global DNA methylation patterns that measured peripheral blood DNA observed both decreases and increases in methylation over time [22]. The observation of both increased and decreased methylation over time may appear inconsistent with other studies, but the method used in this case was an enzyme-based luminometric methylation assay (LUMA) that cleaves $5'$-CCGC-$3'$ sites in a methylation-dependent manner, and the distribution of CCGC sites in the genome is rather different from the distribution of repeat elements and includes more regions of dense CpGs. Nonetheless, a major contribution of this work was an analysis of DNA methylation changes over time among individuals stratified by membership in families, which demonstrated a statistically significant link between change over time in methylation and familiality [22].

Aside from measures of global methylation that represent aggregate measures, recent advances in sequencing and microarray technologies have allowed for highly accurate assessment of DNA methylation levels at individual CpG sites across the genome [23]. Although whole-genome bisulfite sequencing offers a complete characterization of the epigenome, methylation microarray after bisulfite treatment has been the most commonly used genome-wide application. The adoption of array-based technologies, the most popular being the Infinium 450K array and its predecessor the Infinium 27K array, is likely due to their relative cost-effectiveness compared with whole-genome sequencing approaches [24,25]. Additional details on the design, strengths, and limitations of these DNA methylation microarrays are available in Chapter 4.

The emergence of new technologies and production of comprehensive DNA methylation profiles have both helped to confirm prior findings and identify novel characteristics of age-dependent DNA methylation. In support of previous observations, Heyn et al. used whole-genome sequencing approaches to assess the methylomes of both newborn and centenarian CD4+ T cells. The researchers reported that the centenarian DNA had lower DNA methylation content throughout the genome (i.e., at all promoter, exonic, intronic, and intergenic regions), reduced correlation in methylation status of adjacent CpGs, and age-related hypomethylation was particularly pronounced in CpG-poor promoters [26]. The group then validated its findings across newborns, middle-aged individuals, and nonagenarians by using whole-blood DNA methylation profiles from array-based measures and observed that CpG islands had a proclivity for hypermethylation with age. Conversely, new evidence has accumulated to implicate epigenetic alterations as a mechanistic link to the age-dependent deterioration of stem cell function [27]. Three studies that leveraged early microarray technology to interrogate age-dependent methylation reported that age-associated increases in DNA methylation occur preferentially at promoters of developmental genes [6,28,29]. A study by Teschendorff et al. showed that age-dependent hypermethylation is enriched at Polycomb-group target (PCGTs) genes, genes that are collectively repressed to prevent stem cell differentiation [29]. Hypermethylation of stem cell PCGTs was found to occur independent of sex, tissue type, and disease state, providing strong evidence that observed methylation changes were not generated by age-related variation in cell-type composition. Further support for the importance of specific genome-wide hypermethylation events was provided in the report by Rakyan et al., where age-associated hypermethylation was observed at bivalent chromatin domains [28]. Notably, the differentially methylated regions associated with age in this study were replicated in sorted CD4+ T cells and CD14+ monocytes, advocating that these changes arise in precursor or stem cells. Together, these results suggest that age-related hypermethylation is a characteristic of stem cells and that these changes may underlie the observed decline in stem cell function. The conservation of age-associated differentially methylated regions in several tissues suggests that age-dependent methylation changes are not acquired in a random fashion.

It is clear that widespread epigenetic variation arise over the lifetime of an individual. In addition to results reported for blood, consistent observations for genome-wide epigenetic differences have also been found in a variety of tissues that exhibit similar patterns of age-dependent methylation [30−33]. Most of these studies, designed to interrogate patterns of age-dependent methylation, have predominantly used cross-sectional approaches in nonrelated adult populations. For instance, Christensen et al. analyzed 217 nonpathologic tissues from 10 anatomic sites and observed CpG island-dependent correlations between age and methylation [30]. Meanwhile, Grönniger et al. measured DNA methylation in 50 human skin samples from healthy subjects by using the Illumina 27K array and observed that a small proportion of CpGs had age-related increases in DNA methylation of more than 20%. However, more prominent age-related differences in the epidermis compared with the dermis suggested that sun exposure may be a contributing factor to the observed age-related methylation in skin samples [31]. In a larger cohort, Hernandez et al. examined, in human subjects aged 1–102 years, Illumina 27K methylation in 387 brains, in the frontal cortex, temporal cortex, pons, and cerebellum and observed that a majority of CpG loci significantly associated with age were found in CpG islands and exhibited a positive correlation with age [32]. Likewise, Johnson et al. found an enrichment of CpG islands and PCGTs among the CpG loci that exhibited strong correlations with age in a meta-analysis among disease-free breast tissue [33]. Taken together, similar patterns of age-related DNA methylation appear to have emerged across diverse tissues and populations, whereas locus-specific associations with age may vary on a tissue-dependent basis.

To better understand the aging process and more precisely define the time point at which age-related patterning of DNA methylation is initiated, recent research has leveraged twin and pediatric populations. DNA methylation studies in twin cohorts have allowed researchers to evaluate the effect that environmental influences and genetic heritability have on epigenetic variation. For example, a study on twins established that epigenetic discordance between identical siblings increases over time [10]. A more detailed examination of age-dependent DNA methylation dynamics in early life was provided by a recent longitudinal study of buccal epithelium by Martino et al., which showed that CpG sites located in intragenic regions, enhancers, and low-density CpG promoters exhibited rapid methylation changes over the first 18 months of life [34]. Consistent with the notion of differential rates of age-related methylation change based on the developmental period, faster DNA methylation changes in peripheral blood have been observed in adolescents compared with adults [35]. Recently, Bell et al. described results from a study of 172 female twins pointing to the robust association between methylation levels of CpGs with chronological age, but only a small proportion were associated with age-phenotypes, including cholesterol levels [36]. The aging differentially methylated regions were subsequently validated in a younger population of monozygotic twins, generating additional evidence that age-related changes in methylation occur throughout an individual's lifespan. These findings indicate that age-related methylation does not appear to be random and that most age-related methylation events may be neutral or exert a minimal effect on other measures of biologic age. Thus, with the knowledge that DNA methylation is an essential regulator in development and tissue differentiation, it is plausible that age-dependent methylation patterns are established early in life and experience selective pressures over time.

Thus far, published works have demonstrated that age-dependent DNA methylation begins early in life, and the presence of conserved regions of differential methylation across multiple tissue types suggests that a shared mechanism or mechanisms participate in age-dependent methylation.

6.3   CAUSES OF AGE-RELATED DNA METHYLATION CHANGES

Although numerous studies have observed differences in DNA methylation between young and old populations, the mechanisms underlying DNA methylation changes that occur with time remain poorly understood. The deviation away from the normal patterns of DNA methylation marks is likely the result of deregulated processes responsible for maintenance of DNA methylation patterns, a process often referred to as "epigenetic drift." In general, epigenetic drift credited to aging may reflect the sum of reduced methyltransferase fidelity, an accumulation of exposures, stochastic events, and truly age-specific alterations. In trying to understand the mechanisms of epigenetic drift, it is important to consider two unifying features of age-related DNA methylation that occur irrespective of tissue type: (i) that hypermethylation tends to occur at high CpG density promoters, and (ii) loss of methylation more frequently affects CpG sites located in low CpG density regions. There also exist numerous studies that have observed consistent increases at some sites (i.e., PCGTs), suggesting that the process may not be random, or not entirely random. Accordingly, proposals for the mechanisms behind epigenetic drift should be able to account for both the bidirectional nature of age-related changes in DNA methylation patterns and sequence specificity.

DNA methylation is a reversible mark that can serve as an adaptive response to cellular challenges when functioning properly, but deregulation over time can fix these deviations away from normal. Initial explanations about what factors drive epigenetic drift posited that it was the result of changes in the expression levels of the DNA methyltransferases (DNMTs) [12]. Presently, this hypothesis is believed to be overly simplistic, since changes in expression levels of DNMTs alone would fail to account for the complex bidirectional and sequence-specific nature of age-associated DNA methylation changes. In place of this, a plausible explanation that has been put forth is that copying errors of DNA methylation may occur during DNA replication, get propagated to cellular progeny, and accumulate to measurable levels over years [37,38]. The maintenance of DNA methylation marks typically occurs with high fidelity during DNA replication, and it is now appreciated that the Ten-Eleven Translocation (TET) family of proteins, regulators of active DNA demethylation, play a key role in ensuring DNA methylation fidelity [39]. Interestingly, Williams et al. described that most of the TET1 binding sites are located at transcriptional start sites, that TET1 binds a significant proportion of polycomb group target genes, and that TET1 opposes aberrant DNA methylation at CpG-rich sequences [40].

As a consequence, failure of the TET1 enzyme to accurately remove aberrant methylation marks may lead to the age-defining feature of promoter hyper-methylation, areas that are typically devoid of methylation in a normal state. In addition, the so-called *de novo* DNA methyltransferases (DNMT3A and DNMT3B) are proposed to serve a distinct proofreader function in specific regions of DNA that need to maintain methylation [38]. The DNMT3A and DNMT3B enzymes have been shown to associate with methylated CpG islands and repeat elements and are believed to methylate sites that are missed by the maintenance DNA methyltransferase (DNMT1) at the replication fork [41]. The loss of activity by DNMT3A and DNMT3B with age could account for the loss of methylation observed at repeat elements with increasing age [42]. Further, the aforementioned replication error hypothesis also could account for the bidir-ectionality and sequence specificity of age-related DNA methylation changes, since DNA methylation fidelity may be variable across the genome in a way similar to that of higher DNA repair efficiency at actively transcribed genes [43]. Alternatively, it may be that a portion of the observed epigenetic drift represents the accumulation of insults where the epigenetic landscape is eroded over time due to environmental exposures and toxicants. The environmental influences of age-related DNA methylation will be reviewed below, but it is important to indicate that differential accrual of exposures over a lifetime may be partially responsible for observed patterns of age-related methylation.

A major consequence of epigenetic drift is likely to be the compromised function of stem cells. Stochastic variation of methylation aberrations in the stem cell population would create a heterogeneous population. Stem cells in the niche that acquire alterations that provide a survival advantage without trigger-ing apoptosis would be passed on in a clonal fashion to all their progeny. Consequently, methylation errors would accumulate over time in a population of stem cells and be propagated with a directional effect. The implications that age-associated DNA methylation changes have on stem cell plasticity are discussed further below, but the origins of increased variability of DNA methylation among individuals with age are not clear. Thus, additional studies are needed to assay the relative contributions that DNMT fidelity, environmen-tal factors, and stochastic events have on age-associated epigenetic drift.

6.4 TISSUE-SPECIFIC AND TISSUE-INDEPENDENT AGE-ASSOCIATED DNA METHYLATION

It is established that cell-type-specific differential methylation exists [30,44–51]. Cell-type specific DNA methylation patterns naturally led to the postulation that age-dependent DNA methylation may also exhibit tissue spec-ificity. Indeed, early work revealed that gene-specific associations between DNA methylation and age were observed in some tissues and not others [52]. A more comprehensive inspection of tissue-specific methylation subsequently determined that there were significant tissue-specific associations with age across normal human tissues from 11 distinct anatomic sites [30]. Following these preliminary studies, it has become clear that cell type is a potential confounder in studies of DNA methylation, particularly in complex mixtures of cells, such as blood. Thus, the more recent identification of precise differentially methylated regions for particular cell types, as well as methods for accurate classification of cell types using DNA methylation data have been developed [46,53,54]. Typically, the determination of methylation in studies of aging results from measures of a mixture of cell types present in a given tissue or blood sample. In these cases, some of the observed age-related changes in DNA methylation may reflect changes in the cellular composition of tissues that can occur through the normal aging process. Application of methods that account for changes in the proportion of cell types have identified that age-associated skews in underlying blood cell type proportions can have a dramatic impact on statistical inference [46,48,55]. Accordingly, observed and reported DNA methylation signatures of aging confounded by changes in cell proportions will not likely replicate in independent studies, underscoring the importance of adjusting for potential shifts in cell types in studies of DNA methylation, particularly when measuring peripheral blood DNA. At the same time, the impact that cell type heterogeneity may have on the relation between DNA methylation and aging has not been assessed in solid tissues to the extent that it has been in blood, where reference-based cell type adjustment methods were developed. Fortunately, algorithms to address the contribution of various cell types have recently been developed and should ensure more robust results with higher potential for replication [46,48,56].

Although it is known that underlying changes in cell-type composition represent confounding in epigenome-wide association studies, recent evidence that has accounted for cell proportions indicates that age-associated accumulation of DNA methylation changes can also occur independently of cell type [6,28,29,57]. Studies presented in Section 10.1 implicated that age-related hypermethylation of PCGTs occurs independently across diverse groups of tissues. In an attempt to enhance the biologic insights gained from array-based techniques, research has extended beyond methylation analyses of individual CpG dinucleotides to detect methylation modules (clusters of CpGs). Evidence for added interpretability of this analytic approach was shown by a study that identified an age-related co-methylation module that was present in both brain and blood tissues [57]. Another compelling case for the use of module detection algorithms was evident in a study by West et al., which integrated methylation from over 1300 tissues. The study concluded that most age-related methylation modules were tissue specific, although the few modules that were validated independent of tissue type were enriched for PCGTs and transcription factors [58]. A plausible explanation for the consistent preferential methylation targeting of PCGTs across tissues may result from shifts in the number of stem cells, given that changes in stem cell number with age have been observed [59]. Overall, there is ample evidence that both tissue-specific and tissue-independent patterns of age-related DNA methylation occur. Future studies that incorporate emerging statistical techniques for cell type adjustment may help to define the relative contribution that tissue-specific or tissue-independent DNA methylation events have on cellular function and the aging phenotype.

6.5 IMPLICATIONS: AGE-ASSOCIATED DNA METHYLATION AND DISEASE RISK

The contribution of epigenetic alterations to the development of diseases has been extensively studied and is well established (reviewed in [60]). It is less clear how patterns of age-dependent methylation might specifically predispose cells to the development of a disease phenotype. Evidence for a relationship between age-associated drift of DNA methylation and disease initiation has been prominent in studies of human cancers. Seminal work from Teschendorff et al. used genome-wide approaches and observed a highly significant overlap between genes that experience age-associated hypermethylation at promoters and genes that were frequently hypermethylated in cancer [29]. Notably, the promoter regions that underwent hypermethylation during aging and in tumorigenesis were enriched for PCGTs. A separate study that analyzed publicly available genome-wide data also described an enrichment for PCGTs among age-related DNA methylation changes found in normal breast tissues that experienced additional deregulation in breast tumors [33]. Further support for a link between age-related DNA methylation and cancer has come from recent evidence that lifestyle factors can modulate age-associated DNA methylation. Noreen et al. used the Illumina 27K array to profile 178 normal colon samples and demonstrated that age-associated DNA hypermethylation in these tissues can be suppressed by regular aspirin use and hormone replacement therapy (lifestyle factors associated with a lower risk of colorectal cancer (CRC)) but accelerated by long-term smoking and a high body mass index (BMI) (CRC risk factors) [61]. In fact, this study described in greater detail that smoking and high BMI increased the rate of methylation change for both age-related and cancer-related CpG sites, whereas aspirin and hormone replacement therapy reduced the rate of these changes. Thus, factors that are known to be modulators of CRC risk were shown to impact the stability of methylation in aging tissue. In spite of this, not all risk factors for a particular cancer are associated with changes in DNA methylation that are also observed in relation to aging. For example, a study highlighted that aging and chronic sun exposure can result in distinct DNA methylation alteration patterns [31]. Consequently, accumulating evidence has revealed a complex interaction between age-related DNA methylation, cancer, and cancer risk factors. Notably, the intersection of common epigenetic alterations indicates a high probability of shared biology between cancer and aging.

DNA methylation changes that occur with age are likely to be involved in aging-related processes and may increase susceptibility to diseases, where age serves as a major risk factor. Outside the field of cancer epigenomics, few publications exist that have examined the overlap of age-related DNA methylation and methylation patterns in common disease. One such example from Wang et al. concluded that age-specific epigenetic drift represents a substantial contribution to risk of late-onset Alzheimer disease [62]. In a different disease context, premature aging disorders, such as Hutchinson-Gilford Progeria syndrome (HGP) and Werner syndrome [9], exhibit profound DNA methylation changes in the patients with the causal genetic mutations (*WRN* and *LMNA*). Interestingly, patients who do not have genetic mutations in the *WRN* and *LMNA* genes also display marked differences in DNA methylation, which suggests that epigenetic changes can converge to impact disease phenotype that arise from distinct etiologies [63]. Together, these studies have been able to link age-related DNA methylation with disease initiation; however, a few studies are now beginning to assess the causal contribution of DNA methylation to the development of complex diseases. Most convincingly, Liu et al. leveraged a mediation analysis to filter out cell-type-specific effects presumed to be a consequence of disease in a cohort of 354 rheumatoid arthritis cases and identified methylation events that were potential mediators of genetic risk in rheumatoid arthritis [48]. From these early results, it has become clear that age-associated DNA methylation likely has an impact on the phenotype of aging-associated diseases.

The cellular mechanisms underlying age-related methylation events may modulate risk for disease development and are currently under investigation. One proposed mechanism hypothesizes that the aging process leads to the deterioration of stem cell function, which, in turn, manifests as a maladaptive immune response and an increased incidence of disease [64]. Again, support for a decline in stem cell function is largely derived from studies that have shown age-dependent methylation occurs preferentially at PCGTs and bivalent chromatin domains. Changes that occur at PCGTs in a progressive manner over time may then be responsible, in part, for the restriction of stem cell plasticity and exhaustion of stem cell populations. Even in the absence of disease, degradation of the stem cell program may perpetuate the aging process through the generation of an epigenetic field defect if growth or survival advantages are realized. From a public health perspective, critical gaps remain in understanding the relationship between a major risk factor, such as age, and various common diseases. It is plausible that age-dependent changes could be predisposing individuals to common diseases, such as cancer, and that DNA methylation might be useful for risk prediction for patients. Therefore, future studies that more comprehensively characterize those age-related DNA methylation events that raise disease risk are warranted.

6.6 ENVIRONMENTAL FACTORS THAT INFLUENCE DNA METHYLATION PATTERNS OVER TIME

DNA methylation changes with aging can vary by cell type and between individuals and may be affected by both genetic and environmental factors. Several studies have provided evidence that environmental exposures are related with DNA methylation alterations in human subjects, although most of this work has been in samples of diseased tissues. Important early work in normal tissues came from a study on twins, which observed more extensive differences in DNA methylation between identical twins who lived apart compared with twins who lived together and shared a more common environment [10]. The study of exposure-related methylation alterations in normal tissues followed on the observations of exposure-related DNA methylation alterations in diseases, such as cancer, in which tobacco, metals, and air pollutants have been implicated. However, genome-wide measures of normal tissues from disease-free individuals are lacking in target tissues for many exposure types and are more often conducted in surrogates, such as peripheral blood DNA.

Folate, an important nutrient to which humans are exposed in their diet, may affect DNA methylation [65]. Folate is a critical input to the one-carbon metabolism pathway, a major underlying biologic process important to both DNA synthesis and DNA methylation. The one-carbon metabolism pathway is a network of reactions that involves the transfer of one-carbon groups and homocysteine remethylation to methionine occurs using a donor methyl group from folate. Methionine then becomes the methyl donor for cellular methylation reactions, including DNA methylation (*S*-adenosyl methionine), and histone methylation. In an epidemiologic study of breast tumors, which provided detailed dietary data, increased total dietary folate was significantly associated with increased CpG methylation [66]. At the same time, alcohol is a common exposure that interferes with both the uptake of folate in the liver and its redistribution to tissues [67]. In a large study of blood DNA in over 1400 subjects, both alcohol and subject age were inversely related to ALU methylation, whereas BMI and smoking did not exhibit associations with repeat element methylation [68]. Underscoring the importance of potential confounding by cell type and the need for caution in interpreting studies that do not perform adjusted analyses, these authors observed that the proportion of neutrophils and lymphocytes from blood counts in these subjects were related to *LINE-1* methylation. During the *in utero* establishment of DNA methylation patterns in somatic cells, availability of critical nutrients, such as folate, is critical for appropriate development and may have life-long implications for health [69,70]. For example, folate deficiency is related to neural tube defects, and decreased *LINE-1* methylation is related to increased risk of these defects [71,72]. Exposures during critical periods, such as development, early life, and puberty, have the potential to affect DNA methylation patterning and alterations. Shared shifts in the patterns of DNA alterations due to folate and common and widespread exposures, such as alcohol, air pollution, and tobacco, may increase the difficulty in disambiguating exposure and age-related DNA methylation alterations. However, future studies that more comprehensively investigate the potential of nutrients and exposures to alter DNA methylation patterns may assist in identifying clearer patterns of age-related and exposure-related perturbations in DNA methylation. In the meantime, the notion of chronological versus biologic age has emerged in recent studies of DNA methylation and aging.

6.7 DNA METHYLATION AS AN EPIGENETIC/BIOLOGIC CLOCK

It is clear that as cells age, the placement of DNA methylation modifications becomes deregulated and some of these alterations may mark time. The aging process typically has been measured using chronological time owing to its intuitive nature; however, it is now appreciated that biologic age may be a more accurate reflection of an individual's true aging rate. Biologic age represents the intersection of chronological time with the influence that genetic and lifestyle factors have on the pace of aging. Previously, molecular markers, such as telomere length and gene expression, have been used to estimate biologic age, but these methods are currently imprecise and have practical limitations [3,73]. Recently, DNA methylation−based age predictors have emerged and are shown to be reproducible across studies and in a wide number of distinct human tissues, including blood, saliva, and an assortment of solid tissues [74−76]. In contrast to other molecular methods, the consistency of age-associated DNA methylation signatures has been high, with a striking age correlation of 0.96 from Horvath's reported 353 CpG clock [75]. A distinct epigenetic clock, using 71 methylation marks in 656 adults, was developed in blood by Hannum et al. had a similar correlation between chronological age and predicted age at 0.96 with an error of 3.9 years [74]. In the Horvath study, the author demonstrated his multi-tissue age predictor was impressively conserved across 51 normal human tissue types, across species, and also correlated with cell passage number. Clearly, in this context, the DNA methylation changes with age seem to occur independent of tissue type or cell mixture. Although a pathways analysis of the 353 CpGs that tracked to genes in this clock demonstrated an enrichment of genes involved with cell survival, cellular growth, tissue development, and cancer, it is not yet known the exact biochemical processes that underlie the epigenetic clock and alter aging phenotypes [75].

The ability to produce reliable age predictions for an individual tissue as well as specific tissues holds promise for monitoring health, predicting disease risk, and providing insights into modifiable lifestyle factors that promote healthy aging. Indeed, discrepancies found between chronological and biologic age may suggest deregulation in DNA methylation marks and indicate increased disease risk. Horvath et al. demonstrated this phenomenon of age-acceleration in a recent publication in which researchers reported that the epigenetic age of liver was increased by 2.7 years for every 10 units of BMI [77]. Using 450K methylation arrays, the group applied a version of their epigenetic clock algorithm to 141 liver samples and offered the hypothesis that the unexpectedly strong relationship between BMI and epigenetic age may, in part, serve to explain why obese individuals more frequently suffer from age-related liver comorbidities of obesity, such as insulin resistance and liver cancer. Likewise, Hannum et al. and Hovarth independently demonstrated age-acceleration in tumors by applying their respective algorithms to samples from The Cancer Genome Atlas database [74,75]. Interestingly, Horvath further noted that cancer tissues with high age acceleration tended to exhibit fewer somatic mutations. More broadly, these epigenetic clocks have also served to identify general features of aging rates, such as strong gender differences. For example, Hannum et al. used whole blood from 656 individuals to train and test their epigenetic clock and observed that the methylome of men appeared to age 4% faster than that of women [74]. Collectively, epigenetic clocks are promising tools for health assessment and understanding aging processes. At the same time, the clocks described above relied on assumptions of aging linearity by using univariate or multivariate linear models. Periods of rapid growth, such as childhood, can result in age-related methylation changes that occur at a fourfold faster rate. A study by Alisch et al. examined peripheral blood from 398 boys, ages 3−17 years, to demonstrate that age-related DNA methylation changes may not occur in a linear fashion [35]. This represents an important finding that extends beyond to future studies that adjust for age in linear models, suggesting that subjects should be closely matched for age. In spite of potential limitations of precision and modeling, epigenetic clocks that are able to measure potential age acceleration could represent a substantial risk factor for disease as well as mortality. Accordingly, the ability to accurately assess biologic age in blood samples from larger longitudinal studies, such as the Nurses Study or the Framingham Heart study, may aid researchers in the determination of factors that aim to assess and prevent disease.

6.8 SUMMARY AND FUTURE STUDIES

Substantial experimental evidence of genome-wide DNA methylation has accumulated to support a model where normal methylation patterns are deregulated with increasing age. More than 20 years have passed since the early observations of epigenetic drift, yet the exact mechanisms that drive age-dependent methylation remain incompletely defined. Most evidence supports the hypothesis that as the stem cell populations in tissues age epigenetic copying errors occur during the DNA replication process. The propagation of errors would restrict stem cell plasticity through the gradual fixation of methylation errors, thereby more permanently affecting gene expression programs and genomic stability. In its evolutionarily beneficial function, epigenetic drift may serve as an adaptive response to environment challenges to promote a more favorable cellular state. However, over time, this phenotypic flexibility may be detrimental. Ultimately, the results of epigenetic drift have the potential to make significant contributions to the aging phenotype.

Overall, the patterns of age-dependent methylation that have consistently emerged across distinct tissues and different methylation platforms are as follows: (i) High-density CpG promoters may acquire methylation over time, especially at developmental genes; and (ii) CpGs outside these high-density CpG promoters tend to lose methylation with age. Beyond these unifying features, other observations have been too sparse, and the technologies and analytical methods too immature to establish a more detailed portrait of the aging epigenome. Thus, some critical questions about age-related DNA methylation remain and require further investigation. For example, how do age-dependent DNA methylation changes result in meaningful changes to aging-phenotype? Do other epigenetic factors involved in chromatin remodeling play a role in the faithful maintenance of DNA methylation patterns? How does accelerated age in different tissues impact disease risk? Conversely, does the prevention or slowing down of age-related epigenetic drift lead to a reduction in the incidence of diseases associated with aging in humans? Future comprehensive assessments of methylation patterns across tissue-types coupled with appropriate population-based studies are necessary to answer these questions.

REFERENCES

[1] Johnson FB, Sinclair DA, Guarente L. Molecular biology of aging. Cell 1999;96 (2):291−302.

[2] Lopez-Otin C, Blasco MA, Partridge L, Serrano M, Kroemer G. The hallmarks of aging. Cell 2013;153(6):1194−217. 3836174.

[3] de Magalhaes JP, Curado J, Church GM. Meta-analysis of age-related gene expression profiles identifies common signatures of aging. Bioinformatics 2009;25(7):875−81. 2732303.

[4] Deaton AM, Bird A. CpG islands and the regulation of transcription. Genes Dev 2011;25(10):1010−22. 3093116.

[5] Fraga MF, Rodríguez R, Cañal MJ. Genomic DNA methylation−demethylation during aging and reinvigoration of Pinus radiata. Tree Physiol 2002;22(11):813−16.

[6] Maegawa S, Hinkal G, Kim HS, Shen L, Zhang L, Zhang J, et al. Widespread and tissue specific age-related DNA methylation changes in mice. Genome Res 2010;20(3):332−40. 2840983.

[7] Vanyushin BF, Tkacheva SG, Belozersky AN. Rare bases in animal DNA. Nature 1970;225(5236):948−9.

[8] Issa JP, Ottaviano YL, Celano P, Hamilton SR, Davidson NE, Baylin SB. Methylation of the oestrogen receptor CpG island links ageing and neoplasia in human colon. Nat Genet 1994;7(4):536−40.

[9] Issa JP, Vertino PM, Boehm CD, Newsham IF, Baylin SB. Switch from monoallelic to biallelic human IGF2 promoter methylation during aging and carcinogenesis. Proc Natl Acad Sci USA 1996;93(21):11757–62. 38131.

[10] Fraga MF, Ballestar E, Paz MF, Ropero S, Setien F, Ballestar ML, et al. Epigenetic differences arise during the lifetime of monozygotic twins. Proc Natl Acad Sci USA 2005;102 (30):10604–9.

[11] Illingworth RS, Bird AP. CpG islands—'a rough guide'. FEBS Lett 2009;583 (11):1713–20.

[12] Richardson B. Impact of aging on DNA methylation. Ageing Res Rev 2003;2(3):245–61.

[13] Herman JG, Baylin SB. Gene silencing in cancer in association with promoter hypermethylation. N Engl J Med 2003;349(21):2042–54.

[14] Jones PA, Takai D. The role of DNA methylation in mammalian epigenetics. Science 2001;293(5532):1068–70.

[15] Aguilera O, Fernandez AF, Munoz A, Fraga MF. Epigenetics and environment: a complex relationship. J Appl Physiol (1985) 2010;109(1):243–51.

[16] Calvanese V, Fernandez AF, Urdinguio RG, Suarez-Alvarez B, Mangas C, Perez-Garcia V, et al. A promoter DNA demethylation landscape of human hematopoietic differentiation. Nucleic Acids Res 2012;40(1):116–31. 3245917.

[17] Vanyushin BF, Nemirovsky LE, Klimenko VV, Vasiliev VK, Belozersky AN. The 5-methylcytosine in DNA of rats. Tissue and age specificity and the changes induced by hydrocortisone and other agents. Gerontologia 1973;19(3):138–52.

[18] Wilson VL, Smith RA, Ma S, Cutler RG. Genomic 5-methyldeoxycytidine decreases with age. J Biol Chem 1987;262(21):9948–51.

[19] Weisenberger DJ, Campan M, Long TI, Kim M, Woods C, Fiala E, et al. Analysis of repetitive element DNA methylation by MethyLight. Nucleic Acids Res 2005;33 (21):6823–36. 1301596.

[20] Nelson HH, Marsit CJ, Kelsey KT. Global methylation in exposure biology and translational medical science. Environ Health Perspect 2011;119(11):1528–33. 3226501.

[21] Bollati V, Schwartz J, Wright R, Litonjua A, Tarantini L, Suh H, et al. Decline in genomic DNA methylation through aging in a cohort of elderly subjects. Mech Ageing Dev 2009;130(4):234–9. 2956267.

[22] Bjornsson HT, Sigurdsson MI, Fallin MD, Irizarry RA, Aspelund T, Cui H, et al. Intraindividual change over time in DNA methylation with familial clustering. JAMA 2008;299(24):2877–83. 2581898.

[23] Laird PW. Principles and challenges of genomewide DNA methylation analysis. Nat Rev Genet 2010;11(3):191–203.

[24] Lister R, Pelizzola M, Dowen RH, Hawkins RD, Hon G, Tonti-Filippini J, et al. Human DNA methylomes at base resolution show widespread epigenomic differences. Nature 2009;462(7271):315–22. 2857523.

[25] Sandoval J, Heyn H, Moran S, Serra-Musach J, Pujana MA, Bibikova M, et al. Validation of a DNA methylation microarray for 450,000 CpG sites in the human genome. Epigenetics 2011;6(6):692–702.

[26] Heyn H, Li N, Ferreira HJ, Moran S, Pisano DG, Gomez A, et al. Distinct DNA methylomes of newborns and centenarians. Proc Natl Acad Sci USA 2012;109(26):10522–7. 3387108.

[27] Liu L, Rando TA. Manifestations and mechanisms of stem cell aging. J Cell Biol 2011;193(2):257–66. 3080271.

[28] Rakyan VK, Down TA, Maslau S, Andrew T, Yang TP, Beyan H, et al. Human aging-associated DNA hypermethylation occurs preferentially at bivalent chromatin domains. Genome Res 2010;20(4):434−9. 2847746.

[29] Teschendorff AE, Menon U, Gentry-Maharaj A, Ramus SJ, Weisenberger DJ, Shen H, et al. Age-dependent DNA methylation of genes that are suppressed in stem cells is a hallmark of cancer. Genome Res 2010;20(4):440−6. 2847747.

[30] Christensen BC, Houseman EA, Marsit CJ, Zheng S, Wrensch MR, Wiemels JL, et al. Aging and environmental exposures alter tissue-specific DNA methylation dependent upon CpG island context. PLoS Genet 2009;5(8):e1000602. 2718614.

[31] Gronniger E, Weber B, Heil O, Peters N, Stab F, Wenck H, et al. Aging and chronic sun exposure cause distinct epigenetic changes in human skin. PLoS Genet 2010;6(5): e1000971. 2877750.

[32] Hernandez DG, Nalls MA, Gibbs JR, Arepalli S, van der Brug M, Chong S, et al. Distinct DNA methylation changes highly correlated with chronological age in the human brain. Hum Mol Genet 2011;20(6):1164−72. 3043665.

[33] Johnson KC, Koestler DC, Cheng C, Christensen BC. Age-related DNA methylation in normal breast tissue and its relationship with invasive breast tumor methylation. Epigenetics 2014;9(2):268−75. 3962537.

[34] Martino D, Loke YJ, Gordon L, Ollikainen M, Cruickshank MN, Saffery R, et al. Longitudinal, genome-scale analysis of DNA methylation in twins from birth to 18 months of age reveals rapid epigenetic change in early life and pair-specific effects of discordance. Genome Biol 2013;14(5):R42. 4054827.

[35] Alisch RS, Barwick BG, Chopra P, Myrick LK, Satten GA, Conneely KN, et al. Age-associated DNA methylation in pediatric populations. Genome Res 2012;22(4):623−32. 3317145.

[36] Bell JT, Tsai PC, Yang TP, Pidsley R, Nisbet J, Glass D, et al. Epigenome-wide scans identify differentially methylated regions for age and age-related phenotypes in a healthy ageing population. PLoS Genet 2012;8(4):e1002629. 3330116.

[37] Issa JP. Aging and epigenetic drift: a vicious cycle. J Clin Invest 2014;124(1):24−9. 3871228.

[38] Jones PA, Liang G. Rethinking how DNA methylation patterns are maintained. Nat Rev Genet 2009;10(11):805−11. 2848124.

[39] Bell JT, Spector TD. A twin approach to unraveling epigenetics. Trends Genet 2011;27 (3):116−25. 3063335.

[40] Williams K, Christensen J, Pedersen MT, Johansen JV, Cloos PA, Rappsilber J, et al. TET1 and hydroxymethylcytosine in transcription and DNA methylation fidelity. Nature 2011;473(7347):343−8. 3408592.

[41] Jeong S, Liang G, Sharma S, Lin JC, Choi SH, Han H, et al. Selective anchoring of DNA methyltransferases 3A and 3B to nucleosomes containing methylated DNA. Mol Cell Biol 2009;29(19):5366−76. 2747980.

[42] Jintaridth P, Mutirangura A. Distinctive patterns of age-dependent hypomethylation in interspersed repetitive sequences. Physiol Genomics 2010;41(2):194−200.

[43] Haines NM, Kim YI, Smith AJ, Savery NJ. Stalled transcription complexes promote DNA repair at a distance. Proc Natl Acad Sci USA 2014;111(11):4037−42. 3964087.

[44] Bloushtain-Qimron N, Yao J, Snyder EL, Shipitsin M, Campbell LL, Mani SA, et al. Cell type-specific DNA methylation patterns in the human breast. Proc Natl Acad Sci USA 2008;105(37):14076−81. 2532972.

[45] Eckhardt F, Lewin J, Cortese R, Rakyan VK, Attwood J, Burger M, et al. DNA methylation profiling of human chromosomes 6, 20 and 22. Nat Genet 2006;38(12):1378—85. 3082778.

[46] Houseman EA, Accomando WP, Koestler DC, Christensen BC, Marsit CJ, Nelson HH, et al. DNA methylation arrays as surrogate measures of cell mixture distribution. BMC Bioinformatics 2012;13:86. 3532182.

[47] Illingworth R, Kerr A, Desousa D, Jorgensen H, Ellis P, Stalker J, et al. A novel CpG island set identifies tissue-specific methylation at developmental gene loci. PLoS Biol 2008;6(1):e22.

[48] Liu Y, Aryee MJ, Padyukov L, Fallin MD, Hesselberg E, Runarsson A, et al. Epigenome-wide association data implicate DNA methylation as an intermediary of genetic risk in rheumatoid arthritis. Nat Biotechnol 2013;31(2):142—7. 3598632.

[49] Maunakea AK, Nagarajan RP, Bilenky M, Ballinger TJ, D'Souza C, Fouse SD, et al. Conserved role of intragenic DNA methylation in regulating alternative promoters. Nature 2010;466(7303):253—7. 3998662.

[50] Rakyan VK, Down TA, Thorne NP, Flicek P, Kulesha E, Graf S, et al. An integrated resource for genome-wide identification and analysis of human tissue-specific differentially methylated regions (tDMRs). Genome Res 2008;18(9):1518—29. 2527707.

[51] Wu HC, Delgado-Cruzata L, Flom JD, Kappil M, Ferris JS, Liao Y, et al. Global methylation profiles in DNA from different blood cell types. Epigenetics 2011;6(1):76—85. 3052916.

[52] Kwabi-Addo B, Chung W, Shen L, Ittmann M, Wheeler T, Jelinek J, et al. Age-related DNA methylation changes in normal human prostate tissues. Clin Cancer Res 2007;13 (13):3796—802.

[53] Accomando WP, Wiencke JK, Houseman EA, Nelson HH, Kelsey KT. Quantitative reconstruction of leukocyte subsets using DNA methylation. Genome Biol 2014;15(3): R50. 4053693.

[54] Koestler DC, Marsit CJ, Christensen BC, Accomando W, Langevin SM, Houseman EA, et al. Peripheral blood immune cell methylation profiles are associated with nonhematopoietic cancers. Cancer Epidemiol Biomarkers Prev 2012;21(8):1293—302. 3415587.

[55] Langevin SM, Houseman EA, Christensen BC, Wiencke JK, Nelson HH, Karagas MR, et al. The influence of aging, environmental exposures and local sequence features on the variation of DNA methylation in blood. Epigenetics 2011;6(7):908—19. 3154431.

[56] Houseman EA, Molitor J, Marsit CJ. Reference-free cell mixture adjustments in analysis of DNA methylation data. Bioinformatics 2014;30(10):1431—9. 4016702.

[57] Horvath S, Zhang Y, Langfelder P, Kahn RS, Boks MP, van Eijk K, et al. Aging effects on DNA methylation modules in human brain and blood tissue. Genome Biol 2012;13 (10):R97. 4053733.

[58] West J, Beck S, Wang X, Teschendorff AE. An integrative network algorithm identifies age-associated differential methylation interactome hotspots targeting stem-cell differentiation pathways. Sci Rep 2013;3:1630. 3620664.

[59] Jones DL, Rando TA. Emerging models and paradigms for stem cell ageing. Nat Cell Biol 2011;13(5):506—12. 3257978.

[60] Robertson KD. DNA methylation and human disease. Nat Rev Genet 2005;6(8):597—610.

[61] Noreen F, Roosli M, Gaj P, Pietrzak J, Weis S, Urfer P, et al. Modulation of age- and cancer-associated DNA methylation change in the healthy colon by aspirin and lifestyle. J Natl Cancer Inst 2014;106(7). 4112799.

[62] Wang SC, Oelze B, Schumacher A. Age-specific epigenetic drift in late-onset Alzheimer's disease. PLoS One 2008;3(7):e2698. 2444024.

[63] Heyn H, Moran S, Esteller M. Aberrant DNA methylation profiles in the premature aging disorders Hutchinson-Gilford Progeria and Werner syndrome. Epigenetics 2013;8 (1):28–33. 3549877.

[64] Sun D, Luo M, Jeong M, Rodriguez B, Xia Z, Hannah R, et al. Epigenomic profiling of young and aged HSCs reveals concerted changes during aging that reinforce self-renewal. Cell Stem Cell 2014;14(5):673–88. 4070311.

[65] Ly A, Hoyt L, Crowell J, Kim YI. Folate and DNA methylation. Antioxid Redox Signal 2012;17(2):302–26.

[66] Christensen BC, Kelsey KT, Zheng S, Houseman EA, Marsit CJ, Wrensch MR, et al. Breast cancer DNA methylation profiles are associated with tumor size and alcohol and folate intake. PLoS Genet 2010;6(7):e1001043.

[67] Hillman RS, Steinberg SE. The effects of alcohol on folate metabolism. Annu Rev Med 1982;33:345–54.

[68] Zhu ZZ, Hou L, Bollati V, Tarantini L, Marinelli B, Cantone L, et al. Predictors of global methylation levels in blood DNA of healthy subjects: a combined analysis. Int J Epidemiol 2012;41(1):126–39. 3304518.

[69] Li E. Chromatin modification and epigenetic reprogramming in mammalian development. Nat Rev Genet 2002;3(9):662–73.

[70] Gluckman PD, Hanson MA, Cooper C, Thornburg KL. Effect of in utero and early-life conditions on adult health and disease. N Engl J Med 2008;359(1):61–73. 3923653.

[71] Wang L, Wang F, Guan J, Le J, Wu L, Zou J, et al. Relation between hypomethylation of long interspersed nucleotide elements and risk of neural tube defects. Am J Clin Nutr 2010;91(5):1359–67.

[72] Daly LE, Kirke PN, Molloy A, Weir DG, Scott JM. Folate levels and neural tube defects. Implications for prevention. JAMA 1995;274(21):1698–702.

[73] Benetos A, Okuda K, Lajemi M, Kimura M, Thomas F, Skurnick J, et al. Telomere length as an indicator of biological aging: the gender effect and relation with pulse pressure and pulse wave velocity. Hypertension 2001;37(2 Pt 2):381–5.

[74] Hannum G, Guinney J, Zhao L, Zhang L, Hughes G, Sadda S, et al. Genome-wide methylation profiles reveal quantitative views of human aging rates. Mol Cell 2013;49 (2):359–67. 3780611.

[75] Horvath S. DNA methylation age of human tissues and cell types. Genome Biol 2013;14 (10):R115. 4015143.

[76] Weidner CI, Lin Q, Koch CM, Eisele L, Beier F, Ziegler P, et al. Aging of blood can be tracked by DNA methylation changes at just three CpG sites. Genome Biol 2014;15(2): R24. 4053864.

[77] Horvath S, Erhart W, Brosch M, Ammerpohl O, von Schonfels W, Ahrens M, et al. Obesity accelerates epigenetic aging of human liver. Proc Natl Acad Sci USA 2014;111 (43):15538–43. 4217403.

Chapter 7

The Dynamics of Histone Modifications During Aging

Anthony J. Bainor and Gregory David
Department of Biochemistry and Molecular Pharmacology and NYU Cancer Institute,
NYU Langone Medical Center, New York, NY, USA

Chapter Outline

7.1 INTRODUCTION

The age of genome-wide approaches presents a unique opportunity to probe the genomic changes that correlate with aging. The phenotypic outcomes accompanying the aging process are hypothesized to result from the combination of genetics as well as environmental stresses and include telomere shortening, deregulation of transcription, sequence-specific transcription factors, reactive oxygen species (ROS), and genomic damage [1−3]. Given that aging is accompanied by transcriptional changes and increased genomic instability, it is likely that the chromatin fiber plays a pivotal role in this process. One of the current hypotheses linking chromatin and aging is the loss

M. Fraga & A.F. Fernandez (Eds): Epigenomics in Health and Disease.
DOI: http://dx.doi.org/10.1016/B978-0-12-800140-0.00007-8
© 2016 Elsevier Inc. All rights reserved.   **145**

of heterochromatin model of cell aging [4,5]. This theory postulates that the abundance of heterochromatin-rich regions decreases in aging cells as a result of altered replication. Loss of heterochromatin then triggers perturbations in gene expression at the borders of these domains, resulting in aging-associated phenotypes. Although it is unlikely that this model fully explains the array of changes associated with aging, the loss of heterochromatin theory is the first to suggest the direct relationship between aging and the chromatin fiber. The reversible nature of epigenetic modifications supports the notion that therapeutic intervention targeting chromatin modifiers could provide additional treatment options for diseases associated with premature aging. One such example is the combined treatment of progeria cells with the HMG-CoA (3-hydroxy-3-methylglutaryl-coenzyme A) reductase inhibitor mevinolin and the histone deacetylase inhibitor trichostatin A, which leads to heterochromatin reorganization and restoration of normal transcription levels [6]. Although current studies aim to uncover a correlation between specific cellular mechanisms and aging, one of the most pertinent and rather difficult questions to address is which factors drive aging and which factors are merely passengers of aging. In this chapter, we will discuss the studies that investigate the contribution of histone-related processes to aging. More specifically, we will explore nucleosome density, histone variants, and histone modifications, and their influence on the aging process. We will also present studies connecting histone biology to cellular senescence and the DNA damage response (DDR) and explain how each process is associated with aging. Finally, we will discuss how studying progeria syndromes can reveal unexpected links between histones and the aging process.

7.2   PART 1: NUCLEOSOME DENSITY AND AGING

A nucleosome consists of 147 bp of DNA wrapped around a histone octamer. Each octamer consists of two H3/H4 histone heterodimers, and two H2A/H2B histone heterodimers. The positioning of the nucleosome on the chromatin fiber is dynamic; certain histones are constantly being displaced and replaced as a result of transcriptional and nontranscriptional events influencing gene expression. As aging is accompanied by significant transcriptional changes [1,3], it is tempting to postulate that nucleosomes play a pivotal role in the gene expression changes accompanying the aging process.

Studies using micrococcal nuclease (Mnase) digestion in aged human fibroblasts demonstrated that nucleosome spacing increases with age [7,8]. Since this realization, much work has been done to further analyze nucleosome abundance and positioning, and their connection to the aging process. *Saccharomyces cerevisiae*, or budding yeast, represents an ideal organism to study global changes in genome organization as it ages due to its easily defined replicative lifespan (RLS). RLS refers to the number of daughter cells produced by a mother cell prior to senescence and thus provides a model of aging in mitotically active cells [9]. Studies utilizing budding yeast showed an approximately 50% decrease in nucleosome density correlated with RLS aging, resulting in a more open chromatin state [10]. Conversely, ectopic expression of histone H3/H4 leads to a 50% increase in RLS [10]. Interestingly, the experimental depletion of histone H4 resulted in the altered transcription of several genes, including de-repression of those near the subtelomeric regions likely due to loss of heterochromatin; however, differences in RLS were not reported in this study [11]. In spite of the aforementioned studies, the causative role of nucleosome loss in transcriptional deregulation and aging had not been comprehensively identified until recently. The global loss of nucleosomes observed upon replicative aging correlates with de-repression of retrotransposons, elevated levels of chromosomal translocations, and increased levels of γH2A [12]. With age, nucleosome positioning becomes less precise and exhibits reduced periodicity [12]. Nucleosomes specifically located in the promoter–distal gene bodies (450 bp downstream from the transcription start site to 150 bp upstream of the transcription termination site) were found to be displaced approximately 50–100 base-pairs from their original positions [12]. Furthermore, the levels of nucleosomes lost at promoters correlated with gene induction in aging cells. For example, YLR194C, a gene encoding a glycosylphosphatidylinositol (GPI)-anchored protein, exhibits a 14-fold age-dependent increase in expression. Consistently, deletion of YLR194C increased lifespan, whereas its overexpression shortened lifespan [12]. Together, these observations suggest that global alterations of nucleosome density result in gene-specific transcriptional defects, which contribute to aging phenotypes. One important caveat related to studies utilizing genome-wide approaches lies within the methods used to analyze such results. These methods can be difficult to interpret when effects occur on a genome-wide scale, since the identity of the transcripts used as a reference are unknown. This has led to discrepancies which remain to be definitively resolved.

Although studies in yeast have shed some light on the involvement of nucleosome density in the aging process, it is still unclear whether the loss of histones and nucleosome density drives aging in mammalian cells. One study demonstrated that in aged human fibroblasts, telomeric shortening and the subsequent DDR, both hallmarks of cellular aging, lead to a decrease in core histone levels as well as the H3.1/H3.2 chaperones Asf1 and CAF-1. Interestingly, these changes were reversed upon ectopic expression of hTERT [13]. It is possible that decreases in both histone and chaperone abundance prevents cells from restoring the chromatin landscape following replication. This finding may explain how age-associated telomere attrition causes global alterations in gene expression and points to a relationship among telomere shortening, histone regulation, and aging. Surprisingly, although aged cells from various species have reduced histone levels compared with their younger counterparts, the transcription of histones increases with age [14], pointing to possible feedback mechanisms designed to restore histone levels.

Although most studies observe a reduction of histone levels and nucleosome density during the aging process, in *Drosophila melanogaster*, total histone levels do not decrease between 10 and 40 days after eclosion [15]. Therefore, there may be organism-dependent differences in the aging process, which may reflect divergent behavior between dividing and postmitotic cells.

In conclusion, histone levels and nucleosome density decrease with age leading to a more open chromatin landscape and an increase in either global or gene-specific transcription. As replicative aging in yeast can be delayed upon ectopic expression of core histones, the underlying basis for some of the age-related cellular phenotypes may be the increased expression of specific genes resulting from the relaxation of the chromatin fiber. Whether these observations can be extrapolated to more complex organisms remains unclear and will be difficult to address experimentally.

7.3   PART 2: HISTONE VARIANTS

Although the ratios between each core histone are not known to change within each histone octamer, the core histones within the octamer are constantly being exchanged and replaced. Adding to this complexity is the existence of histone variants found incorporated throughout the chromatin fiber. These histone variants can be divided into three subtypes: (i) replication-dependent (RD), (ii) replication-independent and cell cycle phase-independent (RI), and (iii) tissue-specific (TS) [16,17]. In this section, we will present evidence implicating RD and RI histone variants in the aging process.

7.3.1   Histone H3

Five variants of histone H3 exist in *Homo sapiens*: CENP-A, H3.1, H3.2, H3.3, and H3.1t. Of these, H3.1, H3.2, and H3.3 have been implicated in aging. H3.3 is an RI histone that is incorporated into nucleosomes via chaperones Asf1a/b and HIRA throughout the genome. On the other hand, H3.1/H3.2 are RD histones and are incorporated via CAF-1 and Asf1a/b [18].

Both rat brains and human diploid fibroblasts exhibit decreased levels of H3.1 and H3.2 with age, whereas their H3.3 levels increase [19,20]. Similarly, chickens and *Mus musculus* (house mouse) exhibit increased H3.3 levels with age [21,22]. Consistently, Asf1a/b and CAF1 levels decrease in aged human fibroblasts [13], whereas HIRA levels increase in aged primate skin fibroblasts [23]. Although not experimentally demonstrated, global increases in H3.3 with age may result from the inability of aged cells to divide. This would lead to the exchange of RD H3.1 and H3.2 for RI H3.3; however, whether the global increase in H3.3 is causative or correlative in the aging process remains unknown.

7.3.2   Histone H2A

Five variants of H2A exist: H2A.Z, MacroH2A, H2A-Bbd, H2AvD, and H2A.X [24,25]. H2A variants differ at their C-terminal tails in both length and sequence, and occupy different loci in the genome. Of the five H2A variants, H2A.Z, H2A.X, and macroH2A have been implicated in the aging process.

The Swr1 complex deposits H2A.Z at the transcriptional start site of actively expressed genes by replacing an H2A/H2B dimer with an H2A.Z/H2B dimer [26]. Yet H2A.Z has a repressive function when found in gene bodies [27]. Although the levels of H2A.Z do not appear to change during aging, knockdown of H2A.Z in human fibroblasts causes premature senescence. Consistently, knockdown of p400, the human H2A.Z histone chaperone, also induces premature senescence [28,29]. The global levels of H2A.Z do not appear to change with age, but the occupancy of H2A.Z may differ in an age-related manner. This particularly pertains to the notion that nucleosomes, including both H2A.Z and H3.3, are susceptible to disassembly *in vivo* [30].

The histone variant H2A.X is distributed throughout the genome and is essential for double-strand break repair. The ATM (ataxia telangiectasia mutated) and ATR (ataxia telangiectasia and Rad3 related) kinases phosphorylate H2A.X (γH2A.X) at sites of DNA double-strand breaks [31,32]. This event is required for the assembly of specific DNA-repair complexes on damaged DNA, and H2A.X depletion results in chromosomal instability and defective repair [33]. Although the direct involvement of H2A.X in the aging process has yet to be determined, our current understanding suggests a passenger type role. Since the amount of DNA damage present in both *M. musculus* and humans increases with age [34], H2A.X is likely involved in the aging process through the DDR and senescence processes due to its dynamic role in cell cycle exit and DNA damage repair.

MacroH2A, originally found deposited on the inactive X-chromosome, is involved in transcriptional silencing and heterochromatin formation [35]. MacroH2A is also directly involved in the formation of senescence-associated heterochromatic foci or (SAHF). Consistently, levels of macroH2A have been shown to increase in human fibroblasts during replicative senescence, as well as in *M. musculus* and primates with age [36]. Whether this increase is a causative event or merely due to SAHF formation in aged senescent cells remains unclear.

7.3.3   Senescence and Histone Variants

As human cells divide, their telomeres shorten, eventually leading to attrition and cell cycle exit. This phenomenon is described as replicative aging, or cellular senescence, and the contribution of cellular senescence to the aging process has been thoroughly investigated. As mentioned above, the exchange of histone variants necessary for the formation and accumulation of SAHFs, a hallmark of senescence, requires histone chaperones. HIRA and Asf1a aid in the deposition of macroH2A and are required for SAHF formation [37]. Interestingly, the knockdown of HIRA prevents Braf-induced senescence [38]. This presents a counterintuitive result: HIRA is responsible for the exchange of a histone variant H3.3, which correlates with transcriptionally active loci but is also required for the induction of senescence and SAHF formation. Asf1a levels decrease with age, presenting another conflicting result: Asf1a is required for H3.1 and H3.2 exchange but is also required for macroH2A exchange, which increases with age. One explanation is that histone chaperones are promiscuous, which results in the ability of different histone chaperones to act on an array of histone variants. Also, histone chaperones that decrease with age may still be at levels which allow for adequate functionality.

In conclusion, specific variants of histone H3 and H2A are associated with the aging process: H3.1, H3.2 and H3.3, and H2A.X, H2A.Z, and macroH2A. Histone variants H3.1 and H3.2 are preferentially displaced by H3.3 as cells age correlating with active transcription. Conversely, the H2A variants aid in the formation of SAHFs, which correspond to dense chromatin regions. How the balance in chromatin relaxation and condensation influences the aging process remains elusive but is likely to contribute through the transcriptional regulation of specific loci.

7.4 PART 3: HISTONE MODIFICATIONS

Post-translational modifications of histones have been extensively covered in several reviews. Histones can be methylated, acetylated, phosphorylated, and ubiquitinylated, along with other modifications. These covalent modifications can occur on numerous residues of each histone but are mostly concentrated toward the more accessible N-terminal tails [39]. Histone N-terminal tails protrude out of the nucleosome and, through different modifications, generate recruitment interfaces for specific effectors, resulting in specific transcriptional outcomes. As gene transcription varies with age, histone modifications have been directly implicated in the aging process.

The dynamics of histone modifications and their effects on transcriptional outcomes in aging have been extensively studied in aging organisms such as *S. cerevisiae*, *Caenorhabditis elegans* and *M. musculus* [40−42]. Human nucleosomes also exhibit significant and specific differences in histone modifications, such as acetylation and methylation with age [43]. However, it remains unclear whether changes in histone modifications drive aging or are merely a consequence of aging. In this section, we will discuss the effects of activating and repressive histone marks on the aging process and discuss data that suggest either a causative role or a correlative role.

7.4.1   Activating Histone Marks

7.4.1.1   H3K4me3

Trimethylation of histone 3 on lysine residue 4 (H3K4me3) is defined as a mark of active gene transcription and is enriched at the promoters of active genes [44,45]. In *C. elegans*, depletion of ASH2, WDR5, and H3K4me3 methyltransferase SET-2, all components of the H3K4 trimethylation molecular machinery, results in a germline-dependent increase in lifespan due to the silencing of age-related genes. Conversely, when RBR-2, the *C. elegans* homolog of KDM5, a family of H3K4me3 demethylases (HDM), is depleted, global levels of H3K4me3 increase, correlating with a decreased lifespan. Importantly, ectopic RBR-2 expression in the germline is sufficient to extend lifespan [46,47]. It is hypothesized that the balance between RBR-2 and SET2 activity regulates lifespan by modulating global levels of H3K4me3. These results suggest that levels of H3K4me3 inversely correlate with lifespan, in a transgenerational manner [48]. Interestingly, studies have demonstrated that the silencing of the H3K4me1 and −me2 demethylase LSD1 increases lifespan by 25% in *C. elegans*. Similarly, the administration of lithium chloride (LiCl), which decreases LSD1 expression, increases lifespan by 48% [49]. Of note is the peculiar conflicting phenotype elicited by RBR2 and LSD1. This phenomenon could be due, in part, to differential location of action on target genes and in tissues, as well as different temporal regulation. Alternatively, the global amount of H3K4me1/2 may dictate longevity. In *D. melanogaster*, Compass, Trithorax (Trx), and Trx-related complexes are responsible for H3K4 methylation [50]. Silencing of Lid, the RBR-2/KDM5 ortholog, results in an 18% lifespan reduction in males [51]. Surprisingly, Trx, an H3K4me3 histone methyltransferase (HMT) and homolog of Set2, does not seem to affect the lifespan of male *D. melanogaster* [52]. Importantly, until the target loci of the HMT/HDM are identified and their contribution to aging has been elucidated, one cannot rule out the possibility that the KMT and KDM mentioned here target nonhistone proteins to mediate their effects on lifespan.

7.4.1.2   Acetylation

Histone acetylation has been primarily associated with active gene transcription, and its loss has been associated with aging. Global histone acetylation decreases with age in human diploid fibroblasts [53], and in the brains of aged *M. musculus* and *Rattus norvegicus* [54]. Similarly, a decrease in histone acetylation is detected in the cochlea of old *M. musculus* when compared with young *M. musculus* [55]. Finally, in a mouse model of progeria, a premature aging disease, H4K16 was found to be hypoacetylated [56]. However, while much of the *in vivo* mammalian data indicate decreased global histone acetylation with age, data in lower organisms and *in vitro* mammalian culture systems suggest that acetylation of specific lysine residues on histones contributes differently to the aging process.

H4K16ac

H4K16 acetylation is defined as a mark of gene activation [57]. In *S. cerevisiae*, global levels of H4K16ac increase with age, especially at the telomere-proximal sites, which reduce nucleosome–nucleosome interactions [10,42,57]. Depletion of SAS2, an H4K16 histone acetyltransferase (HAT), increases the global levels of H4K16ac and extends the RLS of yeast. Conversely, mutation of lysine 16 to glutamine, a mimic of acetylation, decreases RLS [42]. This collection of studies demonstrates a role for H4K16ac in aging yeast.

The silent information regulator proteins (Sir) are a family of conserved proteins originally discovered in yeast that promote the formation of hetero-chromatin at telomeres, rDNA, and other silent loci through deacetylation of histone N-terminal tails [58]. Consistent with their function in heterochromatin formation, the Sir proteins deacetylate H4K16, and are critical for lifespan extension in yeast, *C. elegans* and *D. melanogaster* [59]. Sir2 levels decrease with age causing an increase in H4K16ac that profoundly affects heterochro-matic regions [42,60,61]. Conversely, the ectopic expression of Sir2 decreases the global levels of H4K16ac, and increases RLS [42,62]. In addition, the shortened lifespan caused by Sir2 mutation is also due to increased recombina-tion within rDNA, which increases the generation of extrachromosomal rDNA circles (ERCs) [62]. Interestingly, loss of histone density at the regions of high H4K16ac near the telomere is observed with age, complementary to the aforementioned age-related loss of nucleosome density [42].

In yeast, Sir3 and Sir4 loss-of-function mutations cause the simultaneous expression of a and α mating type information. This results in increased rDNA recombination and production of ERCs, reducing the lifespan of the organism [62]. Additionally, a specific mutant of Sir4 that renders it unable to be recruited to hidden MAT loci (HM loci) and telomeres promotes a 30% extension of lifespan in *S. cerevisiae* [63]. A follow-up study established that in yeast harboring this mutant, Sir3p and Sir4p were redirected to the nucleolus, concomitant with the relocation of the entire Sir complex from the telomeres to the nucleolus in aged wild-type *S. cerevisiae* [64].

The Sir family of proteins also affect aging in multicellular organisms. One study in *C. elegans* identified that an extra copy of the Sir2 ortholog, Sir-2.1, results in an increase in lifespan by 50% and requires the FOXO transcription factor DAF-16, regulated by the insulin/IGF-1 pathway [65]. Consistently, the ectopic expression of the *D. melanogaster* Sir2 ortholog, dSir2, was reported to extend lifespan in the *D. melanogaster* [66]. It is important to note that the effect of sirtuins on longevity in both *C. elegans* and *D. melanogaster* remain controversial [67]. Mammals have seven homologs of Sir2, SIRT1-7, which catalyze the removal of H4K16 and H3K9 acetylation, silencing gene transcription through the compaction of chromatin [68,69]. One such homolog, SIRT6, when ablated in the mouse, results in genomic instability and aging-like phenotypes [70]. However, SIRT1-deficient mouse embryonic fibroblasts, while refractory to replicative senescence, still undergo both oncogenic and DNA damage-induced senescence [71]. These contrasting results underline the increased complexity of mammalian systems when discussing aging. One question that remains elusive, however, is whether the substrates of SIRTs are histones or nonhistone proteins.

H3K56ac

Yeast Asf1 and the HAT Rtt109 are essential for H3K56ac deposition and expression of histones. Mutation in either gene lowers H3K56ac levels, which reduces both histone transcript levels and histone protein abundance. Consistent with the decrease of H3K56ac in aged yeast, the global depletion of H3k56ac decreases RLS [10,42,72,73]. Surprisingly, too much H3K56ac decreased lifespan as well [10]. This indicates that cells must maintain the ability to both acetylate and deacetylate H3K56 to achieve a normal lifespan. Interestingly, old wild-type yeast have elevated levels of histone transcription [10,14]. This result indicates that cells compensate for the decrease in histone abundance by increasing the transcription of histones. The occupancy of H3K56ac at histone promoters in aged yeast was not explicitly investigated. Rather, the global changes in H3K56ac may influence other age-related gene transcription that remain to be identified [10,42]. Consistent with what was observed in yeast, H3K56ac levels decrease in human cells *in vitro* with age [13].

7.4.2   Repressive Histone Marks

7.4.2.1   H3K27me3

Trimethylation of histone 3 lysine residue 27 (H3K27me3) is a repressive histone mark found enriched in heterochromatin [74,75]. The global levels of H3K27me3 decrease with normal aging in the stroma of *C. elegans* concomitant with an increase in the levels of the H3K27me3 demethylase UTX-1 [76,77]. Strikingly, depletion of UTX-1 extends lifespan in a germline-independent manner [76,77]. This increase is caused, at least in part, by the regulation of insulin-FOXO pathway-related gene expression [76,78,79].

Contrasting to the findings in *C. elegans*, studies in *D. melanogaster* have demonstrated that mutations in two constituents of the PRC2 complex, E(z) and ESC, extend male longevity and correlate with a reduction in global H3K27me3 levels [52]. Strikingly, mutation of Trx, an H3K4me3 methyltransferase, reverts this phenotype [52]. Studies in *C. elegans* and *D. melanogaster* provide conflicting explanations for the role of H3K27me3 in the aging process. In *D. melanogaster*, excessive chromatin compaction leads to a shortened lifespan, whereas in *C. elegans*, chromatin compaction extends lifespan. These findings suggest different regulation of the aging process across species.

7.4.2.2 H3K9me3

Trimethylation of histone 3 at lysine residue 9 (H3K9me3) is found at repressed loci and promotes the compaction of facultative heterochromatin [80]. Global levels of H3K9me3 decrease with passage of cultured human cells [13]. Human progeria cells also exhibit this reduction [81,82]. However, in the *D. melanogaster*, there is an increase of H3K9me3 with age [15]. Much like what has been observed with H3K27me3, there are contradictory findings involving the role of suppressive (H3K27me3 and H3K9me3) marks and the aging process in *C. elegans*, *D. melanogaster*, and humans, pointing to the need for the identification of the target loci responsible for these effects.

7.4.2.3 H4K20me3

Trimethylation of histone 4 at lysine residue 20 (H4K20me3) marks constitutive heterochromatin and increases with age in rat livers [83,84]. It is also increased in human progeria cells [85]. In aged human cycling fibroblasts, however, a decrease of H4K20me3 has been observed [13].

7.4.2.4 H3K36me3

Trimethylation of histone 3 lysine residue 36 (H3K36me3) is enriched in the gene bodies of actively transcribed loci, specifically exons, but evidence also suggests a repressive role [86]. A recent study has implicated H3K36me3 in the regulation of chronological lifespan (CLS), which is the measure of the longevity of a single yeast cell [87]. This study demonstrated that mitochondrial-derived ROS inactivated histone demethylase Rph1p through the activation of the yeast-equivalent DDR kinases ATM and Chk2. This subsequently increased H3K36me3 levels at subtelomeric regions, which enhanced the binding of histone deacetylase Sir3p to this region resulting in an increased telomere stabilization and CLS [88].

7.4.3 Senescence, DNA Damage, and Histone Modifications

Senescence, also termed "replicative cellular aging," was first described by Hayflick who observed that cultured human fibroblasts eventually cease to proliferate and develop an enlarged, flattened morphology [89]. Numerous studies have indicated that cells from aged organisms, ranging from *M. musculus* to baboons to humans, exhibit hallmarks of senescence [23,36,90]. Replicative senescence can result from oxidative stress or telomere shortening, both of which elicit the DDR. Upon DNA damage, H2A.X becomes phosphorylated on serine 139 (γH2A.X) [91,92] H2A.X phosphorylation eventually leads to the recruitment of HP1, which aids in the formation of heterochromatin [93]. With continuous or sustained DNA damage, SAHFs are formed, which sequester damaged DNA into compacted heterochromatin [94,95]. Although speculative, HP1 may act to repress transcription through the stabilization of the nucleosome, thus decreasing the rate of histone exchange. As such, aged human cells have decreased levels of HP1 and exhibit a loss of HP1-recruiting modifications on heterochromatin such as H3K9me3, which complements the hypothesis of loss of heterochromatin aging [82]. Conversely, HP1 increases in senescent human fibroblasts and in 40-day-old *D. melanogaster* compared with 10-day-old *D. melanogaster* [15,36].

It is hypothesized that once the DDR has successfully repaired damaged DNA, the histone marks present before the damaging event are never completely restored to their predamaged state. This leads to alterations in chromatin modifications and gene expression and potentially impacts aging [96]. Studies have demonstrated that aged cells from baboons contain increased levels of DNA damage and heterochromatin marks, such as γ-H2AX, p-Ser 1981-ATM, HP-1β, and HIRA, suggesting that telomere attrition induces the DDR and leads to senescence [97]. Lamin B1 downregulation during senescence is a key trigger of global and local chromatin changes that impact gene expression and aging, specifically, H3K4me3 and H3K27me3 [98]. Importantly, the elimination of senescent cells in a murine model of accelerated aging significantly delays or decreases the associated aging phenotypes, pointing to a direct involvement of senescence in aging [99].

Through the use of formaldehyde-assisted isolation of regulatory elements (FAIRE), the genomic conformations of senescent cells have recently been explored. Dividing cells display features of both open and closed chromatin. In contrast, the profile of replicative senescent cells exhibits a "smoothened" signal. Mainly, this is due to the loss of FAIRE signal in promoters and enhancers of active genes, and gain of signal in heterochromatic-gene-poor regions, such as centromeric or pericentromeric regions [100]. These results indicate that in replicative senescent cells, there is a general closing of chromatin in euchromatic-gene-rich regions, coupled with an overall opening of chromatin in heterochromatic-gene-poor region. Despite these rather striking findings, there are some genes that oppose these trends, demonstrating that this hypothesis is not a comprehensive model for the genomic state of senescent cells.

In senescent cells, γH2AX-positive and H3K27me3-positive chromatin fragments bud off nuclei and are then targeted by lysosomal degradation. Senescent cells also exhibit markers of lysosomal-mediated proteolytic processing of histones, which suggests that the senescence state is maintained, in part, by this mechanism. *In vivo*, depletion of histones correlated with nevus maturation, a well-studied *in vivo* cellular context of senescence [101]. The data presented in this section points to the possibility that as cells age, their chromatin becomes more open. This process coupled with DNA damage and telomere attrition, induces senescence and the formation of SAHFs. These mechanisms compensate one another to protect the cells against transformation. This may explain the discrepancy between the loss of heterochromatin theory and the senescence-related *de novo* heterochromatin formation and suggests that aging corresponds to a convergence of both processes.

7.5 PROGERIA: ACCELERATED AGING DUE TO NUCLEAR ARCHITECTURE DYSFUNCTION

Hutchinson–Gilford progeria syndrome (HGPS, or progeria) is a rare disease where patients display accelerated aging and have a drastically shortened lifespan. HGPS is caused by a Lamin A mutation that causes a C-terminal truncation and the generation of a protein termed "progerin" [102]. Lamin A is associated with the nuclear lamina and provides structural support for the nuclear membrane. Progerin accumulation causes nuclear defects, such as nuclear morphologic abnormalities, increased transcription from pericentric repeats, shortened telomere length, increased DNA damage foci, a decrease of heterochromatin caused by the loss of protein HP1, decreases in global H3K9me3 and H3K27me3, increases in H4K20me3, and a decrease in H3K27 methyltransferase EZH2 [85,102]. Similar to HGPS, naturally aged cells display altered localization of Lamin A, increased levels of progerin, and a decline of HP1 and H3K9me3 levels [82]. Interestingly, the expression of progerin in normal cells is sufficient to induce the abnormalities displayed in HGPS [82,103]. Conversely, if the splicing error resulting in progerin is rectified, the abnormalities are reversed in both progeric and naturally aged cells [81,82].

Both HGPS cells and naturally aged cells have reduced levels of RBBP4/7, HDAC1, and MTA3, all components of the nucleosome remodeling deacetylase (NuRD) chromatin modifying complex [104,105]. Reducing the levels of NuRD complex constituents recapitulates cellular phenotypes characteristic of HGPS and naturally aged cells, such as a reduction in HP1, a decrease in global levels of H3K9me3, and an increase in DNA damage foci [104,105]. Due to the reversible nature of aging through the correction of progerin expression, aging may be driven, at least in part, by dysfunctional nuclear architecture. Progerin could represent a "biologic clock": As levels rise, the nuclear architecture is modulated in such a way that initiates the aging process.

7.6 CONCLUSION

In this chapter, we discussed how epigenetics and histones influence the aging process. We have presented evidence implicating histone density, histone variants, and histone modifications in the aging process. Furthermore, we described how cellular senescence, the DDR, and progeria are influenced by epigenetic processes. It is hypothesized that as we age, chromatin converts to a more open state due, in part, to a loss of histone density and heterochromatin-associated histone modifications. Subsequently, cells divide less frequently, resulting in the substitution of replication-independent histone variants into nucleosomes, which bind DNA less tightly. DNA damage foci also increase with age, which compact chromatin into SAHFs. This then leads to a global increase in heterochromatin and senescence. However, data suggest that the chromatin landscape of senescent cells becomes less distinctive, exhibiting drastic changes in the genomic location of heterochromatin and euchromatin. In addition, senescent cells exhibit lysosomal processing of histones leading to a reduction of histone content. These findings present us with an interesting conundrum: How could the loss of heterochromatin and cellular-senescence-associated *de novo* formation of large heterochromatin foci, both seemingly conflicting processes, converge to contribute to aging? Why is it that senescence, described as natural aging, increases global levels of heterochromatin leading to gene silencing, while aged cells also have a more relaxed chromatin fiber and increased global transcription? Perhaps senescence and the formation of SAHFs is a compensatory mechanism, occurring after the relaxation of the chromatin fiber to protect aged cells from transformation. Despite the speculative nature surrounding cause and consequence of the aging process, the reversibility of histone modifications presents a useful avenue for therapeutic intervention in the treatment of aging diseases, including HGPS. The most provocative studies should aim to elucidate the driving events and understand the passenger events of aging. This information will represent the lexicon required to decipher the puzzle that is aging.

REFERENCES

[1] Maslov AY, Vijg J. Genome instability, cancer and aging. Biochim Biophys Acta 2009;1790(10):963−9.
[2] Hasty P, Campisi J, Hoeijmakers J, van Steeg H, Vijg J. Aging and genome maintenance: lessons from the mouse? Science 2003;299(5611):1355−9.
[3] Busuttil R, Bahar R, Vijg J. Genome dynamics and transcriptional deregulation in aging. Neuroscience 2007;145(4):1341−7.
[4] Villeponteau B. The heterochromatin loss model of aging. Exp Gerontol 1996; 32(4-5):383−94.
[5] Oberdoerffer P, Sinclair DA. The role of nuclear architecture in genomic instability and ageing. Nat Rev Mol Cell Biol 2007;8(9):692−702.

[6] Columbaro M, Capanni C, Mattioli E, et al. Rescue of heterochromatin organization in Hutchinson-Gilford progeria by drug treatment. Cell Mol Life Sci 2005;62(22):2669−78.

[7] Ishimi Y, Kojima M, Takeuchi F, Miyamoto T, Yamada M, Hanaoka F. Changes in chromatin structure during aging of human skin fibroblasts. Exp Cell Res 1987;169(2):458−67.

[8] Macieira-Coelho A, Puvion-Dutilleul F. Evaluation of the reorganization in the high-order structure of DNA occurring during cell senescence. Mutat Res 1989;219(3):165−70.

[9] Robert KM, John RJ. Life span of individual yeast cells. Nature 1959;183(4677):1751−2.

[10] Feser J, Truong D, Das C, et al. Elevated histone expression promotes life span extension. Mol Cell 2010;39(5):724−35.

[11] Wyrick JJ, Holstege FC, Jennings EG, et al. Chromosomal landscape of nucleosome-dependent gene expression and silencing in yeast. Nature 1999;402(6760):418−21.

[12] Hu Z, Chen K, Xia Z, et al. Nucleosome loss leads to global transcriptional up-regulation and genomic instability during yeast aging. Genes Dev 2014;28(4):396−408.

[13] O'Sullivan RJ, Kubicek S, Schreiber SL, Karlseder J. Reduced histone biosynthesis and chromatin changes arising from a damage signal at telomeres. Nat Struct Mol Biol 2010;17(10):1218−25.

[14] Lesur I, Campbell JL. The transcriptome of prematurely aging yeast cells is similar to that of telomerase-deficient cells. Mol Biol Cell 2004;15(3):1297−312.

[15] Wood JG, Hillenmeyer S, Lawrence C, et al. Chromatin remodeling in the aging genome of Drosophila. Aging cell 2010;9(6):971−8.

[16] Hake SB, Allis CD. Histone H3 variants and their potential role in indexing mammalian genomes: the "H3 barcode hypothesis." Proc Natl Acad Sci USA 2006;103(17):6428−35.

[17] Kamakaka RT, Biggins S. Histone variants: deviants? Genes Dev 2005;19(3):295−310.

[18] Burgess RJ, Zhang Z. Histone chaperones in nucleosome assembly and human disease. Nat Struct Mol Biol 2013;20(1):14−22.

[19] Pina B, Suau P. Changes in histones H2A and H3 variant composition in differentiating and mature rat brain cortical neurons. Dev Biol 1987;123(1):51−8.

[20] Rogakou EP, Sekeri-Pataryas KE. Histone variants of H2A and H3 families are regulated during in vitro aging in the same manner as during differentiation. Exp Gerontol 1999;34(6):741−54.

[21] Urban MK, Zweidler A. Changes in nucleosomal core histone variants during chicken development and maturation. Dev Biol 1983;95(2):421−8.

[22] Grove GW, Zweidler A. Regulation of nucleosomal core histone variant levels in differentiating murine erythroleukemia cells. Biochemistry 1984;23(19):4436−43.

[23] Jeyapalan JC, Ferreira M, Sedivy JM, et al. Accumulation of senescent cells in mitotic tissue of aging primates. Mech Ageing Dev 2007;128(1):36−44.

[24] Ausió J, Abbott DW. The many tales of a tail: carboxyl-terminal tail heterogeneity specializes histone H2A variants for defined chromatin function. Biochemistry 2002;41(19):5945−9.

[25] Redon C, Pilch D, Rogakou E, Sedelnikova O, et al. Histone H2A variants h2ax and h2az. Curr Opin Genet Dev 2002;12(2):162−9.

[26] Mizuguchi G, Shen X, Landry J, Wu W-H, Sen S, Wu C. ATP-driven exchange of histone H2AZ variant catalyzed by SWR1 chromatin remodeling complex. Science 2004;303 (5656):343−8.

[27] Coleman-Derr D, Zilberman D. Deposition of histone variant H2A.Z within gene bodies regulates responsive genes. PLoS Genet 2011;8(10):e1002988.

[28] Lee K, Lau ZZ, Meredith C, Park JH. Decrease of p400 ATPase complex and loss of H2A.Z within the p21 promoter occur in senescent IMR-90 human fibroblasts. Mech Ageing Dev 2011;133(11-12):686−94.

[29] Gévry N, Chan HM, Laflamme L, Livingston DM, Gaudreau L. p21 transcription is regulated by differential localization of histone H2A.Z. Genes Dev 2007; 21(15):1869−81.

[30] Jin C, Zang C, Wei G, Cui K, Peng W, Zhao K, et al. H3. 3/H2A. Z double variant−containing nucleosomes mark'nucleosome-free regions' of active promoters and other regulatory regions. Nat Genet 2009;41(8):941−5.

[31] Rogakou EP, Boon C, Redon C, Bonner WM. Megabase chromatin domains involved in DNA double-strand breaks in vivo. J Cell Biol 1999;146(5):905−16.

[32] Paull TT, Rogakou EP, Yamazaki V, Kirchgessner CU, Gellert M, Bonner WM. A critical role for histone H2AX in recruitment of repair factors to nuclear foci after DNA damage. Curr Biol 1999;10(15):886−95.

[33] Celeste A, Petersen S, Romanienko PJ, et al. Genomic instability in *M. musculus* lacking histone H2AX. Science 2002;296(5569):922−7.

[34] Sedelnikova OA, Horikawa I, Zimonjic DB, Popescu NC, Bonner WM, Barrett JC. Senescing human cells and ageing *M. musculus* accumulate DNA lesions with unrepairable double-strand breaks. Nat Cell Biol 2004;6(2):168−70.

[35] Costanzi C, Pehrson JR. Histone macroH2A1 is concentrated in the inactive X chromosome of female mammals. Nature 1998;393(6685):599−601.

[36] Kreiling JA, Tamamori-Adachi M, Sexton AN, et al. Age-associated increase in heterochromatic marks in murine and primate tissues. Aging Cell 2011;10(2):292−304.

[37] Zhang R, Poustovoitov MV, Ye X, et al. Formation of MacroH2A-containing senescence-associated heterochromatin foci and senescence driven by ASF1a and HIRA. Dev Cell 2004;8(1):19−30.

[38] Wajapeyee N, Serra RW, Zhu X, Mahalingam M, Green MR. Oncogenic BRAF induces senescence and apoptosis through pathways mediated by the secreted protein IGFBP7. Cell 2008;132(3):363−74.

[39] Bannister AJ, Kouzarides T. Regulation of chromatin by histone modifications. Cell Res 2011;21(3):381−95.

[40] Lund J, Tedesco P, Duke K, Wang J, Kim SK, Johnson TE. Transcriptional profile of aging in *C. elegans*. Curr Biol 2002;12(18):1566−73.

[41] Bennett-Baker PE, Wilkowski J, Burke DT. Age-associated activation of epigenetically repressed genes in the mouse. Genetics 2003;165(4):2055−62.

[42] Dang W, Steffen KK, Perry R, et al. Histone H4 lysine 16 acetylation regulates cellular lifespan. Nature 2009;459(7248):802−7.

[43] Fraga MF, Ballestar E, Paz MF, et al. Epigenetic differences arise during the lifetime of monozygotic twins. Proc Natl Acad Sci USA 2005;102(30):10604−9.

[44] Bernstein BE, Humphrey EL, Erlich RL, et al. Methylation of histone H3 Lys 4 in coding regions of active genes. Proc Natl Acad Sci USA 2002;99(13):8695−700.

[45] Santos-Rosa H, Schneider R, Bannister AJ, et al. Active genes are tri-methylated at K4 of histone H3. Nature 2002;419(6905):407−11.

[46] Greer EL, Maures TJ, Hauswirth AG, et al. Members of the H3K4 trimethylation complex regulate lifespan in a germline-dependent manner in *C. elegans*. Nature 2010;466 (7304):383−7.

[47] Xiao Y, Bedet C, Robert VJ, et al. Caenorhabditis elegans chromatin-associated proteins SET-2 and ASH-2 are differentially required for histone H3 Lys 4 methylation in embryos and adult germ cells. Proc Natl Acad Sci USA 2011;108(20):8305−10.

[48] Greer EL, Maures TJ, Ucar D, et al. Transgenerational epigenetic inheritance of longevity in *Caenorhabditis elegans*. Nature 2011;479(7373):365−71.

[49] McColl G, Killilea DW, Hubbard AE, Vantipalli MC, Melov S, Lithgow GJ. Pharmacogenetic analysis of lithium-induced delayed aging in *Caenorhabditis elegans*. J Biol Chem 2008;283(1):350−7.

[50] Miller T, Krogan NJ, Dover J, et al. COMPASS: a complex of proteins associated with a trithorax-related SET domain protein. Proc Natl Acad Sci USA 2001;98(23):12902−7.

[51] Li L, Greer C, Eisenman RN, Secombe J. Essential functions of the histone demethylase lid. PLoS Genet 2010;6(11):e1001221.

[52] Siebold AP, Banerjee R, Tie F, Kiss DL, Moskowitz J, Harte PJ. Polycomb Repressive Complex 2 and Trithorax modulate Drosophila longevity and stress resistance. Proc Natl Acad Sci USA 2010;107(1):169−74.

[53] Ryan JM, Cristofalo VJ. Histone acetylation during aging of human cells in culture. Biochem Biophys Res Commun 1972;48(4):735−42.

[54] Peleg S, Sananbenesi F, Zovoilis A, et al. Altered histone acetylation is associated with age-dependent memory impairment in *M. musculus*. Science 2010;328(5979):753−6.

[55] Watanabe K-I, Bloch W. Histone methylation and acetylation indicates epigenetic change in the aged cochlea of *M. musculus*. European archives of oto-rhino-laryngology: official journal of the European Federation of Oto-Rhino-Laryngological Societies (EUFOS): affiliated with the German Society for Oto-Rhino-Laryngology. Head Neck Surg 2013;270(6):1823−30.

[56] Krishnan V, Chow MZ, Wang Z, et al. Histone H4 lysine 16 hypoacetylation is associated with defective DNA repair and premature senescence in Zmpste24-deficient *M. musculus*. Proc Natl Acad Sci USA 2011;108(30):12325−30.

[57] Shogren-Knaak M, Ishii H, Sun J-MM, Pazin MJ, Davie JR, Peterson CL. Histone H4-K16 acetylation controls chromatin structure and protein interactions. Science 2006;311(5762):844−7.

[58] Oppikofer M, Kueng S, Gasser SM. SIR-nucleosome interactions: structure-function relationships in yeast silent chromatin. Gene 2013;527(1):10−25.

[59] Longo VD, Kennedy BK. Sirtuins in aging and age-related disease. Cell 2006;126 (2):257−68.

[60] Imai S, Armstrong CM, Kaeberlein M, Guarente L. Transcriptional silencing and longevity protein Sir2 is an NAD-dependent histone deacetylase. Nature 2000;403 (6771):795−800.

[61] Millar CB, Grunstein M. Genome-wide patterns of histone modifications in yeast. Nat Rev Mol Cell Biol 2006;7(9):657−66.

[62] Kaeberlein M, McVey M, Guarente L. The SIR2/3/4 complex and SIR2 alone promote longevity in Saccharomyces cerevisiae by two different mechanisms. Genes Dev 1999;13 (19):2570−80.

[63] Kennedy BK, Austriaco NR, Zhang J, Guarente L. Mutation in the silencing gene SIR4 can delay aging in S. cerevisiae. Cell 1995;80(3):485−96.

[64] Kennedy BK, Gotta M, Sinclair DA, et al. Redistribution of silencing proteins from telomeres to the nucleolus is associated with extension of life span in S. cerevisiae. Cell 1997;89(3):381−91.

[65] Tissenbaum HA, Guarente L. Increased dosage of a sir-2 gene extends lifespan in Caenorhabditis elegans. Nature 2001;410(6825):227−30.

[66] Rogina B, Helfand SL. Sir2 mediates longevity in the *D. melanogaster* through a pathway related to calorie restriction. Proc Natl Acad Sci USA 2004;101(45):15998−6003.

[67] Burnett C, Valentini S, Cabreiro F, et al. Absence of effects of Sir2 overexpression on lifespan in C. elegans and Drosophila. Nature 2011;477(7365):482−5.

[68]   Pruitt K, Zinn RL, Ohm JE, et al. Inhibition of SIRT1 reactivates silenced cancer genes without loss of promoter DNA hypermethylation. PLoS Genet 2006;2(3):e40.

[69]   Vaquero A, Scher M, Lee D, Erdjument-Bromage H, Tempst P, Reinberg D. Human SirT1 interacts with histone H1 and promotes formation of facultative heterochromatin. Mol Cell 2004;16(1):93–105.

[70]   Mostoslavsky R, Chua KF, Lombard DB, et al. Genomic instability and aging-like phenotype in the absence of mammalian SIRT6. Cell 2006;124(2):315–29.

[71]   Chua KF, Mostoslavsky R, Lombard DB, Pang WW, Saito S, Franco S, et al. Mammalian SIRT1 limits replicative life span in response to chronic genotoxic stress. Cell Metab 2005;2(1):67–76.

[72]   Xu F, Zhang K, Grunstein M. Acetylation in histone H3 globular domain regulates gene expression in yeast. Cell 2005;121(3):375–85.

[73]   Driscoll R, Hudson A, Jackson SP. Yeast Rtt109 promotes genome stability by acetylating histone H3 on lysine 56. Science 2007;315(5812):649–52.

[74]   Cao R, Wang L, Wang H, et al. Role of histone H3 lysine 27 methylation in Polycomb-group silencing. Science 2002;298(5595):1039–43.

[75]   Bernstein BE, Mikkelsen TS, Xie X, et al. A bivalent chromatin structure marks key developmental genes in embryonic stem cells. Cell 2006;125(2):315–26.

[76]   Jin C, Li J, Green CD, et al. Histone demethylase UTX-1 regulates C. elegans life span by targeting the insulin/IGF-1 signaling pathway. Cell Metab 2011;14(2):161–72.

[77]   Maures TJ, Greer EL, Hauswirth AG, Brunet A. The H3K27 demethylase UTX-1 regulates C. elegans lifespan in a germline-independent, insulin-dependent manner. Aging Cell 2011;10(6):980–90.

[78]   Mair W, Dillin A. Aging and survival: the genetics of life span extension by dietary restriction. Annu Rev Biochem 2007;77:727–54.

[79]   Greer EL, Brunet A. Different dietary restriction regimens extend lifespan by both independent and overlapping genetic pathways in C. elegans. Aging Cell 2009; 8(2):113–27.

[80]   Peters AH, Mermoud JE, O'Carroll D, et al. Histone H3 lysine 9 methylation is an epigenetic imprint of facultative heterochromatin. Nat Genet 2001;30(1):77–80.

[81]   Scaffidi P, Misteli T. Reversal of the cellular phenotype in the premature aging disease Hutchinson-Gilford progeria syndrome. Nat Med 2005;11(4):440–5.

[82]   Scaffidi P, Misteli T. Lamin A-dependent nuclear defects in human aging. Science 2006;312(5776):1059–63.

[83]   Schotta G, Lachner M, Sarma K, et al. A silencing pathway to induce H3-K9 and H4-K20 trimethylation at constitutive heterochromatin. Genes Dev 2004;18(11):1251–62.

[84]   Sarg B, Koutzamani E, Helliger W, Rundquist I, Lindner HH. Postsynthetic trimethylation of histone H4 at lysine 20 in mammalian tissues is associated with aging. J Biol Chem 2002;277(42):39195–201.

[85]   Shumaker DK, Dechat T, Kohlmaier A, et al. Mutant nuclear lamin A leads to progressive alterations of epigenetic control in premature aging. Proc Natl Acad Sci USA 2006;103 (23):8703–8.

[86]   Wagner EJ, Carpenter PB. Understanding the language of Lys36 methylation at histone H3. Nat Rev Mol Cell Biol 2012;13(2):115–26.

[87]   Longo VD. Human Bcl-2 reverses survival defects in yeast lacking superoxide dismutase and delays death of wild-type yeast. J Cell Biol 1997;137(7):1581–8.

[88]   Schroeder EA, Raimundo N, Shadel GS. Epigenetic silencing mediates mitochondria stress-induced longevity. Cell Metab 2013;17(6):954–64.

[89] Hayflick L, Moorhead PS. The serial cultivation of human diploid cell strains. Exp Cell Res 1961;25:585−621.

[90] Dimri GP, Lee X, Basile G, et al. A biomarker that identifies senescent human cells in culture and in aging skin in vivo. Proc Natl Acad Sci USA 1995;92(20):9363−7.

[91] Rogakou EP, Pilch DR, Orr AH, Ivanova VS, Bonner WM. DNA double-stranded breaks induce histone H2AX phosphorylation on serine 139. J Biol Chem 1998;273 (10):5858−68.

[92] Burma S, Chen BP, Murphy M, Kurimasa A, Chen DJ. ATM phosphorylates histone H2AX in response to DNA double-strand breaks. J Biol Chem 2001;276(45):42462−7.

[93] Dinant C, Luijsterburg MS. The emerging role of HP1 in the DNA damage response. Mol Cell Biol 2009;29(24):6335−40.

[94] Narita M, Nũnez S, Heard E, et al. Rb-mediated heterochromatin formation and silencing of E2F target genes during cellular senescence. Cell 2003;113(6):703−16.

[95] d'Adda di Fagagna F. Living on a break: cellular senescence as a DNA-damage response. Nat Rev Cancer 2008;8(7):512−22.

[96] Tamburini BA, Tyler JK. Localized histone acetylation and deacetylation triggered by the homologous recombination pathway of double-strand DNA repair. Mol Cell Biol 2005;25(12):4903−13.

[97] Herbig U, Ferreira M, Condel L, Carey D, Sedivy JM. Cellular senescence in aging primates. Science 2006;311(5765):1257.

[98] Shah PP, Donahue G, Otte GL, et al. Lamin B1 depletion in senescent cells triggers large-scale changes in gene expression and the chromatin landscape. Genes Dev 2013; 27(16):1787−99.

[99] Baker DJ, Wijshake T, Tchkonia T, et al. Clearance of p16Ink4a-positive senescent cells delays ageing-associated disorders. Nature 2011;479(7372):232−6.

[100] De Cecco M, Criscione SW, Peckham EJ, et al. Genomes of replicatively senescent cells undergo global epigenetic changes leading to gene silencing and activation of transposable elements. Aging Cell 2013;12(2):247−56.

[101] Ivanov A, Pawlikowski J, Manoharan I, et al. Lysosome-mediated processing of chromatin in senescence. J Cell Biol 2013;202(1):129−43.

[102] Burtner CR, Kennedy BK. Progeria syndromes and ageing: what is the connection? Nat Rev Mol Cell Biol 2010;11(8):567−78.

[103] Scaffidi P, Misteli T. Lamin A-dependent misregulation of adult stem cells associated with accelerated ageing. Nat Cell Biol 2008;10(4):452−9.

[104] Pegoraro G, Kubben N, Wickert U, Göhler H, Hoffmann K, Misteli T. Ageing-related chromatin defects through loss of the NURD complex. Nat Cell Biol 2009;11(10):1261−7.

[105] Pegoraro G, Misteli T. The central role of chromatin maintenance in aging. Aging 2009;1(12):1017−22.

Chapter 8

Epigenomic Studies in Epidemiology

Valentina Bollati[1,2], Valeria Motta[1], Simona Iodice[1] and Michele Carugno[1]

[1]Center of Molecular and Genetic Epidemiology, Department of Clinical Sciences and Community Health, Università degli Studi di Milano, Milan, Italy, [2]Epidemiology Unit, Fondazione IRCCS Ca' Granda Ospedale Maggiore Policlinico, Milan, Italy

Chapter Outline

8.1 INTRODUCTION: FROM CLASSICAL EPIDEMIOLOGY TO EPIGENOMIC EPIDEMIOLOGY

Classical epidemiology investigates disease occurrences and factors associated with disease causation. Classical epidemiologic investigations include incidence and mortality studies in populations characterized by different risk factors. Traditional epidemiology analyzes information derived from surveys,

M. Fraga & A.F. Fernandez (Eds): Epigenomics in Health and Disease.
DOI: http://dx.doi.org/10.1016/B978-0-12-800140-0.00008-X
© 2016 Elsevier Inc. All rights reserved.
163

questionnaires, and registries to identify individuals at high risk of developing diseases, to identify key risk factors, and to inform putative disease prevention measures.

Molecular epidemiology includes the collection of biologic samples, which allows for (i) the quantification of specific elements (e.g., DNA or RNA adducts) and early biologic markers (e.g., somatic mutations), (ii) better characterization of study subjects (e.g., genomic variations in metabolic genes) and risk stratification, and (iii) the use of molecular markers to further classify diseases that are currently categorized by etiology or prognosis.

An additional type of epidemiology is epigenetic epidemiology. Epigenetics describes several molecular mechanisms that alter genome expression and function when under exogenous influences, and thus epigenetics can play a powerful role in linking environment factors to disease development. Since epigenetics usually refers to the study of mechanisms regulating specific, individual genes, epigenome-wide association studies (EWAS) have been developed to investigate epigenetic regulation on a larger scale. Compared with genome-wide association studies (GWAS), EWAS are complicated by the fact that epigenomes are modified by a broad range of stimuli, and thus, defining a reference epigenome is challenging. This chapter addresses factors that should be taken into account when designing an epigenomic epidemiologic study.

8.2   THE CHOICE OF AN APPROPRIATE STUDY DESIGN

An epigenomic epidemiologic study may be used to investigate relationships among exposures to environmental, lifestyle, genetic, and socioeconomic risk factors, the epigenome, and disease development or other specific outcomes. In designing an epigenomic epidemiologic study, it is necessary to understand that the unidirectional linearity of conventional genetic studies does not apply and that epigenetic changes may be an intermediate factor between exposure and disease development or may be a consequence of the disease state. Moreover, an individual's disease status may influence their lifestyle and environmental exposure, and thus the association initially observed in the epigenomc analysis may shift [1].

To account for this potential scenario, in which cause and effect are reversed, termed "reverse causation," it is important to determine *a priori* the temporal sequence of events to be evaluated. Several epidemiologic study designs are available, and each has pros and cons with regard to epigenetic analyses.

8.2.1   Cohort Studies

Cohort (or longitudinal) studies play key roles in human epigenetics because they are able to capture the dynamic nature of epigenetic changes, and they can determine whether epigenetic mechanisms measured over time are related to temporal changes in exposure and disease prevalence [2]. In longitudinal epigenetic epidemiologic studies, groups of healthy individuals are recruited and followed up over a period to record the development of outcomes. In this study design, specific exposure information and biospecimens are collected from all participants at baseline and at follow-ups and are then related to subsequent disease experiences.

The repeated collection of exposure measurements and epigenetic markers at multiple time points throughout the study allows assessment of temporal relationships, strengthens assessments of directions of causality, and increases the ability to detect small effects.

An example of a well-designed longitudinal study is the Normative Aging Study (NAS). The NAS is a longitudinal cohort study established in 1963 in Eastern Massachusetts by the Veterans Administration [3−7] and still ongoing. Community-dwelling men from the greater Boston metropolitan area were screened at entry and enrolled if they had no history of chronic disease. Upon enrollment, participants underwent comprehensive clinical examinations every 3.5 years, which included lung function measurements and collection of extensive medical and lifestyle information. In one of the studies designed on NAS samples, Madrigano et al. found that aging was significantly associated with alterations in DNA methylation, suggesting that DNA hypomethylation, possibly related to decreased genome stability, may mediate the effects of aging over time [8].

The main disadvantage of longitudinal studies is that only limited types of biologic samples can be collected. Since study populations include healthy individuals, invasive sampling cannot be performed (until disease development) and only easily obtained biospecimens, such as blood, buccal cells, nasal cells, saliva, and urine can be collected and stored. Moreover, cohort studies typically include large numbers of participants, which makes biospecimen storage and laboratory analyses expensive.

8.2.2   Birth Cohort Studies

It is believed that the epigenome of an individual, when in the uterus, is more susceptible to exogenous exposures because epigenetic patterns change during the reprogramming waves of embryonic development. Epigenetic marks have been found to undergo erasure and reprogramming during preimplantation leading to tissue-specific gene methylation patterns seen after birth and throughout adulthood. The longitudinal birth cohort is a well-established study design in epigenetic epidemiology, which measures *in utero*, early, late, and end of life exposures and allows for the collection of transgenerational and across-life samples. In addition, birth cohort studies allow for the evaluation of epigenetic mechanisms in placenta, cord, and cord blood, which are easily obtained at delivery from mothers and newborns. In longitudinal birth cohorts, epigenetic changes can be measured over time; can be related to preconceptional, perinatal, or early-life exposures; and may be used to assess health status, disease development, and onset in children from delivery to adulthood.

An ongoing birth cohort study that began in 2012 at Hasselt University in Belgium is called ENVIRONAGE, and its goal is to identify air pollution-susceptible stages during pregnancy [9]. This study enrolls healthy mothers who have delivered healthy infants without complications and evaluates the associations between environmental influences and the molecular mechanisms of aging. After delivery, placental tissues from ENVIRONAGE mothers are collected and evaluated for DNA methylation to determine whether there is an association between particulate matter ($PM_{2.5}$) exposure and methylation patterns in placenta. Janssen et al. evaluated $PM_{2.5}$ levels of air pollution exposure during various prenatal periods ranging from preimplantation to the third trimester and observed a lower degree of placental global DNA methylation in association with exposure to particulate air pollution in early pregnancy, including the critical stages of implantation [9].

8.2.3 Cross-Sectional Studies

An alternative to costly cohort studies is the cross-sectional study design. Like the cohort study design, a cross-sectional study assesses the relationships between exposure, epigenetic changes, and disease, but instead of assessing these relationships over the long term, cross-sectional studies focus on a single time point, and this limits their abilities to establish causality and to analyze dynamic changes in epigenomes. Cross-sectional studies are primarily used to compare the prevalence of epigenetic changes among well-defined groups that differ, for example, in gender, age, or smoking habits.

A cross-sectional study design was used by Nelson et al. to investigate methylation profiles common to non−small cell lung cancer (NSCLC) tumors. Cancerous lung tissues and adjacent nontumor tissues were surgically resected from patients at the Massachusetts General Hospital from 1993 to 1996. Methylation patterns from matched pairs of NSCLC tissue and adjacent nontumor tissue were evaluated using a microarray-based approach (Illumina GoldenGate). To identify the CpG sites most significantly associated with tumor development, odds ratios for each matched tissue pair were calculated at each CpG site [10].

8.2.4 Case-Control Studies

In case-control studies, the study base includes subjects with disease (cases) and subjects without disease (controls) from the same population. "Controls" are defined as individuals who would have become study cases if they had developed the disease. There are multiple types of study bases, including (i) a specific population, such as subjects living in the same geographic boundary, followed up for a precise period, (ii) a selection of cases (if cases are identified in a specific clinic, the corresponding source population is made up of all the people that would attend that clinic if they had the disease [11]), or (iii) a cohort that has already been defined (nested case-control study, see Section 8.2.5). Cases and controls may come from registries (e.g., cancer registries or demographic registries), clinics, and hospitals (monocenter or multicenter) or may be determined at recruitment (cases admitted for first diagnosis, relapse).

The main advantage of this study design, compared with the cohort study design, is that it involves a larger number of readily available and informative cases. Moreover, tissue collection is more feasible, especially if the study population is based in a hospital or clinic. However, this study design is particularly susceptible to misinterpretations due to reverse causation, since case and control individuals often provide biased information regarding their past exposures. In addition, biologic samples are limited to a single time point when the disease is fully developed, and this leads to difficulties in distinguishing whether an epigenetic alteration is a cause or a consequence of a disease.

Langevin et al. performed an epigenome-wide analysis in three population-based case-control studies of bladder cancer, head and neck cancer, and ovarian cancer. The aim of the study was to evaluate the association between cancer status and DNA methylation, taking into account the heterogeneity of cell types within blood and the associated cell lineage specification of DNA methylation signatures.

The first study included bladder cancer cases with available blood samples, identified from the New Hampshire State Cancer Registry from 1994 to 1998, and healthy controls selected from population lists. The second study included head and neck squamous cell carcinoma (HNSCC) cases from nine academic medical facilities in the Boston, Massachusetts, area and controls identified through the Massachusetts town books; case and control individuals were selected randomly by availability. The inclusion criterion was no prior history of cancer, and case and control individuals were grouped according to age, gender, and town of residence. In the third study, methylation data from post-menopausal women diagnosed with primary epithelial ovarian cancer (identified from the UK Ovarian Cancer Population Study [UKOPS]) and from healthy postmenopausal women recruited from the UK Collaborative Trial of Ovarian Cancer Screening (UKCTOCS) were accessed from public data sets and analyzed. The obtained results supported the existence of cancer-associated DNA methylation profiles in the blood of solid tumor patients that are independent of alterations in normal leukocyte distributions [12].

8.2.5   Nested Case-Control Studies

A nested case-control study is typically embedded within a well-defined cohort study to identify predictive epigenetic biomarkers of disease. Cases are defined as events that occur within cohorts during follow-ups, and controls are randomly selected within the cohort among the noncases.

The risk estimate is thus calculated on a small sample of the entire cohort population. Such an approach allows for obtaining more information than what is already available in the cohort and makes the study less time consuming, less costly, and more efficient. Given the efficiency of the nested case-control design, it is well suited to assess rare diseases and outcomes and to assess diseases that are characterized by long induction and latent periods and is ideal for studies in which exposure data are difficult or expensive to obtain. Moreover, biologic samples might have been collected at the time of subject enrollment in the original cohort, thus prior to developing the disease of interest.

A major limitation of the nested case-control design is that multiple outcomes cannot be investigated because controls are selected from noncases when an event occurs.

The European Prospective Investigation into Cancer and Nutrition (EPIC) cohort contains 520,000 subjects from the general population, and blood samples were collected from each individual during enrollment. Shenker et al. performed a nested case-control study on the Italian subcohort (EPIC-Turin), in which data on female patients with breast and colon cancers were collected prospectively and matched individually with controls for gender, the season when blood was drawn, and the time between follow-ups. All individuals were healthy at the time of blood collection, and cases were selected among individuals who subsequently developed either breast cancer or colon cancer. Using an epigenome-wide approach (Infinium HumanMethylation450 BeadChip), this study assessed the effects of smoking on DNA methylation [13].

8.2.6 Twin Studies

The advantage of twin studies, both in genetics and epigenetics, is the ability to study complex diseases and outcomes without the complications that arise from individual genetic variation. Examining the epigenetic patterns of monozygotic (MZ) twins may be useful to identify epigenetic differences that are driven by exposure to environmental factors. Bell et al. investigated the effects of aging on DNA methylation patterns in female twins (ages 32–80 years) with age-related phenotypes using an epigenome-wide approach (Infinium HumanMethylation27 BeadChip). The participants were healthy female MZ and dizygotic (DZ) twins from the TwinsUK cohort (St. Thomas' UK Adult Twin Registry). The obtained results suggest that in a small set of genes, DNA methylation may be a candidate mechanism of mediating not only environmental effects but also genetic effects on age-related phenotypes [14].

8.3   ENVIRONMENTAL EPIGENETICS

Gender, age, and lifestyle factors, including diet (folate, phytoestrogens, green tea), smoking habits, and physical activity, have been shown to influence epigenetic patterns. Environmental epigenetics is an emerging field that focuses on the effects of environmental factors on epigenetic mechanisms. The WHO report that more than 7 million deaths in 2012 occurred as a result of exposure to air pollution demonstrates the necessity and relevance of environmental epigenetics.

Environmental factors, such as metals, chemicals, and air pollutants, have been linked to aberrant changes in epigenetic pathways both in experimental and epidemiologic studies. Most of the human studies conducted thus far have focused on DNA methylation [15,16] and micro-RNAs (miRNAs) [17], whereas only few studies have investigated the *in vitro* effects of environmental chemicals on histone modifications [18].

There are a number of well-studied toxicants that have been found to be associated with epigenetic modifications (i.e., DNA methylation, histone modification, and miRNA expression), and they include arsenic, cadmium, nickel, chromium, methylmercury, benzene, asbestos, bisphenol A, diethylstilbestrol, persistent organic pollutants (POPs), and air pollution (fine and ultra-fine particulate matter) [19].

It is often difficult to identify the cause-and-effect relationships among exogenous stimuli, epigenetic changes, and disease development in environmental exposure studies. Even though they are cumulative, epigenetic changes induced by environmental exposure are usually small and difficult to detect. In spite of this limitation, current evidence demonstrates that epigenetics has great potential to further our understanding of the molecular mechanisms of environmental toxicants and for predicting health-related risks due to environmental exposure and individual susceptibility [20].

8.4   VALIDATION OF RESULTS

Given the complexity of data produced by epigenomic epidemiologic investigations, it is challenging to separate robust signals from noise. Selection of an appropriate study design and validation of results are necessary to minimize false-positive and false-negative findings. Methods to validate results may be technical (on the same study subjects but using a different technique) or biologic (on different study subjects). Technical validation is desirable to avoid reproducing the problems and statistical biases that were in the original assay.

A discovery-only study is a single study without a validation step. Because a large number of associations are investigated simultaneously, discovery-only studies often produce false findings. To overcome this limitation, a number of study design strategies have been proposed. In the split-sample design, the study population is divided into two groups. One group is analyzed by using an epigenome-wide approach, and the second group is analyzed by using candidate target analyses. In the single-study cross-validation design, subsets of the study population are assayed multiple times to evaluate result reproducibility. The design has a stage one, which is discovery, and a stage two, which is replication, and the investigation is conducted as two independent studies to ensure validity of findings. In the meta-analysis design, no new biologic results are produced, but results from multiple studies are analyzed. Meta-analyses allow for large sample sizes and data sets, and because of this, results are applicable to large, general populations rather than to small, focused populations.

8.5 BIOLOGIC SAMPLE SELECTION

Since epigenetic marks are highly tissue specific, the ideal tissues to analyze in an epidemiologic study are target tissues, but this may not always be possible.

Although diseased tissues are easily obtained for cancer studies, some complications do arise. First, if the goal of a study is to investigate risk factors for cancer development, one must be aware that cancerous tissues are not reflective of the predisease state. Moreover, cancer tissues are often stored as formalin-fixed and paraffin-embedded (FFPE) tissues that may be difficult to work with. Although case tissues are easily obtained, control tissues are often impossible to obtain. Instead, some studies compare cancer tissue and tumor-adjacent noncancer tissue obtained from the same patient. Nevertheless, this approach also has risks, since adjacent tissue may harbor the same genetic abnormalities and epigenetic changes as those of cancer tissue [21,22]. In addition, only retrospective studies can utilize target tissues. Prospective studies can only collect noninvasive biospecimens, such as blood, urine, saliva, and buccal cells. Determination of which biospecimens to collect is not a trivial matter, and for some diseases, the simplest approach is to collect biospecimens that are most proximal to the target tissue. For example, when studying bladder cancer, cells in urine sediments would be collected, and when studying leukemia, peripheral white blood cells would be collected. For many diseases, such as brain diseases, this approach is not applicable, and three alternative approaches have been proposed. The "embryo layer" approach involves collecting cells derived from the same embryo layer as the target tissue (e.g., brain and buccal cells both derive from neuroectoderma). This approach is well suited for studying epigenetic changes induced *in utero*. The "uniform effects" approach suggests that even if the baseline levels of epigenetic marks vary among tissues, all tissues may be equally impacted by exposure. But, it should be kept in mind that different tissues also have different levels of exposure due to the distinct distribution of toxicants throughout the body. The "highest dose, first target" approach involves collecting biospecimens from the first target of exposure because the exposure effect will presumably be greatest in those biospecimens (e.g., nasal mucosae would be collected when studying inhalable pollutants).

Nevertheless, most epidemiologic studies are based on existing cohorts, in which the only available biologic samples are blood or buffy coat, buccal cells, or urine. Furthermore, archived biospecimens are often a collection of different cell types—for example, blood is a heterogeneous tissue, and epigenetic levels may vary as a result of variations in blood cell compositions among samples. For blood, normalizing samples using blood cell counts can solve this problem. Other tissues are more complicated—for example, cancer tissues rarely contain 100% cancer cells, and different ratios of cancer cells to normal cells may significantly affect results obtained from epigenetic analyses.

When determining which samples are suitable for epigenetic investigations, such parameters as sample storage, nucleic acid extraction, and the availability of whole cells should be considered (Table 8.1). Fresh tissue and blood are the gold standards and allow for analysis of multiple epigenetic marks, including DNA methylation, histone modifications, and miRNAs. They also allow for the separation of different cell types, such as the separation of lymphocytes from total blood. Cell separations cannot be performed on frozen tissue and blood samples, but $-80\,°C$ storage is recommended for analysis of RNA markers, and $-20\,°C$ storage is recommended for analysis of DNA markers.

FFPE tissues have many advantages, including ease of handling and long-term, inexpensive storage. However, formalin cross-links proteins and decreases polymerase chain reaction (PCR) efficiency, and nucleic acids extracted from FFPE are highly degraded and of low yield. This low quality and low yield make experiments that require large amounts of nucleic acids, such as microarray analysis, impossible.

Buccal or nasal cells have been sampled for decades because they are easy to collect and because nucleic acids can be extracted from them for molecular epidemiology studies. Since they are considered respiratory cells, they are useful for evaluating exposures to inhalable agents. A considerable advantage of sputum and saliva collection is that they can be self-collected, but since the nucleic acid extraction yields are low, saliva and sputum samples are not ideal for EWAS. Since urine contains a small number of erythrocytes, leukocytes, and squamous cells and since these cells are routinely examined in the clinical laboratory, they are useful biospecimens. In addition, urinary sediments can be used to extract nucleic acids for epidemiologic studies and whole urine can be used to analyze noncellular nucleic acids, such as extracellular miRNAs and free DNA.

If properly stored at $-20\,°C$, purified DNA can be used to assess DNA methylation, but histone modifications cannot be assessed. To assay for histone modifications, DNA must be extracted after cross-linking (usually by formalin treatment) to ensure that histone proteins are not removed during DNA purification.

TABLE 8.1 Commonly Available Sample for Epigenetic Investigations

Source	Biospecimen description	Storage	DNA methylation		Histone modifications		miRNA expression	
			Genome-wide	Target-specific	Genome-wide	Target-specific	Genome-wide	miRNA-specific
TISSUE	Biopsy	Fresh	X	X	X	X	X	X
	Surgical specimen	Frozen −20 °C	X	X	X	X	X	X
	Microdissected	Frozen −80 °C	X	X	X	X	X	X
		Frozen in liquid nitrogen	X[a]	X	X	X	X	X
		Formalin-fixed	X[a]	X	X[a]	X	X[a]	X
BLOOD	Whole blood	Fresh	X	X	X	X	X	X
	Buffy coat	Frozen −20 °C	X	X	X	X	X	X
	Lymphocytes/Monocytes/Erythrocytes	Frozen −80 °C	X	X	X	X	X	X
	Cell sorting (Ab-conjugated beads)	Frozen in liquid nitrogen	X	X	X	X	X	X
BLOOD CARDS	Dried blood spot	Room temperature	X[a]	X				
BUCCAL CELLS	Buccal swab	Fresh	X	X			X	X
	Mouthwash	Frozen −20 °C	X	X			X	X
NASAL CELLS	Nasal brushing	Frozen −80 °C	X	X			X	X
	Nasal scrub	Frozen in liquid nitrogen	X	X			X	X

SPUTUM	Induced sputum	Frozen −20 °C	X[a]		X		X	X
	Expectorated sputum	Frozen −80 °C	X[a]		X		X	X
SALIVA	Saliva	Room temperature (with preservatives)	X[a]		X		X	X
		Fresh			X			X
URINE	Whole urine	Frozen −20 °C			X		X	X
	Urinary sediment	Frozen −80 °C			X		X	X
		Frozen in liquid nitrogen			X		X	X
DNA	DNA extracted by organic extraction	Frozen −20 °C	X		X			
	DNA extracted by salting out	Frozen −80 °C	X		X			
	DNA extracted by Silica-based methods	Frozen in liquid nitrogen	X		X			
		Frozen −20 °C					X	X
RNA	Total RNA[b]	Frozen −80 °C					X	X
	miRNA-enriched RNA	Frozen in liquid nitrogen					X	X

[a]In principle possible, but strongly depending on nucleic acid yield and degradation.
[b]When approaching a new study on previously extracted RNA is highly recommended to verify miRNA presence since several extraction kits do not preserve miRNA fraction.

If properly stored at $-80\,°C$, purified RNA can be used to assess RNA markers, such as miRNA expression. In addition, RNA samples may be treated with reagents that hinder degradation.

8.6   METHODS SELECTION

Examination of the epigenetic status of many different loci has begun to replace specific targeted approaches to study the associations between epigenetic marks and health and disease. Epidemiologic studies usually involve large sample sizes, and therefore, selecting a cost-effective method of analysis is critical. Next-generation sequencing (NGS) platforms for high-throughput sequencing are available for studying methylation patterns, histone modifications, and miRNAs, but NGS has many limitations, including high costs, time-intensive labor, potentially high error rates, and relative inefficiency for assessing large numbers of samples.

Table 8.2 lists the primary techniques used and highlights the issues associated with each, which should be considered when designing an epigenomic epidemiologic study.

8.6.1   DNA Methylation

Cytosine methylation is a stable DNA modification, thus methylation analyses can be performed on frozen and FFPE specimens, as well as on specimens that are spotted on cards. In this section, we summarize the common techniques to analyze methylation in epidemiologic studies.

8.6.1.1   Bisulfite-Conversion-Based Assays

Sodium bisulfite DNA treatment allows for discrimination between methylated and unmethylated cytosines. Bisulfite conversion was one of the first techniques developed for methylation analysis and is still considered the gold standard [21] because it allows for quantitative comparisons of methylation levels at single-base resolution. The main limitation of this assay is the inability to discriminate between hydroxymethylcytosine from methylcytosine.

Whole-genome bisulfite sequencing (WGBS) allows for single CpG site resolution and precise mapping of specific loci. Bisulfite treatment is known to degrade DNA, and inefficient cytosine conversion may occur with long sequences. Reduced-representation bisulfite sequencing (RRBS) [22] reduces sequencing costs by focusing analyses on CpG dense regions that make up approximately 15% of total CpGs in the genome.

Use of genome-wide microarray-based platforms, such as the Illumina Infinium HumanMethylation450 BeadChip, is the preferred method for EWAS because of cost, coverage, and labor advantages. This array-based

TABLE 8.2 Techniques That Might Be Applied to Epigenomic Epidemiologic Studies

Epigenetic mark	Technique	Chemistry	Specificity	Sensitivity	Input material	Coverage	Cost
DNA METHYLATION	RRBS	Bisulfite + Restriction enzymes	++	+++	10–300 ng gDNA[a]	15% of CpGs	€€€
	Infinium 450K	Bisulfite	+	+	500 ng bisDNA[b]	485,764 CpGs	€
	Pyrosequencing	Bisulfite	++	++	500 ng gDNA[a]	Locus specific	€€
	MeDIP-seq	Immunoprecipitation	+	+	500 ng gDNA[a]	Genome wide	€
	MBD-seq	MBD-magnetic bead	++	++	>500 ng gDNA[a]	~27 million CpGs	€
HISTONE MODIFICATION	ChIP-Seq	Immunoprecipitation	+	++	$1–5 \times 10^5$ cells	Genome wide	€€€
	ChIP-chip	Immunoprecipitation	+	+	5×10^7 cells	Genome wide	€€€
microRNA	High-throughput qRT-PCR		++	+++	10–500 ng total RNA	~800 human miRNAs	€
	Microarray chips		++	+++	100 ng–1 µg total RNA	1200–1700 human miRNAs (miR-base 18.0)	€
	NGS		+++	++	500 ng–10 µg total RNA	De-novo miRNA identification	€€€

[a]gDNA stands for genomic DNA.
[b]bisDNA stands for bisulfite DNA.

technique is commonly used during initial discovery stages, and although it is highly sensitive and specific, it is often followed by a locus-specific validation step, such as pyrosequencing.

8.6.1.2   Enrichment-Based Methods

To enrich for and to sequence select bisulfite-converted loci, various purification strategies are utilized and combined with NGS techniques. MeDIP-seq [23] is a methylation analysis method that utilizes monoclonal antibodies against 5-methylcytosine to immunoprecipitate DNA and is compatible with NGS. MBD-seq [24] is a method that enriches for methyl-binding domains of methyl-binding proteins, such as MBD2 or MBD3L1, using monoclonal antibodies against methyl-binding domains [25].

8.6.2   Histone Modifications

Genome-wide analyses of histone modifications are performed with immunoprecipitated DNA by microarray hybridization (ChIP-chip) [25] or by high-throughput sequencing (ChIP-seq) [26] to identify and characterize histone-DNA interactions. The specificities and sensitivities of immunoprecipitation-based techniques depend on the specificity and sensitivity of the antibody chosen for the epigenetic mark under investigation. The limitations of ChIP-chip include probe design, cross-hybridization, and background noise. ChIP-seq generally produces profiles with higher resolution, requires less DNA, and is used to analyze the entire genome, and thus ChIP-seq has largely replaced ChIP-chip for genome-wide studies.

8.6.3   miRNAs

miRNA expression profiles can be generated using real-time quantitative PCR (qRT-PCR), hybridization-based methods, and NGS. The method to be selected depends upon the type of sample to be analyzed and the RNA preparation protocol used. qRT-PCR methods based on high-throughput profiling (TaqMan OpenArray Human MicroRNA Panel, miScript miRNA PCR Arrays) allow for the quantification of limited amounts of miRNAs. Further, qRT-PCR is the preferred method because of its sensitivity, specificity, accuracy, and simple protocols.

Several miRNA microarray chip platforms that differ in probe design and detection stringency are commercially available (Affymetrix GeneChip miRNA 3.0 array, Agilent Human microRNA Microarrays system, Exiqon miRCURY LNA microarray). The limitation of this method is the availability and stringency of probes on the chip platform that pair with miRNAs of interest. The newly developed Nanostring nCounter [27] uses two sequence-specific capture probes to allow for discrimination between similar variants of a single miRNA. NGS technologies (Illumina/Solexa, GA Roche/454 GS

FLX Titanium, ABI/SOLID) allow for complete miRnomes to be sequenced and allow for the discovery of novel miRNAs and isoforms. In addition, NGS will likely become the gold standard for miRNA analysis because of its ability to sequence short fragments in a high-throughput mode.

8.7   EXTRACELLULAR NUCLEIC ACID MARKERS

In addition to regulating the intracellular environment, epigenetic mechanisms may function in the extracellular environment. Most nucleic acids are located within cells, but a small quantity of nucleic acids are found circulating in extracellular fluids, and they are termed "circulating nucleic acids" (CNA). CNA analysis may be a viable, noninvasive approach to early disease diagnosis. Circulating DNA can be found in serum and plasma, and its methylation status can be an early indicator of cancer. The main limitation of CNA analysis is the low quantity and quality (it is highly degraded) of DNA in serum and plasma.

Circulating DNA can be also found as nucleosomes complexed with histone proteins as a result of apoptosis or active secretion by inflammatory cells [28]. Deligezer et al. has shown that specific, post-translational modifications in plasma samples were independent of the quantity of circulating nucleosomes [29]. It has been suggested that specific histone modifications in plasma function to selectively activate some immune pathways and inflammatory responses [30]. Like cellular histone modification analyses, extracellular histone analyses are time consuming and problematic for large epidemiologic studies.

Extracellular vesicles (EVs) are $0.05-1$ μm membrane vesicles, which are actively released by human cells into the bloodstream or into other bodily fluids. They shuttle bioactive molecules from their tissues of origin to recipient cells [31]. The fate of EVs after binding to the surface of recipient cells is currently unknown, but recent evidence suggests that they may fuse with recipient cell membranes and deliver their contents directly into the cytoplasm of recipient cells. It has been suggested that after being internalized within target cells, EVs may transport miRNAs [34,35], thereby facilitating intercellular and interorgan communication [32]. Moreover, miRNA expression in circulating EVs has been detected in plasma samples from normal subjects, and a predictive role for peripheral blood miRNA signatures in human disease has been hypothesized [32].

Analysis of EV miRNA content is challenging because normalizing the samples is difficult. miRNA expression data are typically normalized to commonly used endogenous reference miRNAs. But it is unknown whether miRNAs are packaged into EVs randomly or whether miRNAs are packaged into EVs by an active signaling process. If miRNAs are packaged randomly, standard endogenous reference miRNAs are suitable for normalization. If miRNAs are packaged nonrandomly, the use of standard endogenous reference miRNAs is arbitrary and inappropriate. To solve this problem, two strategies have been developed. Using the global mean normalization strategy, miRNAs are normalized to the average Quantification Cycle (Cq) or Threshold Cycle (Ct) value for all miRNAs measured. This strategy is useful for epigenomics studies that assay large numbers of miRNAs [33]. Alternatively, miRNAs may be normalized to an external spike-in RNA control introduced during the phenolic phase of the RNA extraction protocol [34]. However, the best method for normalization of miRNAs in EVs is currently being debated.

8.8 SAMPLE SIZE SELECTION AND STATISTICS

Selecting the appropriate sample size is important for all experimental studies but is more critical for large epigenome-wide studies. The complexity of these experiments and the large number of targets to be evaluated make determining sample sizes a challenge. When evaluating large numbers of targets, the potential for false findings is high. In addition, if a study is conducted with too few subjects, small differences will not be detected, and reliable statistical inferences will not be possible. The classical approach to sample size estimation involves testing a hypothesis by selecting a sample size and a significance threshold that control for the type I error rate, which is the probability of incorrectly rejecting a true null hypothesis or the probably of concluding falsely that a relationship exists. This approach is well suited to discovery studies as well as studies in which information from other similar experiments and prior knowledge of the reliability of the measurements is known. But, this may not apply to EWAS.

Microarray experiments are common to EWAS as they identify differences in gene expression among several covariance groups (e.g., age, gender, diet habits). In EWAS, statistical tests determine whether a particular gene is not differentially expressed across groups (a null hypothesis). A standard method to reduce error rates is the family-wise error rate (FWER), which is the probability of having at least one type I error over the course of the entire study; the Bonferroni correction is one type of FWER. However, FWER may be too conservative and may fail to identify significant differences [35]. The false discovery rate (FDR) method proposed by Benjamini and Hochberg [36] and Storey [37] may be more appropriate for EWAS. FDR controls for the expected proportion of rejected null hypotheses that are actually true. FDR methods are not only less conservative than FWER methods and have more power to identify significant differences but also have a higher probability of type I errors. Since EWAS vary substantially, the sample size effect and variance must be estimated for each study [38–40]. Sample size effects may be determined by assessing fold changes between differentially expressed genes, by assessing mean differences between cases and controls or by assessing odds ratios for binary responses, but none of these methods measures the precision of estimated differences or ratios.

To estimate variance, data from similar previous experiments should be evaluated. If larger variances are identified, larger sample sizes are required. Further, when being conservative, variance estimates from genes that display the greatest variation should be selected. If sample size is limited, then variance may not be estimated reliably [41]. For complex multivariate models, sample calculations may be difficult and may require ad hoc methods, such as simulation-based power estimates, or Bayesian analyses. However, currently there is little information on Bayesian analyses of size effects.

8.9 CONFOUNDING FACTORS AND EFFECT MODIFIERS: DEALING WITH COMPLEX SYSTEMS

Since a wide range of factors, such as environmental exposures, ethnicity, age, gender, and lifestyle habits, influences epigenetic markers, confounding factors must be considered in EWAS. A confounding factor is defined as a variable that correlates with both dependent and independent variables, and this confounding variable may obscure the relationship between the independent and dependent variables (Figure 8.1A). For example, the methylation state of lung cancer tissues may be altered relative to healthy tissues. Further, many patients with lung cancer are also smokers, and smoking is known to have direct effects on DNA methylation and on lung cancer development. If smoking is not taken into account as the adjustment variable in the model, we could observe a distortion in our study results.

Some other variables might exert their role as effect modifiers. Effect modification (or effect-measure modification) refers to the "situation in which a measure of effect changes over values of some other variable" [11]; in other words, effect modifiers act varying the relationship between the independent variable and the dependent variable. (Figure 8.1B). For example, smoking and asbestos exposure are both risk factors for lung cancer. Nonsmokers exposed to asbestos have a fivefold increased risk of lung cancer, and many studies demonstrate that smoking increases the risk of lung cancer 20-fold. The contemporary presence of both risk factors, however,

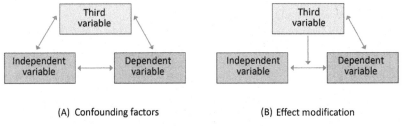

(A) Confounding factors   (B) Effect modification

FIGURE 8.1 Confounding factors (panel A) and effect modifiers (panel B).

increases the risk of developing lung cancer up to 64 times. This synergy shows that the presence or absence of asbestos exposure is somehow able to modify the effect of smoke on lung cancer risk.

8.10 CONCLUSIONS AND PERSPECTIVES

Investigations of epigenetic markers in epidemiologic studies address a critical gap in the current knowledge of the molecular mechanisms that link environmental exposures, lifestyle habits, and modifiable factors to disease development. Despite the above-mentioned limitations, applying epidemiologic analyses to epigenomic studies will improve the identification and classification of disease risks and will facilitate the development of new pharmacologic targets.

REFERENCES

[1] Relton CL, Davey Smith G. Is epidemiology ready for epigenetics? Int J Epidemiol 2012;41(1):5−9.

[2] Ng JW, Barrett LM, Wong A, Kuh D, Smith GD, Relton CL. The role of longitudinal cohort studies in epigenetic epidemiology: challenges and opportunities. Genome Biol 2012;13(6):246.

[3] Lepeule J, Baccarelli A, Motta V, Cantone L, Litonjua AA, Sparrow D, et al. Gene promoter methylation is associated with lung function in the elderly: the Normative Aging Study. Epigenetics 2012;7(3):261−9.

[4] Wilker EH, Alexeeff SE, Poon A, Litonjua AA, Sparrow D, Vokonas PS, et al. Candidate genes for respiratory disease associated with markers of inflammation and endothelial dysfunction in elderly men. Atherosclerosis 2009;206(2):480−5.

[5] Hunninghake GM, Cho MH, Tesfaigzi Y, Soto-Quiros ME, Avila L, Lasky-Su J, et al. MMP12, lung function, and COPD in high-risk populations. N Engl J Med 2009; 361(27):2599−608.

[6] He JQ, Foreman MG, Shumansky K, Zhang X, Akhabir L, Sin DD, et al. Associations of IL6 polymorphisms with lung function decline and COPD. Thorax 2009;64(8):698−704.

[7] Ren C, Fang S, Wright RO, Suh H, Schwartz J. Urinary 8-hydroxy-2′-deoxyguanosine as a biomarker of oxidative DNA damage induced by ambient pollution in the Normative Aging Study. Occup Environ Med 2011;68(8):562−9.

[8] Madrigano J, Baccarelli A, Mittleman MA, Sparrow D, Vokonas PS, Tarantini L, et al. Aging and epigenetics: longitudinal changes in gene-specific DNA methylation. Epigenetics 2012;7(1):63−70.

[9] Janssen BG, Godderis L, Pieters N, Poels K, Kici Ski M, Cuypers A, et al. Placental DNA hypomethylation in association with particulate air pollution in early life. Part Fibre Toxicol 2013;10(1):22.

[10] Nelson HH, Marsit CJ, Christensen BC, Houseman EA, Kontic M, Wiemels JL, et al. Key epigenetic changes associated with lung cancer development: results from dense methylation array profiling. Epigenetics 2012;7(6):559−66.

[11] Rothman KJ. Epidemiology an introduction. 2nd ed. New York: Oxford University Press, New York; 2012.

[12] Langevin SM, Houseman EA, Accomando WP, Koestler DC, Christensen BC, Nelson HH, et al. Leukocyte-adjusted epigenome-wide association studies of blood from solid tumor patients. Epigenetics 2014;9(6):884−95.

[13] Shenker NS, Polidoro S, van Veldhoven K, Sacerdote C, Ricceri F, Birrell MA, et al. Epigenome-wide association study in the European Prospective Investigation into Cancer and Nutrition (EPIC-Turin) identifies novel genetic loci associated with smoking. Hum Mol Genet 2013;22(5):843−51.

[14] Bell JT, Tsai PC, Yang TP, Pidsley R, Nisbet J, Glass D, et al. Epigenome-wide scans identify differentially methylated regions for age and age-related phenotypes in a healthy ageing population. PLoS Genet 2012;8(4):e1002629.

[15] Tarantini L, Bonzini M, Apostoli P, Pegoraro V, Bollati V, Marinelli B, et al. Effects of particulate matter on genomic DNA methylation content and iNOS promoter methylation. Environ Health Perspect 2009;117(2):217−22.

[16] Pilsner JR, Liu X, Ahsan H, Ilievski V, Slavkovich V, Levy D, et al. Folate deficiency, hyperhomocysteinemia, low urinary creatinine, and hypomethylation of leukocyte DNA are risk factors for arsenic-induced skin lesions. Environ Health Perspect 2009; 117(2):254−60.

[17] Motta V, Angelici L, Nordio F, Bollati V, Fossati S, Frascati F, et al. Integrative Analysis of miRNA and inflammatory gene expression after acute particulate matter exposure. Toxicol Sci 2013;132(2):307−16.

[18] Zhou X, Sun H, Ellen TP, Chen H, Costa M. Arsenite alters global histone H3 methylation. Carcinogenesis 2008;29(9):1831−6.

[19] Baccarelli A, Bollati V. Epigenetics and environmental chemicals. Curr Opin Pediatr 2009;21(2):243−51.

[20] Bollati V, Baccarelli A. Environmental epigenetics. Heredity 2010;105(1):105−12.

[21] Clark SJ, Statham A, Stirzaker C, Molloy PL, Frommer M. DNA methylation: bisulphite modification and analysis. Nat Protoc 2006;1(5):2353−64.

[22] Meissner A, Gnirke A, Bell GW, Ramsahoye B, Lander ES, Jaenisch R. Reduced representation bisulfite sequencing for comparative high-resolution DNA methylation analysis. Nucleic Acids Res 2005;33(18):5868−77.

[23] Harris RA, Wang T, Coarfa C, Nagarajan RP, Hong C, Downey SL, et al. Comparison of sequencing-based methods to profile DNA methylation and identification of monoallelic epigenetic modifications. Nat Biotechnol 2010;28(10):1097−105.

[24] Aberg KA, Xie LY, Nerella S, Copeland WE, Costello EJ, van den Oord EJ. High quality methylome-wide investigations through next-generation sequencing of DNA from a single archived dry blood spot. Epigenetics 2013;8(5):542−7.

[25] Huebert DJ, Kamal M, O'Donovan A, Bernstein BE. Genome-wide analysis of histone modifications by ChIP-on-chip. Methods 2006;40(4):365−9.

[26] Barski A, Cuddapah S, Cui K, Roh TY, Schones DE, Wang Z, et al. High-resolution profiling of histone methylations in the human genome. Cell 2007;129(4):823−37.

[27] Geiss GK, Bumgarner RE, Birditt B, Dahl T, Dowidar N, Dunaway DL, et al. Direct multiplexed measurement of gene expression with color-coded probe pairs. Nat Biotechnol 2008;26(3):317−25.

[28] Jahr S, Hentze H, Englisch S, Hardt D, Fackelmayer FO, Hesch RD, et al. DNA fragments in the blood plasma of cancer patients: quantitations and evidence for their origin from apoptotic and necrotic cells. Cancer Res 2001;61(4):1659−65.

[29] Deligezer U, Akisik EE, Erten N, Dalay N. Sequence-specific histone methylation is detectable on circulating nucleosomes in plasma. Clin Chem 2008;54(7):1125−31.

[30] Dieker JW, Fransen JH, van Bavel CC, Briand JP, Jacobs CW, Muller S, et al. Apoptosis-induced acetylation of histones is pathogenic in systemic lupus erythematosus. Arthritis Rheum 2007;56(6):1921−33.

[31] Thery C, Ostrowski M, Segura E. Membrane vesicles as conveyors of immune responses. Nat Rev Immunol 2009;9(8):581−93.

[32] Hunter MP, Ismail N, Zhang X, Aguda BD, Lee EJ, Yu L, et al. Detection of microRNA expression in human peripheral blood microvesicles. PLoS ONE 2008;3(11):e3694.

[33] Mestdagh P, Van Vlierberghe P, De Weer A, Muth D, Westermann F, Speleman F, et al. A novel and universal method for microRNA RT-qPCR data normalization. Genome Biol 2009;10(6):R64.

[34] Jiang L, Schlesinger F, Davis CA, Zhang Y, Li R, Salit M, et al. Synthetic spike-in standards for RNA-seq experiments. Genome Res 2011;21(9):1543−51.

[35] Storey JD, Tibshirani R. Statistical significance for genomewide studies. Proc Natl Acad Sci USA 2003;100(16):9440−5.

[36] Benjamini Y, Hochberg Y. Controlling the false discovery rate: a practical and powerful approach to multiple testing. J R Stat Soc Ser B 1995;85:289−300.

[37] Storey JD. A direct approach to false discovery rates. J R Stat Soc Ser B Stat Methodol 2002;64:479−98.

[38] Hwang D, Schmitt WA, Stephanopoulos G. Determination of minimum sample size and discriminatory expression patterns in microarray data. Bioinformatics 2002;18(9):1184−93.

[39] Liu P, Hwang JT. Quick calculation for sample size while controlling false discovery rate with application to microarray analysis. Bioinformatics 2007;23(6):739−46.

[40] Yeung SH, Liu P, Del Bueno N, Greenspoon SA, Mathies RA. Integrated sample cleanup-capillary electrophoresis microchip for high-performance short tandem repeat genetic analysis. Anal Chem 2009;81(1):210−17.

[41] Dobbin K, Simon R. Sample size determination in microarray experiments for class comparison and prognostic classification. Biostatistics 2005;6(1):27−38.

Chapter 9

The DNA Methylomes of Cancer

Renée Beekman, Marta Kulis and José Ignacio Martín-Subero
Institut d'Investigacions Biomèdiques August Pi i Sunyer, Barcelona, Departamento de Anatomía Patológica, Farmacología and Microbiología, Universitat de Barcelona, Spain

Chapter Outline

9.1 INTRODUCTION TO THE EPIGENETIC LANGUAGE

The term "epigenetics" was used for the first time in 1942 by Conrad Hal Waddington, who defined it as the mechanisms by which genotypes give rise to phenotypes during development [1]. This link with developmental biology is no longer used and currently, epigenetics is frequently defined as the study of changes in gene expression that occur independently of changes in the primary DNA sequence. The epigenetic language comprises various regulatory layers, such as DNA methylation, chromatin marks (histone modifications or variants), nucleosome positioning, and nucleosome accessibility. By far, the most widely studied epigenetic mark in the context of human diseases is

M. Fraga & A.F. Fernandez (Eds): Epigenomics in Health and Disease.
DOI: http://dx.doi.org/10.1016/B978-0-12-800140-0.00009-1
© 2016 Elsevier Inc. All rights reserved.
183

DNA methylation, but histone modifications are also essential to understand the role of DNA methylation. In fact, recent studies are starting to analyze multiple epigenetic marks to obtain a more precise characterization of epigenome in the context of normal differentiation and various diseases. In the next paragraph, we will present a succinct overview on histone marks, and then we will proceed with a more detailed explanation of DNA methylation.

Chromatin has the structural role of packing DNA into the cell nucleus. However, this is not an inert structure but, rather, a highly dynamic scaffold essential in regulating gene expression. Multiple post-translational modifications of the N-terminal tails of histones are involved in this process, such as methylation, acetylation, and phosphorylation, among others, and the number of modifications keeps on increasing [2]. Different combinations of these marks form a specific "histone code" that is associated with gene expression or silencing [3]. With the recent development of whole-genome approaches to characterize the distribution of histone marks across the genome, a more precise insight into the regulatory role of chromatin has been made available. Specific histone modifications not only seem to be associated with activation or silencing of gene expression but can also correlate with other genomic functions. Currently, the genome can be segmented into various functional categories called "chromatin states," such as active, weak, or poised promoters; active or weak enhancers; insulators (mainly sites of CTCF binding); transcribed regions; repressed regions (Polycomb-repressed); and heterochromatic regions [4]. These chromatin patterns vary among different cell types, and therefore, chromatin modulation and activity of distinct regions of the genome are cell-type specific.

9.2   DEFINITION AND CLASSICAL ROLES OF DNA METHYLATION

DNA methylation is one of the most intensely studied epigenetic modifications in mammals and it has an important impact on normal cell physiology. At the biochemical level, DNA methylation consists on the covalent addition of a methyl group ($-CH_3$) to cytosine, generally within the context of CpG dinucleotides. Typically, these dinucleotides are concentrated in clusters, called "CpG islands" (CGIs) that are enriched in promoter and first exon regions. In the human genome, nearly 60% of all human promoters contain CGIs [5]. Cytosine methylation is mediated by a class of enzymes called "DNA methyltransferases." Five members of the DNMT family have been identified in mammals: DNMT1, DNMT2, DNMT3A, DNMT3B, and DNMT3L. DNMT1 appears to be involved in restoring the parental DNA methylation pattern in the newly synthesized daughter strand during cell division, thereby ensuring the maintenance of methylation patterns during multiple cell generations. *De novo* DNA methylation is carried out by DNMT3A and DNMT3B. Their activity is critical for developmental stages, as shown by studies in which the inactivation of each of these genes leads to severe phenotypes [6]. DNMT2 and DNMT3L are not thought to function as cytosine methyltransferases. However, DNMT3L was shown to stimulate *de novo* DNA methylation by DNMT3A [7,8]. In contrast to DNA methylation, the exact mechanisms leading to DNA demethylation still remain controversial. DNA demethylation may occur passively through lack of maintenance during cell division or actively though the function of Ten-Eleven Translocation (TET) family of proteins or activation-induced cytidine deaminase (AID) followed by base-excision repair that introduces an unmethylated cytosine [9].

According to the first studies in the field, gene silencing seemed to be the main function of DNA methylation [10,11]. A variety of DNA methylation analysis, mostly centered on CGI methylation, showed that methylation of promoter region results in inhibition of gene expression [12]. This hypothesis was also confirmed by experiments in which DNA methylation was reversed, either by demethylating agents or disruption of the DNA methylation machinery (e.g., deletions of DNMTs) [13,14]. Upon demethylation, the studied genes became re-expressed. This inverse correlation between gene expression and DNA methylation was confirmed by several studies throughout the years, becoming finally a generally accepted model. Based on accumulated evidence, it is now widely accepted that DNA methylation is involved in multiple physiologic processes, such as organismal development and cell differentiation, genomic imprinting, X-chromosome inactivation, suppression of repetitive elements, and genomic stability.

9.3 DNA METHYLATION IN CANCER: A HISTORICAL PERSPECTIVE

As DNA methylation is associated with processes essential for cell physiology, it is not surprising that alterations in DNA methylation levels or patterns may be linked to various diseases, most notably in cancer. Already in 1983, Feinberg and Vogelstein observed a reduction of DNA methylation of specific genes in human colon cancer cells as compared with normal tissues [15]. In the same year, Gama-Sosa et al. described a global reduction of the 5-methylcytosine content of DNA from tumor samples [16]. Since these initial findings, we have significantly broadened our knowledge in the field [12]. Overall, it has been observed that carcinogenesis is accompanied by a global decrease of CpG methylation. This phenomenon was mainly suggested to be related to genomic instability, particularly by demethylation of repetitive genomic elements and transposable elements and, less frequently, to activation of silenced oncogenes [17]. Although the mayor effect seen in cancer is global loss of DNA methylation, until recently, the majority of the studies in different types of cancer were focused on another generally observed phenomenon cancer, DNA methylation gain in promoter regions, mostly containing CGIs. As DNA methylation mediates gene silencing, gain of DNA methylation at CGIs has been repeatedly shown to be involved in tumor suppressor genes inactivation in virtually all cancer types [12,18]. Hence, CGI hypermethylation represents today one of the major hallmarks of cancer and affects dozens of genes from the main cellular pathways, such as DNA repair (i.e., *MGMT*), Ras signaling (*RASSFIA*), cell cycle control (*p16INK4a, RB*), the p53 network (*TP73*), and apoptosis (*DAPK1*). A more detail analysis of tumor suppressor gene silencing in cancer can be found elsewhere [19−21]. The classical model of the role of DNA methylation in cancer, as described above, is illustrated in Figure 9.1A.

More recent analyses on broader, genome-wide scale, however, have invited us to revise these initial models of cancer epigenetics. It is becoming increasingly clear from these unbiased studies that the role of DNA methylation changes in cancer is more complex than initially thought; different layers of genetic and epigenetic information as well as the cell of origin from which the cancer originates have to be taken into account to interpret those data in a biologically and clinically meaningful way. These new concepts derived from genome-wide approaches will be further addressed in more detail in this chapter.

9.4 HIGH-THROUGHPUT APPROACHES TO DETECT DNA METHYLATION CHANGES

DNA methylation has been the subject of intense research over the last decades. Therefore, a wide range of methods have been developed to detect and quantify this epigenetic mark (reviewed by Laird, 2010) [22]. Furthermore, an enormous progress has also been made toward completing whole-genome DNA methylomes. 5mCs can be detected by three general strategies based on (i) bisulfite conversion of DNA, (ii) methyl-sensitive restriction enzymes, and (iii) immunoprecipitation (affinity enrichment) assays. Bisulfite treatment of DNA converts unmethylated cytosines into uracil, whereas methylated cytosines stay unchanged. Thus, sequencing after bisulfite treatment allows us to detect and quantify DNA methylation levels in individual CpG residues. Although bisulfite treatment is a reliable method to distinguish methylated and unmethylated cytosines, the recent discovery of 5mC derivatives, such as 5-hydroxymethylcytosine (5hmC), call for more careful use of this technique, as it does not distinguish between 5mC and 5hmC. This implies that a proportion of genomic loci identified as methylated may actually be hydroxymethylated [23]. Another approach to detect methylated cytosines is based on methylation-sensitive restriction enzymes that can digest the target sequence only if it is not methylated. A comparison between digestion patterns of these restriction enzymes and their isoschizomer or neoschizomer, that is, the enzymes that digest DNA independent of the DNA methylation patterns, gives information on the

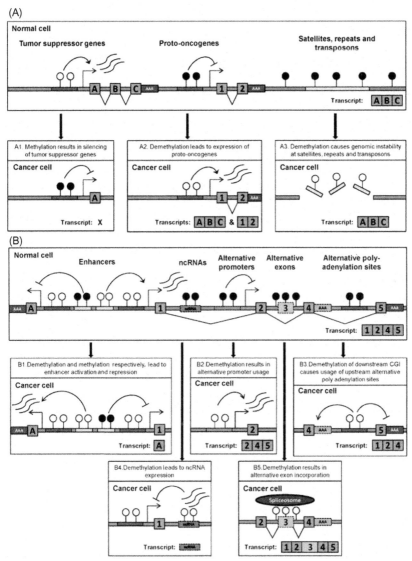

FIGURE 9.1 Classical (A) and new roles (B) of DNA methylation in cancer. In each model, the DNA methylation in normal cells is depicted in the upper panel, followed by the changes in cancer in the lower panels. Closed lollipops indicate methylated CpGs, and open lollipops represent unmethylated CpGs. Color coding: promoters/CGIs (red), alternative promoter/CGI (pink), exons (green), alternative exon (light green), enhancers (yellow), polyadenylation sites (purple), alternative polyadenylation site (lila), ncRNA (orange), transposons, satellites, and repeats (light blue). In each case, the expected expressed transcript is depicted, considering that all other DNA methylation sites as shown in the normal cells remain unaltered. In the classical model, three processes were considered to be of relevance: (A1) aberrant methylation of promoters (red) of

(Continued)

DNA methylation status. Although this kind of strategies can often be successfully used to evaluate global DNA methylation patterns, one of the disadvantages is that they only recognize those CpGs that are localized withing the enzyme restriction site. A third method for distinguishing between methylated and unmethylated cytosines is related to immuno-precipitation. The development of antibodies against 5mC provides an opportunity to detect these epigenetic marks across the genome. Moreover, the use of specific antibodies against MBD proteins (which specifically recognize 5mC) allows to immunoprecipitate the methylated fragments of DNA that can be subjected to further identification analysis [24,25]. This approach is limited by its resolution, which is restricted to the size of the immunoprecipitated DNA fragments and the CpG density. All methods of DNA methylation detection can be coupled with different downstream techniques and platforms to properly measure DNA methylation. Depending on the methodology, the number of samples and CpG nucleo-tides that can be analyzed as well as the resolution of the technique is different [22].

The application of microarray systems has facilitated the identification of differentially methylated sites at a genome-wide scale. Since their development, microarrays used for epigenomic purposes have mainly covered promoter regions and/or CGIs. In the recent years, microarray platforms have improved their coverage, and nowadays, the most widely applied DNA methylation microarray in the context of cancer is the Infinium HumanMethylation450k BeadChip, which allows us to quantify CpG methylation across the genome at a single CpG resolution [26]. Overall, the 450k BeadChip array is able to measure the methylation levels of 485,512 CpG sites located in 5′ regions (comprising promoter regions up to 1500 bp from the TSS, 5′UTR and the first exon) as well as in the gene body and 3′UTR regions of nearly all RefSeq genes, noncoding RNAs (ncRNAs), microRNAs (miRNAs), and intergenic regions. Moreover, it covers

---

◀ tumor suppressor genes, leading to their downregulation, (A2) aberrant demethylation of promoters (red) of proto-oncogenes, leading to their upregulation and (A3) aberrant demethyla-tion of transposons, satellites, and repeats, leading to genome instability; the depicted genes are not affected. Insights from whole genome studies show that many processes might be deregu-lated by DNA methylation changes in cancer as depicted in B. The depicted examples involve gene expression regulation and alternative isoform expression: (B1) DNA methylation changes in enhancers (yellow) may lead to their activation or repression, (B2) DNA methylation may lead to alternative promoter (pink) usage, leading to skipping of exon 1, (B3) demethylation of a nearby CGI (red) may lead to use of an alternative polyadenylation site (lila), leading to skipping of exon 5, (B4) DNA demethylation may result in expression of intragenic (or intergenic, not shown) ncRNAs (orange) that affect gene expression, the ncRNA represses the expression of the depicted transcript, (B5) DNA demethylation may lead to alternative splicing, leading to assembly of the splicing machinery (the spliceosome) and incorporation of alternative exon 3 (light green).

96% of all known CGIs. Furthermore, the 450k array requires quite low sample input (as low as 500 ng of DNA). Thus, coupled with its relatively low cost, it represents a potent tool for high-throughput methylation profiling of large sample cohorts.

In spite of the power of microarrays to detect methylation levels, they are still far from characterizing the whole DNA methylome at a single base-pair resolution. To solve this limitation, over the past years, several methods have been developed to examine DNA methylation profiles using next-generation sequencing approaches. All strategies of DNA methylation detection can be coupled with different downstream techniques and platforms to properly measure DNA methylation, which has led to the development of multiple methods [22]. Reduced-representation bisulfite sequencing is a widely used method to quantify DNA methylation levels of around 1 million CpGs [27], and even an enhanced method is available to measure approximately 2 million CpGs [28]. However, the most robust and comprehensive technique for DNA methylation analysis at a single-base-pair resolution is whole-genome bisulfite sequencing (WGBS) [29]. It allows for obtaining an unbiased representation of DNA methylation maps throughout the genome and constant improvements in this technology help increase its accuracy. The methylation status of over 90% of all cytosines in the genome can be measured. The analysis of large amounts of data generated by WGBS demand sophisticated statistical methods and bioinformatic techniques [22]. A considering development of the computational tools have been made during the last few years, but the data analysis is still a bottleneck of the WGBS method. Furthermore, one of the disadvantages of WGBS is its economical cost. It is believed that to achieve accurate results, at least 30-times coverage is needed, which makes this technique still rather expensive. Therefore, the analysis of DNA methylation in large numbers of samples still relies on other sequencing strategies and array-based approaches. Important insights originated from different studies using mostly WGBS or the 450k BeadChip array will be further discussed in this chapter.

9.5 THE GENOME-WIDE DNA METHYLOME OF CANCER CELLS: OVERVIEW AND GENERAL INSIGHTS

The application of unbiased methods to detect DNA methylation changes in cancer cells have started to provide new insights into the potential roles of DNA methylation in different types of tumors. So far, thousands of cases have been analyzed using methylation arrays, mostly in the context of the International Cancer Genome Consortium (ICGC) and The Cancer Genome Atlas (TCGA). At the time of writing (November 2014), these consortia had data available from 8451 (array-based DNA methylation, not otherwise specified) and 1135 (450k Beadchip array) cases, respectively [30,31]. Considering WGBS studies, significantly fewer studies were available in the literature at the time of writing: ependymoma ($n = 6$), pediatric high-grade glioma ($n = 13$), medulloblastoma ($n = 34$), chronic lymphocytic leukemia (CLL) ($n = 2$), and colon cancer ($n = 4$) [32−37]. In this review, instead of summarizing the data generated for each individual tumor type, we will present an integrated overview on the DNA methylome of cancer by distilling general concepts from the current available literature in the field.

As mentioned previously, the two major DNA methylation changes described in the literature are the hypermethylation of CGIs leading to gene silencing and global hypomethylation associated with chromosomal instability. However, the impact of these two epigenetic modifications needs to be revisited in the light of recent findings. First, global hypomethylation is a generally observed phenomenon not only in cancers with genomic instability but also in cancers with rather stable genomes, such as CLL [34]. Hence, extensive hypomethylation does not lead to genomic instability *per se*. Second, although the presence of DNA methylation at CGIs seems to ensure a repressive chromatin environment, its absence is not necessarily associated with gene activation [38]. Several other mechanisms have been shown to inhibit gene expression even if the CGIs are unmethylated. For example, it is known that the polycomb repressive complex, which mediates the trimethylation of H3K27 (H3K27me3), can lead to a closed chromatin state in the absence of DNA methylation [39]. However, even though DNA methylation has frequently been considered a mechanism inducing *de novo* gene silencing, different studies show that DNA methylation may occur in the promoters of already silenced genes. For instance, cancer-associated hypermethylation frequently targets genes already silenced in nontumoral cells by H3K27me3 [40−42]. Thus, although it is true that some tumor suppressor genes become *de novo* methylated and *de novo* silenced in cancer, hypermethylation affects mostly genes already silenced in normal cells [43,44]. This can suggest that DNA methylation at promoter regions could be a secondary event playing a role in achieving stable gene inactivation. Hence, in many cases, CGI methylation may be more a consequence than a cause of gene repression.

Furthermore, whole-genome studies are starting to reveal the importance of DNA methylation not only within promoter regions and CGIs but also in low CpG density regions, in gene bodies, and in intergenic regions [45]. It was seen that in those contexts, the classical association between gene expression and DNA methylation was no longer sustainable. Within these regions, both positive and negative correlations between transcription and DNA methylation have been described. However, of important note is that an association between gene expression and DNA methylation does not necessarily indicate that alterations in DNA methylation cause changes in gene expression levels. Instead, it is becoming increasingly clear that the activity of a gene is determined by a complex interplay among different epigenetic layers, including DNA methylation, histone modifications, and chromatin accessibility.

9.6   DNA METHYLATION CHANGES OUTSIDE PROMOTERS IS A MAJOR FINDING IN CANCER

From genome-wide DNA methylation studies, it is becoming increasingly clear that DNA methylation changes in cancer frequently occur in gene bodies and intergenic regions. Many different studies have addressed the potential role of DNA methylation changes within these regions. These studies generally suggest that DNA methylation far away from promoter regions is also, in part, associated with gene expression by regulating processes related to transcription. First, a high number of transcripts are transcribed form alternative promoters [46]; therefore, we may speculate that DNA methylation at alternative promoters may play a general role in expression of alternative transcripts. This hypothesis was confirmed by a study in which 61 alternative promoters showed a correlation between DNA methylation and the expression of their corresponding alternative transcripts [47]. Furthermore, in brain tumors, alternative promoters showed a high correlation between DNA methylation, occupancy of trimethylation at H3K4 (H3K4me3), a chromatin mark characterizing active promoters, and expression of alternative transcripts [48], suggesting, again, an important role of DNA methylation in alternative promoters. Methylation of intragenic CGIs may not only regulate transcription of alternative transcript but may affect expression of the regular transcript as well. An example is the *RB1* locus that harbors an imprinted intragenic CGI at the second exon [49]. In CLL, low methylation levels of this CGI result not only in increased expression of the alternative transcript but also in decreased expression of the main transcript [34]. This is possibly regulated by the binding of the transcription initiation complex to the alternative promoter, affecting the elongation speed of the regular transcript. Hence, positive and negative correlation with intragenic DNA methylation can go hand in hand, dependent on whether the alternative or the regular transcript is considered.

Another mechanism by which DNA methylation and other epigenetic mechanisms are suggested to regulate transcription is via alternative splicing and polyadenylation [50]. This hypothesis originates from the observation that nucleosome positioning, histone modifications and DNA methylation are unequally distributed between introns and exons [51]. CpG density was found to be higher in splice donor than in acceptor sites and in introns surrounding alternatively spliced exons in comparison with constitutively used exons [52]. These data provided the first evidence that DNA methylation may play a role in alternative splicing. Additionally, CpGs surrounding acceptor sites usually have increased DNA methylation levels, whereas CpGs in donor sites are usually unmethylated. The CD45 gene is a mechanistic example of how DNA methylation and alternative splicing can be linked [53]. In this study, it was shown that methylation of its fifth exon results in exon skipping, whereas its demethylation results in exon incorporation in the final transcript. The authors reported that CTCF binding to the unmethylated fifth exon may lead to RNA polymerase II pausing, allowing time for the splicing machinery to include the exon in the transcript. It is generally known that splicing factors are mutated in various tumors, including CLL [54]. The fact that in CLL hundreds of differentially methylated CpGs are close to alternative splice sites clearly suggest that DNA methylation may add another level of regulating splicing in cancer [34]. However, to what extent DNA methylation influences splicing remains to be elucidated because an association between DNA methylation and alternative splicing could be found for only 2 out 16 studied genes [34]. Likely, the interaction between genetic and other epigenetic layers plays an important role in splicing regulation. DNA methylation has also been linked to the usage of alternative polyadenylation sites, another mechanism by which transcripts of different length can be produced [55]. Hence, different studies show that epigenetic mechanisms, including intragenic DNA methylation, regulate RNA processing, thus providing evidence for another role of this epigenetic mark in gene expression.

DNA methylation has also been linked to the expression of ncRNAs, which are known key regulators in such processes as proliferation, differentiation, apoptosis, and cell development [56]. One of the best studied classes of ncRNAs comprises miRNAs, which are short RNA fragments (18−25 nucleotides) known to downregulate their target gene by messenger RNA (mRNA) degradation or inhibiting translation [57]. Hence, by regulating miRNA expression, DNA methylation can indirectly affect gene expression. For example, hypermethylation of mir-124 in colon cancer and mir-127 in bladder cancer leads to their down regulation. As a consequence, their oncogenic targets, CDK6 and BCL-6, respectively, are upregulated [58,59]. Another family of ncRNAs are long ncRNAs (lncRNAs), with a length of 200−2000 nucleotides. They are known to act post-transcriptionally like miRNA, but they may further act by competing for transcription factor (TF)-binding sites in regulatory elements, targeting chromatin modifiers or enhancing transcription [60,61]. For example, the lncRNA AFAP1-AS1 was shown to be hypomethylated in esophageal carcinoma, leading to its overexpression. Its mechanism of action remains unknown, as its inhibition did not lead to changes in the coding the *AFAP1* gene on mRNA or protein level. However, knockdown of AFAP1-AS1 reduced proliferation, invasiveness and programed cell death, suggesting an important role in carcinogenesis [62]. Besides ncRNAs, it is agreed that DNA methylation also has a role in the expression of transposable elements, such as LINEs (long-interspersed nuclear elements), SINEs (short-interspersed nuclear elements), and retroviruses [63,64]. Many of them are located within gene bodies; however, how DNA methylation affects their transcription remains unclear. A possible mechanism was suggested by Lorincz et al., who showed in an *in vitro* model that high DNA methylation levels at intragenic transposable elements leads to chromatin condensation, possibly resulting in lower RNA polymerase II elongation efficiencies [65].

Although the examples of how DNA methylation outside promoter regions may affect gene expression are numerous, many CpGs are not associated with alternative promoters, alternative exons, alternative polyadenylation sites, ncRNAs, or transposable elements [66]. This suggests that DNA methylation might influence transcription via other mechanisms. Recently, different studies have suggested another role of transcriptional regulation by DNA methylation outside promoter regions, that is, by targeting both intragenic and intergenic enhancer regions. Generally, DNA methylation seems to be associated with enhancer repression, whereas low methylation levels correspond to their activation. The role of DNA methylation in enhancer regions will be extensively discussed in the next section. All examples given above on how DNA methylation may play a role in cancer, as is becoming increasingly clear from whole-genome methylation studies, are shown in Figure 9.1B.

9.7 ALTERED DNA METHYLATION IN CANCER IS BIASED TOWARD PARTICULAR CHROMATIN STATES

As mentioned previously, the presence of different histone modification and features, such as CTCF binding determine the chromatin state of genomic regions, broadly separating promoters, enhancers, insulators, transcribed regions, polycomb-repressed regions, and heterochromatin. Among different cell types, a high variation in chromatin states exists, representing the execution of different gene expression programs necessary for proper cell functioning [4]. An extensive study of 42 WGBS profiles, covering 30 different human cell and tissue types, shows that differences in DNA methylation among cell types target specific chromatin states [67]. Globally, in this study, Ziller et al. observed that embryonic stem (ES) cells have the highest level of DNA methylation, followed by primary cells. In contrast, colon cancer cells show a sharp decrease in DNA methylation. Furthermore, these authors observed that the differential DNA methylated regions among ES cells, ES-cell derived cell types, and primary cells are small (75% smaller than 1 Kb) and located distal of transcriptional start sites. Interestingly, these differentially methylated regions show a high co-localization with features characteristic for regulatory elements (DNAseI hypersensitivity and H3K27ac) and transcription-factor-binding sites. Furthermore, low DNA methylation levels at regulatory sites in all investigated cell types by Ziller et al. are enriched for cell-type-specific TF binding sites, suggesting that DNA methylation patterns are associated with the expression of cell type specific TFs. When comparing colon cancer with a matched control, they found that 40% of the differentially methylated CpGs overlap with the previously identified differentially methylated CpGs among ES-cell derived cell types and primary cells that were enriched for regulatory elements.

The data presented above suggest that not only in normal differentiation but also in cancer cells regulatory elements outside promoters are subject to DNA methylation changes. Only few examples of this phenomenon are known so far. First of all, in a large CLL cohort compared with its normal counterparts, DNA hypomethylation was most frequently observed in enhancer regions [34]. Furthermore, in this study the correlation between enhancer methylation and gene expression was stronger at enhancer regions than at promoter regions. This was also observed by Blattler et al., who compared the colon cancer cell lines HCT116 and the DNMT1- and DNMT3B-deficient DKO1, and the latter showed a reduction of 95% of DNA methylation levels [68]. In this study, these authors observe that DNA methylation loss at promoter in DKO1 cells does not cause activation of gene expression on a large scale. Interestingly however, loss of DNA methylation at distant regions was associated with the presence of the active enhancer mark H3K27ac in DKO1 cells and was correlated with higher expression levels of neighboring genes. In a recent study, it was furthermore shown that aberrant enhancer activity was obtained in the breast cancer cell line MCF7 and the prostate cancer cell line PC3, which was accompanied by loss in DNA methylation [69]. However, in the same study, it was shown that the magnitude of regulatory elements aberrantly losing enhancer activity ($n = 6356$) in the breast cancer cell line MCF7 in comparison with HMEC (human mammary epithelial) cells, was threefold higher than the aberrant gain of enhancer activity ($n = 2142$). This phenomenon of enhancer "decommissioning" was associated with high methylation levels at these regulatory sites. Hence, the examples described above show that both loss and gain of DNA methylation at regulatory elements may correlate with changes in enhancer activity in cancer cells.

Two other chromatin states that are particularly prone to acquire DNA methylation changes in cancer are polycomb-repressed regions and heterochromatin [29,32,33,41,70]. Polycomb-repressed regions are highly likely to become hypermethylated in cancer, possibly through the recruitment of DNMTs to these regions by enhancer of zeste homolog 2, a component of the polycomb repressor 2 complex [71,72]. The role of DNA hypermethylation in polycomb-repressed regions is described above in the section "The genome-wide DNA methylome of cancer cells: overview and general insights." In brief, DNA methylation at these regions seems not to affect gene expression, but affects already silenced genes. However, it may block the capability of a cell to reactivate genes, which may be essential for malignant transformation. In contrast to polycomb-repressed regions, heterochromatin, which shows a strong overlap with late-replicating regions and association with the nuclear lamina, is particularly prone to losing DNA methylation during malignant transformation [29,32,33,70]. It is thought that this phenomenon may be the result of passive methylation loss upon replication. Activation of gene expression has not been observed in heterochromatic regions losing DNA methylation in cancer, and some studies suggest that even the opposite phenomenon occurs, that is, gene repression [29,33,70].

In summary, DNA methylation changes in cancer mainly affect enhancers, polycomb-repressed regions and heterochromatin. As the potential effect of DNA methylation changes in different chromatin states may be different, it is important to place the observed DNA methylation changes in light of the proper chromatin architecture in order to understand the potential functional role of differential DNA methylation in cancer.

9.8   NORMAL REFERENCE SAMPLES OF THE CANCER EPIGENOME

It is generally accepted that DNA methylation patterns depend on cell type and differentiation stage. Hence, the methylation pattern of a cell is a footprint of its cell type and maturation status. This phenomenon is challenging to study because it requires extensive cell sorting, usually of small cell populations, and high cell purity to be able to analyze the differentiation stages of a lineage in a biologically meaningful way. Similar to normal cells, cancer cells have a methylation footprint as well. This footprint is the result of at least two different factors; it harbors characteristics of the cell of origin where the tumor originated from and cancer-specific changes, such as those affecting promoters, enhancers, polycomb-repressed regions, and heterochromatin. The hematopoietic system is a paradigmatic model to study DNA methylation footprints in cancer in light of their cell of origin, as it well known that different hematopoietic neoplasms develop from particular cell lineages at specific maturation stages. In fact, the WHO classification of hematopoietic tumors is based both on their cell lineage and maturation stage [73]. Furthermore, subpopulations of hematopoietic lineages can be isolated in sufficient numbers and purity to analyze their DNA methylome. Multiple studies have shown that various differentiation stages of early hematopoiesis, lineage commitment, and maturation of specific cell subtypes are accompanied by a clear modulation of the DNA methylome [74−79].

In line with previous findings, in a study of a large CLL cohort, a massive remodeling of the DNA methylome from naive B cells toward class-switch memory B-cells was observed by WGBS [34]. Interestingly, in this study, it was shown that the methylation patterns of CLL cases harbored a small fraction of the footprint of their cell of origin, which seemed to be hidden within the massive DNA methylation changes occurring in these tumors. By comparing two mayor subgroups of CLL, that is, cases with unmutated and mutated immune globulin heavy chains (IGHV), a DNA methylation signature was identified, showing that CLLs with unmutated IGHV resemble naive B cells, whereas CLLs with mutated IGHV clustered with memory B cells. Interestingly, in an unsupervised cluster analysis, both CLL subtypes resembled more memory B cells, suggesting that during malignant transformation, the CLLs with unmutated IGHV have undergone many DNA methylation changes that also occur during normal differentiation from naive B cells to memory B cells. These changes were massive and masked the true cell of origin of the unmutated CLLs.

Based on the above studies, an essential issue in cancer epigenetics is to determine the correct normal counterpart of a cancer cell. Comparing with one reference may suggest that a region is differentially methylated, whereas comparing with another reference might not show differences. The same may hold true when comparing chromatin states. Hence, using the wrong reference samples may lead to incorrect interpretations of data. How can we find the proper reference for a cancer cell? First, we have to know the underlying clinical and biologic characteristics—for example, in what organ did it originate, is it blocked in differentiation and, if so, in which cellular stage? Second, surface markers may help define the cell of origin even further. Third, similarities in epigenetic make-up with normal counterparts may guide us in this process. All parameters given above should be considered with care, although, as during malignant transformation, the cell may partially progress in differentiation, as was shown for the CLLs with unmutated IGHV [34]. Consequently, the cancer cell may resemble a more differentiated cell rather than its cell of origin. Hence, the best clues to determine the proper cell of origin may probably come from computational dissection of the DNA methylation profiles in different fractions, representing (i) cell of origin, (ii) differentiation related changes that have occurred during transformation, and (iii) cancer specific changes.

Although the example above focuses on a hematopoietic neoplasm, in solid tumors, it will likely be as important to study the tumors in light of their proper cell of origin. However, in these cases, we are challenged by technical difficulties in terms of the presence of mixed cell populations in tumors and the proper isolation of normal subpopulations. This is likely one of the reasons why the differentiation of nonhematopoietic lineages is not very well described from the DNA methylation point of view yet. Nevertheless, given the fact that the epigenome is influenced by stage-specific and tissue-specific changes, it seems essential for any type of tumor to analyze both the cancer cells and their normal counterparts as highly pure cell populations as mixed populations influence the epigenomic profiles and may lead to inaccurate, and even wrong, conclusions.

9.9 DNA METHYLATION CHANGES: CAUSE OR CONSEQUENCE OF CANCER?

The changes in the DNA methylation footprints of cancer that also occur during normal differentiation or that are reminiscent of the cell of origin should not be characterized as a cause or consequence of cancer but are inherent to the cellular lineage and differentiation stage of the tumor. Hence, to understand whether DNA methylation changes in cancer are a cause or a consequence of malignant transformation, it first has to be determined which DNA methylation changes are cancer specific. The next issue we are facing to answer the question listed above, is that it is still not clear at which stage of tumor transformation DNA methylation alterations take place. A major problem is that for many cancers, we do not have precancerous stages that might give us additional clues on when cancer-specific DNA methylation changes occur. Some studies, however, provide some insights into this issue. For example, in colon cancer, high levels of aberrant DNA methylation are observed already in polyps, which may suggest that it is an early event in carcinogenesis [43]. It was also proposed that DNA methylation related with aging may predispose for neoplastic transformation. In a study of whole-blood samples from older individuals, it was shown that the age-related methylation of polycomb target genes is a preneoplastic condition and may lead to gene expression changes associated with carcinogenesis [80]. Additionally, immortalization of B cell cells by Epstein-Barr virus, which is associated with lymphoma and nasopharyngeal cancer, leads to extensive hypomethylation. The large blocks of hypomethylation seen in these cells are highly similar to the ones observed in cancer, suggesting that these methylation changes are an early event in malignant transformation [81].

We may draw some conclusions by considering the major DNA methylation changes observed in cancer, that is, hypomethylation of enhancers and heterochromatin and hypermethylation of polycomb-repressed regions. As stated previously, hypomethylation of enhancers is associated with increased enhancer activity at affected loci. However, whether DNA hypomethylation is a cause or a consequence of enhancer activation during either differentiation or malignant transformation remains unclear. Possibly, TF occupancy at enhancers may affect the maintenance of DNA methylation via interfering with DNMT binding, leading to gradual loss of DNA methylation during cell division. In this case, enhancer demethylation is a consequence, rather than a cause, of enhancer activation [82]. A different scenario is that enhancers actively lose methylation via mechanisms mediated by TETs, AID, and the apolipoprotein B mRNA-editing enzyme, catalytic polypeptide-like proteins [9,83,84]. This model is supported by the fact that enhancers that are modulated during differentiation are frequently targeted by TET1 and are enriched for hydroxymethylated cytosines, an important step in DNA demethylation [85,86]. Although the causes and consequences of DNA methylation changes in enhancers remain to be elucidated, hypomethylation of heterochromatin seems to be without functional impact, and DNA methylation at polycomb-repressed regions seems to occur at promoters of already silenced genes. These data clearly suggest that most DNA methylation changes in cancer do not lead to deregulated gene expression. But then, which are the consequences of these apparently "inert" epigenetic changes? Compared with the versatile gene silencing mediated by histone modifications, hypermethylation of polycomb-repressed regions may be associated with stable gene silencing [87]. Due to the stability of DNA methylation, hypermethylation of already silenced genes has been proposed to be associated with a reduction of epigenetic plasticity in cancer cells [88]. Hypomethylation of heterochromatin is a general phenomenon, also seen in cells that go into senescence and at the final steps of normal differentiation [80,89,90]. Hence, it seems to be a general physiologic feature. We may hypothesize that a complex interplay of different biologic characteristics that are shared between cancer cells and differentiated cells, such as proliferative history, senescence, and terminal differentiation, may cause this physiologic phenomenon without imposing functional impact.

In summary, the majority of DNA methylation changes in cancer likely does not have a functional effect but should be seen as either dependent on the cell of origin and differentiation status of the tumor cells or as a consequence of malignant transformation. Of all DNA methylation changes, however, it has to be hypothesized that some have functional impact, for example, in terms of changing processes related to gene expression. The challenge for the future will be to identify these DNA methylation changes and to understand their functional role in carcinogenesis.

9.10   CLINICAL USE OF DNA METHYLATION IN CANCER

DNA methylation is a stable mark that can rather easily and reliably be measured and implemented as a standard laboratory test in the clinics. Hence, it can be applied as a diagnostic or a prognostic tool and can also be used in clinical decision making. Many examples of the usefulness of this mark in the clinics can be found [91−93]. A specific application for example is that DNA methylation biomarkers may be able to distinguish cancer subgroups with different prognosis, thereby having predictive value. Additionally, as DNA methylation may occur early in disease, it can be used for early disease detection and monitoring of disease progression. Furthermore, it can be used to identify primary tumors in the case of cancers with unknown primary origin [90]. Some examples of how DNA methylation marks can be used in clinical practice will be discussed in further detail below.

Previously, our own group described that a DNA methylation signature of 1649 CpGs defined in 139 CLL cases has been found that can distinguish two major clinical subgroups of CLL based on their cell of origin [34]. These two subgroups furthermore differ significantly in response to therapy and prognosis. Interestingly, the same DNA methylation signature was able to distinguish a third CLL subgroup that had an intermediate phenotype in terms of DNA methylation pattern and prognosis. Although this signature harbored prognostic potential, in clinical practice, it is generally not appropriate to implement a 1649-CpGs signature as a prognostic tool. Therefore, we extracted a CpG signature containing five CpGs of which DNA methylation levels could accurately be measured by bisulfite pyrosequencing. DNA methylation levels of these five CpGs reliably divided CLL cases in the three subgroups previously identified. These results were validated in an extended CLL dataset from the same group ($n = 211$), and in an independent validation cohort ($n = 97$) [94]. In a different study of 295 CLL patients, the methylation status of even a single CpG located in the *ZAP70* gene, enabled to distinguish two CLL subgroups with different prognostic outcome [95]. Furthermore, Sandoval et al. presented a similar example of prognostic significance of a DNA methylation signature in non−small cell lung cancer (NSCLC) [96]. In this study DNA methylation patterns of 444 NSCLC patients were analyzed by the 450k BeadChip array (Illumina). The 10,000 most variable DNA methylation sites enabled identification of a subgroup of NSCLC patient that had a shorter recurrence time and worse prognosis. By statistical analysis and DNA pyrosequencing in an additional validation cohort of 143 NSCLC patients, they identified five CpGs that were highly correlated with cancer recurrence. These were located in five genes: *HIST1H4F*, *PCDHGB6*, *NPBWR1*, *ALX1*, and *HOXA9*.

In rare cases, when cancer patients present with metastatic disease, the primary tumor cannot be identified. However, for clinical decision making, it is very important to know the origin of a tumor in order to assign the right treatment and to increase the chances of survival for these patients. In these cases, DNA methylation can be used as a diagnostic marker. This was exemplified in an extensive study by Fernandez et al., in which they reliably measured DNA methylation levels of 1322 CpG sites in 1628 human samples, covering 424 normal tissue samples, 1054 tumorigenic samples, and samples from 150 other disorders [90]. First, the authors showed that one-third of the analyzed CpGs ($n = 511$) show differential DNA methylation between the normal tissue samples. Additionally, they showed that a substantial higher number (1003) of CpGs, were significantly different among the different tumor types. Interestingly, by defining the DNA methylation patterns of tumors with unknown primary origin and comparing them with the DNA methylation patterns of the 1054 tumor samples, the authors were able to assign primary tumors in 29 out of 42 cancers with unknown primary origin. The other 13 cases appeared not to harbor a tumor that was analyzed in the initial data set. By detailed pathologic analysis, 7 out of 9 of the predicted primary tumors were later confirmed. To be able to use DNA methylation patterns as a clinical tool, a requisite is that these CpGs show low individual heterogeneity. This is exactly what the authors observed for the 1322 analyzed CpGs among 180 leukocyte samples from different individuals. Although for each CpGs this may be dependent on the tissue type and should therefore always be assessed when defining new DNA methylation biomarkers. It has to be said that with the fast publication of genome-wide methylation studies, we may expect that the potential of detecting DNA methylation markers relevant for clinical use is rapidly increasing. However, studies in large and moreover in independent patient cohorts are essential to validate their applicability in clinical practice.

9.11   CONCLUSIONS AND FUTURE DIRECTIONS

Whole-genome DNA methylation studies have invited us to revise the initial hypotheses on DNA methylation in cancer carefully. Furthermore, the ample examples presented in this chapter clearly show that the delineation of the precise role of DNA methylation in cancer calls for integrative approaches, whereby different layers of information, such as genetic context, the DNA methylome, histone modifications, epigenetic states, genome accessibility, three-dimensional chromatin structure, and the transcriptome, have to be taken into account. Other layers that were not discussed in this chapter but are of importance in understanding the role of DNA methylation in cancer include the metabolome and the proteome. An example of how the metabolome can influence the DNA methylation machinery comes from studies investigating the role of isocitrate dehydrogenase (IDH1 and IDH2) mutations in myeloid malignancies [97]. These studies clearly suggest that aberrant production of 2-hydroxyglutarate instead of α-ketoglutarate by mutant IDH1/2 results in inhibition of hydroxylation of 5mC by TET2 and its subsequent DNA demethylation. Of important note is that besides all the potential layers of information listed above, other layers, of which we are not aware of or that are difficult to investigate, likely play a role. Recent evidence shows that, for example, DNA methylation at cytidines is not the only epigenetic mark occurring at 5mCs. Other potential events at CpGs are 5hmC and its derivates. At the time of this writing, a lot of effort had been put into developing tools to measure these derivatives. Hence, in the near future, data for these additional epigenetic marks will become available, and their role in cancer will become clearer.

The systems biology field is a rapidly growing discipline that focuses on complex interactions within biologic systems and will prove to be essential for data integration of different layers of information. Its strength lies in the fact that (i) it offers mathematical models to understand complex systems in biology, (ii) it provides tools to filter out experimental and biologic noise, (iii) it enables dissection of different layers of biologically relevant information, and (iv) it can be applied to any biologic data set of good quality. Currently, a major field of interest where systems biology can be applied is the mapping of the three-dimensional chromatin structure to understand enhancer−promoter interactions. For this purpose, different adaptations of the chromatin conformation capture technique are used [98], followed by computational modeling of the contact frequencies between different genomic loci. Understanding these interactions enables us to study the effect of DNA methylation in enhancers in a more precise way, as we can predict their proper target genes. Another application in which systems biology is useful is in modeling cellular heterogeneity when analyzing different integrative layers in cellular pools. To understand cellular heterogeneity even better, however, single-cell analysis is necessary, but this is technically and computationally challenging. However, after filtering out experimental and biologic noise by means of systems biology approaches, single-cell data sets may provide us the means to investigate the one-to-one correlation between informative layers, such as DNA methylation, histone modifications, three-dimensional chromatin structure, and gene expression, which is more informative than correlating population means in cellular pools.

The role of DNA methylation and other epigenetic mechanisms mainly has been studied in the context of intracellular processes, for example, by focusing on its role in transcriptional regulation. However, we may also look at it from other perspectives. For example, we may ask ourselves how the DNA methylome may aid in establishing the developmental and environmental needs of the cancer cell. Does it provide selection, growth, and/or survival advantages? To answer this question, we might look at similarities between DNA methylation patterns in cancer cells and those associated with normal differentiation, aging, and senescence. Differentiation and aging are discussed in more detail in Chapters 11 and 9, respectively. Strikingly, during aging and senescence, cells broadly obtain a similar DNA methylation profile as cancer cells in terms of gaining DNA methylation in polycomb-repressed regions and losing methylation in heterochromatin [80,89,90]. First, we might hypothesize that specific DNA methylation patterns may be involved in escaping apoptosis, a feature important in cellular senescence as well. Second, we may hypothesize that locking gene expression by stable DNA methylation might save energy by eliminating the necessity to repress these genes by other mechanisms. This energy may be used by the cancer cell for other processes, such as proliferation. Long-lived cells and senescent cells might be using the same strategy to reduce unnecessary waist of energy by the body. Third, we may speculate that other processes guided by DNA methylation changes may be related to inducing proliferation by upregulating cell cycle regulators or escaping the immune system by hiding specific surface molecules.

Finally, based on the data presented in this chapter, we provide guidelines that have to be taken into account when studying whole genome methylation in cancer, which include (i) the proper reference sample should be used to study the cancer epigenome; (ii) both the cancer and the reference samples that are studied should be of high cell purity; (iii) integration of different layers of information is essential in understanding the role of DNA methylation in cancer; and (iv) the role of DNA methylation should not only be explored to explain intracellular processes but should also be studied in the context of the developmental and environmental necessities of the cancer cell. Although the number of whole-genome DNA methylation studies is growing rapidly, the number of samples analyzed by WGBS still remains rather small. Nevertheless, WGBS studies are essential in understanding the role of DNA methylation changes throughout the complete genome and to integrate this layer of information with other genome-wide data sets. Major efforts, however, are currently underway to analyze the cancer epigenome from different perspectives, including WGBS. These efforts mainly come from large consortia, such as the ICGC, the TCGA, the International Human Epigenome Consortium, and the Cancer Epigenome Consortium [30,31,99,100]. Extensive collaborations, analysis of many different samples and data integration of multiple layers of information, as done in these large international consortia, is the key to achieve a deeper understanding of DNA methylation in cancer and will ultimately result in better treatment strategies and better quality of life for patients.

ACKNOWLEDGMENTS

The authors' studies on epigenomics are currently funded by the European Union's Seventh Framework Programme through the Blueprint Consortium (grant agreement 282510), the Spanish Ministry of Economy and Competitivity (MINECO) (project SAF2009-31138), Fundació La Marató de TV3 (project 20132130), and the European Hematology Association. R.B. is supported by a fellowship from the EU (Marie Curie) and J.I.M.-S. is a Ramon y Cajal researcher of the MINECO.

REFERENCES

[1] Waddington CH. The epigenotype. 1942. Int J Epidemiol 2012;41(1):10−13.

[2] Kouzarides T. Chromatin modifications and their function. Cell 2007;128(4):693−705.

[3] Jenuwein T, Allis CD. Translating the histone code. Science 2001;293(5532):1074−80.

[4] Ernst J, Kheradpour P, Mikkelsen TS, Shoresh N, Ward LD, Epstein CB, et al. Mapping and analysis of chromatin state dynamics in nine human cell types. Nature 2011;473 (7345):43−9.

[5] Saxonov S, Berg P, Brutlag DL. A genome-wide analysis of CpG dinucleotides in the human genome distinguishes two distinct classes of promoters. Proc Natl Acad Sci USA 2006;103(5):1412−17.

[6] Okano M, Bell DW, Haber DA, Li E. DNA methyltransferases Dnmt3a and Dnmt3b are essential for *de novo* methylation and mammalian development. Cell 1999;99(3):247−57.

[7] Chedin F, Lieber MR, Hsieh CL. The DNA methyltransferase-like protein DNMT3L stimulates *de novo* methylation by Dnmt3a. Proc Natl Acad Sci USA 2002;99(26):16916−21.

[8] Deplus R, Brenner C, Burgers WA, Putmans P, Kouzarides T, de Launoit Y, et al. Dnmt3L is a transcriptional repressor that recruits histone deacetylase. Nucleic Acids Res 2002;30(17):3831−8.

[9] Bhutani N, Burns DM, Blau HM. DNA demethylation dynamics. Cell 2011;146 (6):866−72.

[10] Holliday R, Pugh JE. DNA modification mechanisms and gene activity during development. Science 1975;187(4173):226–32.

[11] Riggs AD. X inactivation, differentiation, and DNA methylation. Cytogenet Cell Genet 1975;14(1):9–25.

[12] Esteller M. Epigenetics in cancer. N Engl J Med 2008;358(11):1148–59.

[13] Christman JK. 5-Azacytidine and 5-aza-2′-deoxycytidine as inhibitors of DNA methylation: mechanistic studies and their implications for cancer therapy. Oncogene 2002;21(35):5483–95.

[14] Rhee I, Bachman KE, Park BH, Jair KW, Yen RW, Schuebel KE, et al. DNMT1 and DNMT3b cooperate to silence genes in human cancer cells. Nature 2002; 416(6880):552–6.

[15] Feinberg AP, Vogelstein B. Hypomethylation distinguishes genes of some human cancers from their normal counterparts. Nature 1983;301(5895):89–92.

[16] Gama-Sosa MA, Slagel VA, Trewyn RW, Oxenhandler R, Kuo KC, Gehrke CW, et al. The 5-methylcytosine content of DNA from human tumors. Nucleic Acids Res 1983;11(19):6883–94.

[17] Ehrlich M. DNA methylation and cancer-associated genetic instability. Adv Exp Med Biol 2005;570:363–92.

[18] Jones PA, Baylin SB. The epigenomics of cancer. Cell 2007;128(4):683–92.

[19] Esteller M. Epigenetic gene silencing in cancer: the DNA hypermethylome. Hum Mol Genet 2007;(16 Spec No 1):R50–9.

[20] Boultwood J, Wainscoat JS. Gene silencing by DNA methylation in haematological malignancies. Br J Haematol 2007;138(1):3–11.

[21] Jacinto FV, Esteller M. Mutator pathways unleashed by epigenetic silencing in human cancer.. Mutagenesis 2007;22(4):247–53.

[22] Laird PW. Principles and challenges of genomewide DNA methylation analysis. Nat Rev Genet 2010;11(3):191–203.

[23] Huang Y, Pastor WA, Shen Y, Tahiliani M, Liu DR, Rao A. The behaviour of 5-hydroxymethylcytosine in bisulfite sequencing. PLoS ONE 2010;5(1):e8888.

[24] Ballestar E, Paz MF, Valle L, Wei S, Fraga MF, Espada J, et al. Methyl-CpG binding proteins identify novel sites of epigenetic inactivation in human cancer. EMBO J 2003;22(23):6335–45.

[25] Lopez-Serra L, Ballestar E, Fraga MF, Alaminos M, Setien F, Esteller M. A profile of methyl-CpG binding domain protein occupancy of hypermethylated promoter CpG islands of tumor suppressor genes in human cancer. Cancer Res 2006;66(17):8342–6.

[26] Bibikova M, Barnes B, Tsan C, Ho V, Klotzle B, Le JM, et al. High density DNA methylation array with single CpG site resolution. Genomics 2011;98(4):288–95.

[27] Meissner A, Gnirke A, Bell GW, Ramsahoye B, Lander ES, Jaenisch R. Reduced representation bisulfite sequencing for comparative high-resolution DNA methylation analysis. Nucleic Acids Res 2005;33(18):5868–77.

[28] Akalin A, Garrett-Bakelman FE, Kormaksson M, Busuttil J, Zhang L, Khrebtukova I, et al. Base-pair resolution DNA methylation sequencing reveals profoundly divergent epigenetic landscapes in acute myeloid leukemia. PLoS Genet 2012;8(6):e1002781.

[29] Lister R, Pelizzola M, Dowen RH, Hawkins RD, Hon G, Tonti-Filippini J, et al. Human DNA methylomes at base resolution show widespread epigenomic differences. Nature 2009;462(7271):315–22.

[30] International Cancer Genome C, Hudson TJ, Anderson W, Artez A, Barker AD, Bell C, et al. International network of cancer genome projects. Nature 2010;464(7291):993–8.

[31] Heng HH. Cancer genome sequencing: the challenges ahead. Bioessays 2007;29(8):783−94.

[32] Berman BP, Weisenberger DJ, Aman JF, Hinoue T, Ramjan Z, Liu Y, et al. Regions of focal DNA hypermethylation and long-range hypomethylation in colorectal cancer coincide with nuclear lamina-associated domains. Nat Genet 2012;44(1):40−6.

[33] Hansen KD, Timp W, Bravo HC, Sabunciyan S, Langmead B, McDonald OG, et al. Increased methylation variation in epigenetic domains across cancer types. Nat Genet 2011;43(8):768−75.

[34] Kulis M, Heath S, Bibikova M, Queiros AC, Navarro A, Clot G, et al. Epigenomic analysis detects widespread gene-body DNA hypomethylation in chronic lymphocytic leukemia. Nat Genet 2012;44(11):1236−42.

[35] Hovestadt V, Jones DT, Picelli S, Wang W, Kool M, Northcott PA, et al. Decoding the regulatory landscape of medulloblastoma using DNA methylation sequencing. Nature 2014;510(7506):537−41.

[36] Bender S, Tang Y, Lindroth AM, Hovestadt V, Jones DT, Kool M, et al. Reduced H3K27me3 and DNA hypomethylation are major drivers of gene expression in K27M mutant pediatric high-grade gliomas. Cancer Cell 2013;24(5):660−72.

[37] Mack SC, Witt H, Piro RM, Gu L, Zuyderduyn S, Stutz AM, et al. Epigenomic alterations define lethal CIMP-positive ependymomas of infancy. Nature 2014;506(7489):445−50.

[38] Bergman Y, Cedar H. DNA methylation dynamics in health and disease. Nat Struct Mol Biol 2013;20(3):274−81.

[39] Cao R, Wang L, Wang H, Xia L, Erdjument-Bromage H, Tempst P, et al. Role of histone H3 lysine 27 methylation in Polycomb-group silencing. Science 2002; 298(5595):1039−43.

[40] Ohm JE, McGarvey KM, Yu X, Cheng L, Schuebel KE, Cope L, et al. A stem cell-like chromatin pattern may predispose tumor suppressor genes to DNA hypermethylation and heritable silencing. Nat Genet 2007;39(2):237−42.

[41] Schlesinger Y, Straussman R, Keshet I, Farkash S, Hecht M, Zimmerman J, et al. Polycomb-mediated methylation on Lys27 of histone H3 pre-marks genes for *de novo* methylation in cancer. Nat Genet 2007;39(2):232−6.

[42] Widschwendter M, Fiegl H, Egle D, Mueller-Holzner E, Spizzo G, Marth C, et al. Epigenetic stem cell signature in cancer. Nat Genet 2007;39(2):157−8.

[43] Keshet I, Schlesinger Y, Farkash S, Rand E, Hecht M, Segal E, et al. Evidence for an instructive mechanism of *de novo* methylation in cancer cells. Nat Genet 2006; 38(2):149−53.

[44] Martin-Subero JI, Kreuz M, Bibikova M, Bentink S, Ammerpohl O, Wickham-Garcia E, et al. New insights into the biology and origin of mature aggressive B-cell lymphomas by combined epigenomic, genomic, and transcriptional profiling. Blood 2009;113(11):2488−97.

[45] Jones PA. Functions of DNA methylation: islands, start sites, gene bodies and beyond. Nat Rev Genet 2012;13(7):484−92.

[46] Kimura K, Wakamatsu A, Suzuki Y, Ota T, Nishikawa T, Yamashita R, et al. Diversification of transcriptional modulation: large-scale identification and characterization of putative alternative promoters of human genes. Genome Res 2006;16(1):55−65.

[47] Cheong J, Yamada Y, Yamashita R, Irie T, Kanai A, Wakaguri H, et al. Diverse DNA methylation statuses at alternative promoters of human genes in various tissues. DNA Res 2006;13(4):155−67.

[48] Maunakea AK, Nagarajan RP, Bilenky M, Ballinger TJ, D'Souza C, Fouse SD, et al. Conserved role of intragenic DNA methylation in regulating alternative promoters. Nature 2010;466(7303):253−7.

[49] Kanber D, Berulava T, Ammerpohl O, Mitter D, Richter J, Siebert R, et al. The human retinoblastoma gene is imprinted. PLoS Genet 2009;5(12):e1000790.

[50] Brown SJ, Stoilov P, Xing Y. Chromatin and epigenetic regulation of pre-mRNA processing. Hum Mol Genet 2012;21(R1):R90–6.

[51] Chodavarapu RK, Feng S, Bernatavichute YV, Chen PY, Stroud H, Yu Y, et al. Relationship between nucleosome positioning and DNA methylation. Nature 2010;466 (7304):388–92.

[52] Malousi A, Maglaveras N, Kouidou S. Intronic CpG content and alternative splicing in human genes containing a single cassette exon. Epigenetics 2008;3(2):69–73.

[53] Shukla S, Kavak E, Gregory M, Imashimizu M, Shutinoski B, Kashlev M, et al. CTCF-promoted RNA polymerase II pausing links DNA methylation to splicing. Nature 2011;479(7371):74–9.

[54] Scott LM, Rebel VI. Acquired mutations that affect pre-mRNA splicing in hematologic malignancies and solid tumors. J Natl Cancer Inst 2013;105(20):1540–9.

[55] Elkon R, Ugalde AP, Agami R. Alternative cleavage and polyadenylation: extent, regulation and function. Nat Rev Genet 2013;14(7):496–506.

[56] Guil S, Esteller M. Cis-acting noncoding RNAs: friends and foes. Nat Struct Mol Biol 2012;19(11):1068–75.

[57] Deng S, Calin GA, Croce CM, Coukos G, Zhang L. Mechanisms of microRNA deregulation in human cancer. Cell Cycle 2008;7(17):2643–6.

[58] Lujambio A, Ropero S, Ballestar E, Fraga MF, Cerrato C, Setien F, et al. Genetic unmasking of an epigenetically silenced microRNA in human cancer cells. Cancer Res 2007;67 (4):1424–9.

[59] Saito Y, Liang G, Egger G, Friedman JM, Chuang JC, Coetzee GA, et al. Specific activation of microRNA-127 with downregulation of the proto-oncogene BCL6 by chromatin-modifying drugs in human cancer cells. Cancer Cell 2006;9(6):435–43.

[60] Mercer TR, Dinger ME, Mattick JS. Long non-coding RNAs: insights into functions. Nat Rev Genet 2009;10(3):155–9.

[61] Ponting CP, Oliver PL, Reik W. Evolution and functions of long noncoding RNAs. Cell 2009;136(4):629–41.

[62] Wu W, Bhagat TD, Yang X, Song JH, Cheng Y, Agarwal R, et al. Hypomethylation of noncoding DNA regions and overexpression of the long noncoding RNA, AFAP1-AS1, in Barrett's esophagus and esophageal adenocarcinoma. Gastroenterology 2013;144 (5):956–966.e4.

[63] Hoffmann MJ, Schulz WA. Causes and consequences of DNA hypomethylation in human cancer. Biochem Cell Biol 2005;83(3):296–321.

[64] Ehrlich M. DNA methylation in cancer: too much, but also too little. Oncogene 2002;21 (35):5400–13.

[65] Lorincz MC, Dickerson DR, Schmitt M, Groudine M. Intragenic DNA methylation alters chromatin structure and elongation efficiency in mammalian cells. Nat Struct Mol Biol 2004;11(11):1068–75.

[66] Jjingo D, Conley AB, Yi SV, Lunyak VV, Jordan IK. On the presence and role of human gene-body DNA methylation. Oncotarget 2012;3(4):462–74.

[67] Ziller MJ, Gu H, Muller F, Donaghey J, Tsai LT, Kohlbacher O, et al. Charting a dynamic DNA methylation landscape of the human genome. Nature 2013;500(7463):477–81.

[68] Blattler A, Yao L, Witt H, Guo Y, Nicolet CM, Berman BP, et al. Global loss of DNA methylation uncovers intronic enhancers in genes showing expression changes. Genome Biol 2014;15(9):469.

[69] Taberlay PC, Statham AL, Kelly TK, Clark SJ, Jones PA. Reconfiguration of nucleosome-depleted regions at distal regulatory elements accompanies DNA methylation of enhancers and insulators in cancer. Genome Res 2014;24(9):1421–32.

[70] Hon GC, Hawkins RD, Caballero OL, Lo C, Lister R, Pelizzola M, et al. Global DNA hypomethylation coupled to repressive chromatin domain formation and gene silencing in breast cancer. Genome Res 2012;22(2):246–58.

[71] Vire E, Brenner C, Deplus R, Blanchon L, Fraga M, Didelot C, et al. The Polycomb group protein EZH2 directly controls DNA methylation. Nature 2006;439(7078):871–4.

[72] O'Hagan HM, Wang W, Sen S, Destefano Shields C, Lee SS, Zhang YW, et al. Oxidative damage targets complexes containing DNA methyltransferases, SIRT1, and polycomb members to promoter CpG Islands. Cancer Cell 2011;20(5):606–19.

[73] Harris NL, Jaffe ES, Diebold J, Flandrin G, Muller-Hermelink HK, Vardiman J, et al. World Health Organization classification of neoplastic diseases of the hematopoietic and lymphoid tissues: report of the Clinical Advisory Committee meeting-Airlie House, Virginia, November 1997. J Clin Oncol 1999;17(12):3835–49.

[74] Calvanese V, Fernandez AF, Urdinguio RG, Suarez-Alvarez B, Mangas C, Perez-Garcia V, et al. A promoter DNA demethylation landscape of human hematopoietic differentiation. Nucleic Acids Res 2012;40(1):116–31.

[75] Martin-Subero JI, Ammerpohl O, Bibikova M, Wickham-Garcia E, Agirre X, Alvarez S, et al. A comprehensive microarray-based DNA methylation study of 367 hematological neoplasms. PLoS ONE 2009;4(9):e6986.

[76] Ji H, Ehrlich LI, Seita J, Murakami P, Doi A, Lindau P, et al. Comprehensive methylome map of lineage commitment from haematopoietic progenitors. Nature 2010;467(7313):338–42.

[77] Shaknovich R, Cerchietti L, Tsikitas L, Kormaksson M, De S, Figueroa ME, et al. DNA methyltransferase 1 and DNA methylation patterning contribute to germinal center B-cell differentiation. Blood 2011;118(13):3559–69.

[78] Lee ST, Xiao Y, Muench MO, Xiao J, Fomin ME, Wiencke JK, et al. A global DNA methylation and gene expression analysis of early human B-cell development reveals a demethylation signature and transcription factor network. Nucleic Acids Res 2012;40(22):11339–51.

[79] Lai AY, Mav D, Shah R, Grimm SA, Phadke D, Hatzi K, et al. DNA methylation profiling in human B cells reveals immune regulatory elements and epigenetic plasticity at Alu elements during B-cell activation. Genome Res 2013;23(12):2030–41.

[80] Teschendorff AE, Menon U, Gentry-Maharaj A, Ramus SJ, Weisenberger DJ, Shen H, et al. Age-dependent DNA methylation of genes that are suppressed in stem cells is a hallmark of cancer. Genome Res 2010;20(4):440–6.

[81] Hansen KD, Sabunciyan S, Langmead B, Nagy N, Curley R, Klein G, et al. Large-scale hypomethylated blocks associated with Epstein-Barr virus-induced B-cell immortalization. Genome Res 2014;24(2):177–84.

[82] Stadler MB, Murr R, Burger L, Ivanek R, Lienert F, Scholer A, et al. DNA-binding factors shape the mouse methylome at distal regulatory regions. Nature 2011; 480(7378):490–5.

[83] Popp C, Dean W, Feng S, Cokus SJ, Andrews S, Pellegrini M, et al. Genome-wide erasure of DNA methylation in mouse primordial germ cells is affected by AID deficiency. Nature 2010;463(7284):1101–5.

[84] Tahiliani M, Koh KP, Shen Y, Pastor WA, Bandukwala H, Brudno Y, et al. Conversion of 5-methylcytosine to 5-hydroxymethylcytosine in mammalian DNA by MLL partner TET1. Science 2009;324(5929):930−5.

[85] Pastor WA, Pape UJ, Huang Y, Henderson HR, Lister R, Ko M, et al. Genome-wide mapping of 5-hydroxymethylcytosine in embryonic stem cells. Nature 2011;473 (7347):394−7.

[86] Serandour AA, Avner S, Oger F, Bizot M, Percevault F, Lucchetti-Miganeh C, et al. Dynamic hydroxymethylation of deoxyribonucleic acid marks differentiation-associated enhancers. Nucleic Acids Res 2012;40(17):8255−65.

[87] Jones PA. The DNA methylation paradox. Trends Genet 1999;15(1):34−7.

[88] Gal-Yam EN, Egger G, Iniguez L, Holster H, Einarsson S, Zhang X, et al. Frequent switching of Polycomb repressive marks and DNA hypermethylation in the PC3 prostate cancer cell line. Proc Natl Acad Sci USA 2008;105(35):12979−84.

[89] Cruickshanks HA, McBryan T, Nelson DM, Vanderkraats ND, Shah PP, van Tuyn J, et al. Senescent cells harbour features of the cancer epigenome. Nat Cell Biol 2013; 15(12):1495−506.

[90] Fernandez AF, Assenov Y, Martin-Subero JI, Balint B, Siebert R, Taniguchi H, et al. A DNA methylation fingerprint of 1628 human samples. Genome Res 2012;22(2):407−19.

[91] Heyn H, Esteller M. DNA methylation profiling in the clinic: applications and challenges. Nat Rev Genet 2012;13(10):679−92.

[92] Mikeska T, Bock C, Do H, Dobrovic A. DNA methylation biomarkers in cancer: progress towards clinical implementation. Expert Rev Mol Diagn 2012;12(5):473−87.

[93] Bock C. Epigenetic biomarker development. Epigenomics 2009;1(1):99−110.

[94] Queiros AC, Villamor N, Clot G, Martinez-Trillos A, Kulis M, Navarro A, et al. A B-cell epigenetic signature defines three biologic subgroups of chronic lymphocytic leukemia with clinical impact. Leukemia 2014;Aug:25.

[95] Claus R, Lucas DM, Ruppert AS, Williams KE, Weng D, Patterson K, et al. Validation of ZAP-70 methylation and its relative significance in predicting outcome in chronic lymphocytic leukemia. Blood 2014;124(1):42−8.

[96] Sandoval J, Mendez-Gonzalez J, Nadal E, Chen G, Carmona FJ, Sayols S, et al. A prognostic DNA methylation signature for stage I non-small-cell lung cancer. J Clin Oncol 2013;31(32):4140−7.

[97] Figueroa ME, Abdel-Wahab O, Lu C, Ward PS, Patel J, Shih A, et al. Leukemic IDH1 and IDH2 mutations result in a hypermethylation phenotype, disrupt TET2 function, and impair hematopoietic differentiation. Cancer Cell 2010;18(6):553−67.

[98] de Wit E, de Laat W. A decade of 3C technologies: insights into nuclear organization. Genes Dev 2012;26(1):11−24.

[99] Bae JB. Perspectives of international human epigenome consortium. Genomics Inform 2013;11(1):7−14.

[100] Beck S, Bernstein BE, Campbell RM, Costello JF, Dhanak D, Ecker JR, et al. A blueprint for an international cancer epigenome consortium. A report from the AACR Cancer Epigenome Task Force. Cancer Res 2012;72(24):6319−24.

Chapter 10

Genome-Wide Epigenetic Studies in Neurologic Diseases

Ashwin Woodhoo[1,2]
[1]CIC bioGUNE, Centro de Investigación Biomédica en Red de Enfermedades Hepáticas y Digestivas, Bizkaia, Spain, [2]IKERBASQUE, Basque Foundation for Science, Bilbao, Spain

Chapter Outline

10.1 INTRODUCTION

Epigenetics is a fascinating and burgeoning field of biology that is making a significant impact on reshaping our understanding of the physiology and pathology of both the central nervous system (CNS) and the peripheral nervous system (PNS), a field that is now commonly referred to as "neuroepigenetics" [1,2]. Epigenetics refers to stable and heritable, and yet dynamic and reversible, changes in gene expression, which do not result from alterations in DNA sequence. The epigenome, the complex pattern of different epigenetic modifications in the genome, plays an essential role in instructing the unique gene expression profile in each cell type, together with its genome. The epigenome is broadly used to describe the global, comprehensive interplay of different sequence-independent processes, including DNA methylation, histone modification, and regulation by noncoding RNAs (ncRNAs) that modulate the transcriptome of a cell [3−5].

It is now clear that epigenetic mechanisms play a key role in many fundamental neurobiologic processes, ranging from neural stem cell maintenance and differentiation, to learning and memory. Not surprisingly, it is now being

M. Fraga & A.F. Fernandez (Eds): Epigenomics in Health and Disease.
DOI: http://dx.doi.org/10.1016/B978-0-12-800140-0.00010-8
© 2016 Elsevier Inc. All rights reserved.

increasingly recognized that epigenetics is intrinsically linked to neurologic diseases, ranging from neurodegenerative and neurodevelopmental disorders to neuropsychiatric diseases [3,6]. There is also an emerging conceptual framework that links the epigenetic machinery and processes to neurologic disease states [7]. The most studied paradigm is, not surprisingly, mutations in epigenetic factors that lead to neurologic syndromes. To date, more than a dozen single gene mutations in these factors, which have been shown to be associated with brain disorders, including mental retardation, Rubinstein-Taby syndrome, and autism, have been described [1]. Rett syndrome (OMIM 312750), which is an X-linked neurologic disease caused by genetic defects in a methyl-CpG binding protein (MeCP2), is one of the most salient examples [8].

Genetic variation in genes encoding epigenetic enzymes, including single-nucleotide polymorphisms (SNPs), is also associated with neurologic disorders. For example, it was recently shown that the risk of ischemic stroke was associated with an SNP in the intron of the *HDAC9* gene (odds ratio (OR) 1.38; 95% confidence interval (CI) 1.22−1.57), in the large-scale Genome-wide Association Study (GWAS) study by the International Stroke Genetics Consortium [9]. Alteration of expression, localization, or function of epigenetic factors caused by abnormal disease-associated pathways and proteins is another evolving paradigm that has ascribed to the pathophysiologic defects in neurologic diseases. For example, deficiency of ataxia telangiectasia mutated protein indirectly leads to nuclear HDAC4 accumulation in neurons, which contributes to the pathogenesis of the neurodegenerative disease [10].

10.2 NEUROEPIGENETICS IN THE "OMICS" ERA

The exponential growth of epigenomic datasets and publications in the last few years have helped to identify thousands of long intervening ncRNA (lincRNA) genes, *cis*-regulatory sequences, including promoters and enhancers, and annotation of long-range chromatin interactions [4,5]. In the nervous system, epigenomic profiling has helped understand major processes related to normal brain function and plasticity as well as neurologic diseases [5]. Here, we describe the use of these epigenomic maps in understanding neurologic disease mechanisms, in both the CNS and the PNS, ranging from neurodegenerative and neurodevelopmental disorders to neuropsychiatric diseases.

10.2.1 Alzheimer Disease

Alzheimer disease (AD) is the leading cause of dementia in older adults worldwide, afflicting 11% of people aged 65 years or older, and 32% of people aged 85 years or older. It is expected that by 2050, about 100 million people will be affected by this disease worldwide [11]. The disease is characterized by progressive neurodegeneration in selected brain regions, including temporal and parietal lobes that lead to memory loss and a decline in the patient's cognitive and functional abilities, severely affecting quality of life and being ultimately fatal [12].

The classical hallmarks of AD include the accumulation of extracellular amyloid b-peptide (Aβ) plaques and intracellular neurofibrillary tangles, as well as neuronal loss. In the rare early-onset form of AD (EOAD), genetic mutations of the amyloid-β precursor protein (*APP*) and the presenilin genes (*PSEN1, PSEN1*) have a clear connection to the disease because they directly lead to Aβ plaque accumulation [12]. However, the late-onset form of AD (LOAD), the more common form of the disease, demonstrates a number of non-Mendelian features, and there is increasing evidence that epigenetic mechanisms could play a potential role in the pathogenesis of the disease.

One important piece of evidence suggesting that epigenetic mechanisms could be involved in the disorder comes from a study on monozygotic twins discordant for AD [13]. The authors showed significantly reduced levels of DNA methylation in the temporal neocortex neuronal nuclei of the twin with AD, compared with the twin without dementia. Other studies that focused on examining global and gene-specific DNA methylation patterns also suggested that DNA methylation could play a role in AD [14].

In small-scale studies, the genome-wide DNA methylation patterns in the cerebral cortex were identified in postmortem examination of AD patients and mice models of AD [15,16]. Both studies showed relatively modest changes in DNA methylation levels on a global scale. In the first study, in which the DNA methylome was interrogated in 12 LOAD and 12 age-matched and gender-matched controls using an Illumina Infinium HumanMethylation27 BeadArray, the authors found that 948 CpG sites, representing 918 unique genes, were differentially methylated in AD patients, although the mean methylation difference was only relatively modest (2.9%; interquartile range (IQR) = 0.88−4.2) [15]. Nevertheless, they also found that methylation of a CpG site in the promoter region of *TMEM59*, a gene involved in post-translational modification of APP and has been implicated in AD pathogenesis [17], was most strongly associated with AD status and was associated with functional changes in RNA and protein expression.

In the second study, the DNA methylome of the cerebral cortex in two mice models of AD was examined using custom-designed DNA methylation arrays targeted at 384 genes related to sensory perception, cognition, neuroplasticity, and mental diseases [16]. Notably, the authors found DNA hypermethylation of four genes in both mice models, which were accompanied by a reduction in the expression of the corresponding RNA transcripts and protein, a pattern broadly confirmed in AD patients. These genes have key functions in molecular signaling pathways and cellular structures, including CREB-activation and the axon initial segment, highlighting a potential role in AD pathogenesis. Out of these four genes, SORBS3 is particularly interesting since another study also showed DNA hypermethylation at its promoter regions [18]. It encodes for a cell adhesion molecule in both neurons and glia and has been implicated in synapsis [19].

Recently, in two major studies, the methylome profile of human brain was characterized by using large independent cohorts of autopsied samples [20,21]. In both studies, the methylation levels of 415,8484 CpG dinucleotides were examined by using the Illumina 450K HumanMethylation array. In the first study, the methylation level was measured in a sample of dorsolateral prefrontal cortex in 708 subjects [21]. The authors identified 71 CpGs, whose methylation levels significantly correlated with the burden of neuritic amyloid plaques, a quantitative measure of AD neuropathology. Interestingly, the authors found that the altered methylation profiles occurred early in the development of AD pathology and at presymptomatic stages, suggesting that altered DNA methylation could contribute to the pathologic process, although much remains to be learned in this key area. Notably, 12 of the 71 CpGs identified in the primary cohort of samples were also found to be significantly associated with AD pathology in an independent cohort of 117 subjects, in which Braak staging was used as a quantitative measure of AD pathology. Using another independent cohort consisting of 202 AD and 197 non-AD individuals, several differentially expressed genes were found to be associated with the differentially methylated regions. Strikingly, several of these genes, including *ANK1, DIP2A, RHBDF2, RPL13, SERPINF1*, and *SERPINF2* were found to be connected to an AD susceptibility network derived from genetic studies.

In the second study, methylation profiling was also performed in large independent autopsied human brain cohorts by using the Illumina 450K HumanMethylation array [20]. However, in this case, these analyses were done using different parts of the brain with different vulnerability to AD. Using a primary cohort of 122 subjects, the authors found that a hypermethylated region in the ankyrin gene (*ANK1*) was associated with the AD-related neuropathology in three different cortical regions, including the entorhinal cortex (EC), which is the primary and early area of AD neuropathology. Strikingly, no significant disease-associated DNA hypermethylation was observed neither in the cerebellum, which is largely protected from neurodegeneration in AD, nor in the whole blood samples. Notably, the authors showed that neuropathology-associated DNA hypermethylation in the *ANK1* gene was also found in an independent cohort of 144 cortical samples by using Illumina 450K HumanMethylation array and in another cohort of 62 samples by using bisulfite pyrosequencing. Importantly, DNA hypermethylation of this gene was also detected in the other large-cohort study mentioned above [21].

A number of studies have shown that regulation of gene expression by histone modifications could be implicated in AD. Differential expressions of histone modifications and histone modifying enzymes have been observed in AD patients and in mice models. Particularly, in monozygotic twins discordant for AD, increased levels of H3K9me3 were found in the hippocampus and temporal cortex of the twin with AD [22]. In another study, HDAC2 was found to be overexpressed in a mouse model of AD and in small cohort of AD patients [23]. Interestingly, using targeted chromatin immunoprecipitation (ChIP) analyses of the hippocampus from the animal AD model, the authors found that HDAC2 was significantly enriched at the promoter regions of several genes related in memory and synaptic plasticity, which had been shown to be downregulated in AD patients. Further ChIP analyses revealed hypoacetylation of several histone residues at these genes. However, so far, genome-wide mapping of histone modifications or histone-modifying diseases has not been performed, as for DNA methylation, and this remains a key challenge for future work. Nevertheless, there seems be an ongoing effort in characterizing different histone modifications in different brain regions from AD patients [24].

Similar to histone modifications, studies done on the expression patterns of ncRNAs in AD have been limited mostly to targeted assays in mice models of AD and in postmortem brain specimens from AD subjects by using a small sample size. A number of microRNAs (miRNAs) and long ncRNAs (lncRNAs) have been found to be deregulated in AD, several of which have been implicated in regulation of key AD-associated genes [22]. Nevertheless, in a recent study, miRNA profiling of the hippocampus from a cohort of 41 LOAD cases and 23 age-matched controls was carried out using the nCounter system, covering about 641 human miRNAs [25]. Out of these, 35 miRNAs were found be differentially expressed. Similarly, nCounter profiling of an independent cohort of 49 prefrontal cortex from Braak-stratified patients and seven control samples showed 41 miRNAs to be differentially expressed. To overcome the limitation of the nCounter analysis, which covers only about 30% of all human miRNAs found in the miRBase database, the authors also deep sequenced the miRNAs in a modest cohort of prefrontal cortex samples (4 controls, 2 early-stage AD patients, and 6 late-stage AD patients), where they found 85 differentially expressed miRNAs. Notably, the authors found a significant downregulation of miR-132-3p in the three distinct analyses performed, irrespective of tissue origin or profiling assay, and went on to demonstrate that the deregulation of this miRNA appeared to be present mainly in neurons with hyper-phosphorylated Tau. Downregulation of miR-132-3p in AD was also found in other independent studies [22], suggesting that this miRNA could have a potentially important role in AD.

In a small-scale study, involving 10 patients, miRNA profiling was performed using the Exiqon LNA arrays [26]. The authors showed that there were more extensive changes in miRNA profiles in gray matter than in white matter. In another array-based study, in which levels of both miRNAs and mRNAs were simultaneously profiled from the parietal lobe of 4 AD patients and 4 controls, a robust relationship was found between miRNA levels and their mRNA in the brain [27]. In an NGS-based study [28], there were no significant differences in miRNA profiles in postmortem brain tissue obtained from controls and AD patients, although this was likely due to the small sample size (5 AD patients and 2 controls).

MiRNA profiling has also been performed in cerebrospinal fluid (CSF) or blood from AD patients to identify potential biomarkers. Using array-based technology, in which a total of 462 miRNAs were analyzed, miR34a and miR181b were found to be increased in mononuclear blood cells from a cohort of 16 AD patients compared with 16 controls [29]. Similarly, using array-based nanostring technology, in which a total of 462 miRNAs were analyzed in the CSF of small cohort of alive patients (8 controls and 8 AD patients), miR-27a-3p was found to be reduced in the AD patients [30]. In an NGS-based study, Leidinger et al. [31] found differential expression of 140 unique miRNA in peripheral blood from 22 healthy controls and 48 AD patients. Strikingly, using a signature of 12 miRNAs, they were able to differentiate with high diagnostic accuracies between AD patients and control individuals.

10.2.2 Parkinson Disease

Parkinson disease (PD) (OMIM 168699) is the second most common progressive neurodegenerative disorder, estimated to affect $1-2\%$ of population over 60 years and rising to $3-5\%$ in those older than 85 years in industrialized countries. The major clinical symptoms include resting tremor, muscle rigidity, bradykinesia, and postural instability. Neuropathologically, PD is characterized by severe loss of dopaminergic neurons in the substantia nigra, and presence of cytoplasmic inclusions of fibrillary protein aggregates, known as Lewy bodies, composed of misfolded α-synuclein [32].

More than 90% of cases can be interpreted as sporadic PD and are thought to rise from a combination of susceptibility genes and environmental factors. The remaining 10% of cases have a strict familial etiology, and in recent years, mutations in six genes have been identified as causes of PD: *SNCA* (which encodes α-synuclein), *PARK2* (parkin), *PINK1* (PTEN-induced kinase protein 1), *DJ1* (protein DJ-1), *UCHL1* (ubiquitin carboxyl-terminal hydrolase isozyme L1), and *LRRK2* (leucine-rich repeat serine/threonine-protein kinase 2) [6].

The current view of the etiology of PD is that it is multifactorial and involves a complex interplay between a number of gene networks and the environment. Although less well studied than AD, there is emerging evidence that epigenetic mechanisms, including DNA methylation, histone modifications and ncRNAs, could contribute to the onset and development of both the sporadic and familial forms of PD [14,33]. Postmortem studies of brain tissue have shown a significant reduction in global DNA methylation, specifically in the substantia nigra of PD patients compared with that of healthy individuals [34]. Subsequent studies have shown a decrease in nuclear DNMT1 levels, associated with the interaction and cytoplasmic sequestration of DNMT1 by α-synuclein, could be responsible for the decreased DNA methylation in PD [35]. A recent methylome profiling study on postmortem prefrontal cortex samples from a small cohort of PD ($n = 5$) and healthy individuals ($n = 6$) broadly confirmed the reduced global DNA methylation in PD subjects [36]. The authors, using the Illumina 450K HumanMethylation array to interrogate DNA methylation of individual CpGs in these samples, found close to 3000 differentially methylated CpGs, with about 90% being hypomethylated. Strikingly, a similar trend was seen when DNA methylation levels in peripheral blood leukocytes was examined in the same study. Importantly, the authors found that about 75% of the top differentially methylated loci were reported to be deregulated in a GWAS study, including at least four genes highly associated with PD risk (*HLA-DQA1, GFPT2, MAPT*, and *MIR886*). Another more limited study, in which the authors used the Illumina HumanMethylation27 BeadChip, which covers only 275,000 individual CpG sites, to examine DNA methylation profiles in the putamen and cortex of 12 healthy individuals and PD patients, showed only the *CYP2E1* gene to be hypomethylated in the PD samples [37].

Little is currently known about the role of histone modifications in PD patients, although some studies with cell cultures or animal models have shown a likely role in PD [33,38]. In particular, it has been shown that nuclear-targeted α-synuclein can interact with histones and inhibit histone acetylation. Furthermore, several histone deacetylase inhibitors can rescue α-synuclein-mediated toxicity [33,38]. By contrast, the role of miRNAs is a very active area of research, and several seminal studies have shown the role of specific miRNAs in targeting PD-related genes and how PD-related genes act to regulate the miRNA processing machinery [39].

Global microRNA profiling has been performed for different PD-related model systems and in PD patients. Asikainen et al. [40] used NCode miRNA microarrays to evaluate expression levels of 115 annotated miRNAs in different *Caenorhabditis elegans* PD models. They found up to 12 specific miRNAs that were differentially expressed in the models. Similarly, Gillardon et al. [41] used a microarray-based platform to interrogate expression levels of 266 unique miRNAs in the brainstem tissue of the α-synuclein (A30P)-transgenic mouse model for PD and found a consistent decrease of several miRNAs in mutant mice. Differential miRNA profiling has also been performed in human brains with important implications [42,43]. In the study, the authors used Exiqon miRCURY arrays, which cover all miRNAs in the miRBase registry (v 8.1), to analyze miRNA profiles in 6 control individuals and 11 PD patients. The authors found consistent downregulation of miR-34 b/c in different brain regions of PD patients, including the substantia nigra, occurring at early stages of the disease. Subsequent experiments showed that this downregulation could underlie mitochondrial dysfunction and oxidative stress. Global miRNA profiling of peripheral tissues have also been carried. Using the Exiqon miRCURY arrays, Martins et al. [43] identified 18 differentially regulated miRNAs in peripheral blood mononuclear cells from 13 control and 19 PD patients. In another study, the authors examined miRNA profiles of leukocytes from 6 control and 7 PD patients by using NGS and found 16 differentially expressed miRNAs in PD patients compared with controls [44].

10.2.3   Huntington Disease

Huntington disease (HD) (OMIM 143100) is a dominantly inherited neurodegenerative disorder with a prevalence of 5−10 cases per 100,000 people worldwide. Clinical symptoms of HD include motor, psychiatric, and cognitive abnormalities that lead to a gradual loss of functional capacity and shortened lifespan. Neurodegeneration is most pronounced in medium spiny gamma-aminobutyric acid (GABA)-ergic neurons of the striatum, although several neuronal types and brain areas, including the cortex, are also affected to a lesser degree [45]. HD is caused by dominant mutations that abnormally expand a CAG trinucleotide repeat sequence at the 5′ terminal of the Huntington gene (*HTT*), leading to a polyglutamine expansion in the HTT protein. Neuronal degeneration in HD could be due to either depletion of the normal HTT protein, which has important cellular function, including endocytosis and vesicle trafficking, or gain of toxic function of the mutant HTT protein [46].

Several studies have implicated distinct layers of epigenetic dysregulation in the molecular pathophysiology of HD [45]. In a recent study, the DNA methylome of mouse striatal cells lines, expressing wild-type or mutant HTT, was examined by reduced-representation bisulfite sequencing (RRBS) [47]. The authors found that expression of the mutant HTT induced significant changes in DNA methylation patterns, which showed substantial overlap with expression of genes. Among these, several key genes that encode for proteins involved in neurogenesis were identified, including *Sox2, Pax6* and *Nes*, which were characterized by increased methylation and reduced expression. Since impaired hippocampal neurogenesis has been reported in animal models of HD, these data argue for a potential role of DNA methylation in controlling gene expression and the pathologic process in HD.

In another study, the genome-wide distribution of DNA hydroxymethylation (5 hmC) was examined in striatum and cortex of a mouse model of HD by capture-based enrichment and NGS [48]. The authors found a global decrease in 5 hmC levels in both brain regions in the HD model, and gene ontology analyses of the differentially hydoxymethylated regions showed

TABLE 10.1 Different Histone Modifications Profiles in Huntington Disease

Platform	Histone mark analyzed	Model	References
ChIP-chip	H3K9/14ac	Striatum of the transgenic R6/2 mouse model of HD	[49]
ChIP-seq	H3K9/14ac H4K12Ac	Hippocampus of transgenic N171-HD82Q mouse model for HD	[50]
ChIP-seq	H3K4me3	Cortex and striatum of the transgenic R6/2 mouse model of HD	[51]
ChIP-seq	H3K9me3	Mouse striatal lines expressing wild-type or mutant HTT	[52]

enrichment for a number of HD-related categories, including neuronal development, differentiation, function, and survival.

There is ample evidence to suggest that mutant HTT can lead to abnormal post-translation modifications of histones, including acetylation, methylation, phosphorylation, and ubiquitination, either directly or indirectly in HD, arguing for a critical role in the pathogenesis of the disease [45,46]. Genome-wide distribution of several histone modifications marks have been performed in different models of HD, which have provided a more complete and, at the same time, complex perspective of how the influence of altered epigenetic marks on transcriptional dysregulation in HD (Table 10.1). Histone acetylation has been one of the better studied histone marks, and two genome-wide studies have been performed. McFarland et al. [49] used ChIP-chip to interrogate the genome-wide distribution of the acetylated histone H3 mark H3K9/14ac in striatum of the transgenic R6/2 mouse model of HD. They found that the total number of histone sites occupied by the histone mark was significantly decreased in the HD model, although they did not find a strong correlation between the changes in this mark with changes in gene expression in the HD model compared with wild-type mice.

In another study, the more refined ChIP followed by NGS (ChIP-seq) technology was used to analyze this histone mark, in addition to the H4K12Ac mark, in the hippocampus of transgenic N171-HD82Q mouse model for HD [50]. Similar to above, the authors found hypoacetylation for the H3K9/14 at hundreds of loci, and poor correlation between the global changes in altered transcript levels and hypoacetylation in the mutant mice. Hypoacteylation of the H4K12 mark was more severe than for the H3K9/14 mark in the HD model, although similarly, there was poor correlation between gene expression and level of acetylation. However, the authors found a small subset of genes with concurrent changes in acetylation and expression levels, which were affected in different brain areas, mouse models, and patients. They suggest that these genes could play a role in the etiology of the disease.

Histone methylation is another mark that has been studied on a genomewide, and two studies have focused on the following marks: H3K4me3 [51] and H3K9me3 [52]. In the first study, the authors examined H3K4me3 occupancy across the genome in the R6/2 model for both the cortex and the striatum by using ChIP-seq. They identified a specific H3K4me3 signature at promoters of wild-type mice that predicted with high likelihood that these would be transcriptionally downregulated in the mutant mice [51]. In the second study, genomic distribution of the H3K9me3 mark was examined by ChIP-seq in mouse striatal cells lines, expressing wild-type or mutant HTT [52]. The authors found that genes involved in pathways related to HD pathology, including synaptic transmission, cell motility and neuronal pathways were significantly correlated with H3K9me3 occupancy at their promoter regions.

Similarly, there is now increasing evidence that a group of neuronalspecific miRNAs display reduced expression in the brains of HD patients and mouse models. miRNA profiling at a genomic scale has been carried out for mice models and for human HD samples. Using the Agilent Mouse miRNA Microarray 8×15 kit, which detects 567 mouse miRNAs, Lee et al. [53] analyzed the miRNA profiles in the striata of two transgenic models of HD (YAC128 and R6/2). Both models showed deregulation of many miRNAs, several of which were commonly deregulated in both models. Similarly, miRNA microarray profiling, using the Illumina Universal 12 BeadChip arrays, was carried out in the cerebral cortex of the N171-82Q mouse HD model [54]. They found several deregulated miRNAs in the mutant mice, among which alteration of miR200 family members proved particularly interesting, given their predicted role in targeting genes regulating synaptic function, neurodevelopment, and neuronal survival.

miRNA profiling in HD patients has also been carried out by independent groups with different technologic platforms, which all suggest an important role of miRNA dysfunction in human pathology. Using the TaqMan Array Human miRNA Panel, which detects 365 mature miRNAs, Packer et al. [55] showed that about 12 miRNAs were dysregulated in the cortex of early stage HD patients ($n = 13$) compared with control individuals ($n = 7$). In another study, the more sensitive technique of NGS was carried out on small RNA samples derived from the frontal cortex and striatum of two HD patients and two control individuals [56]. The authors found 160 and 121 dysregulated miRNAs in the frontal cortex and striatum, respectively, from the HD samples, with a high proportion commonly regulated in both brain tissues. These data were broadly validated using the Agilent Human miRNA microarrays, using another cohort of 4 control and 4 HD patients. The authors found that the altered expression of the deregulated miRNAs could contribute to the aberrant expression of their putative targets. In another NGS study in the prefrontal cortex of 12 HD and 9 control individuals, several miRNAs were found to be deregulated in the HD samples [57]. The authors found that several of the miRNAs were located in intergenic regions of Hox clusters, and their data argue for a neuroprotective role of these Hox-related miRNAs in HD. There is also evidence that lncRNAs could also be deregulated in HD, as shown by reanalysis of previously described microarray profiles [58]. However, so far, no genome-wide profiling study for this class of ncRNAs has been carried out.

10.2.4   Multiple Sclerosis

Multiple sclerosis (MS) (OMIM 126200) is a chronic inflammatory and neurodegenerative disease of the brain and spinal cord, and its cause is not known. It is characterized by autoimmune-mediated loss of myelin and axons that lead to progressive neurologic deterioration. The most common symptoms of this disorder include muscular weakness and rigidity, visual alterations, cognitive abnormalities, and urinary, intestinal, and sexual dysfunction. The heterogeneous clinical course and the low concordance rate in monozygotic twins indicate the involvement of complex heritable and environmental factors in the pathogenesis of the disease. Emerging data suggest that MS pathophysiology could be regulated by epigenetic mechanisms [59,60].

There is now increasing evidence that DNA methylation patterns are altered in MS. In a targeted analysis, Liggett et al. [61] examined the methylation levels of 56 genes in a custom-designed array in blood plasma obtained from 30 control and 30 MS patients. Strikingly, they found significant differences in methylation levels in 15 out of the 56 genes analyzed in the two groups. In a more recent study, the DNA methylome pattern was examined in postmortem brain tissue from 28 MS-affected and 19 control individuals by using Illumina 450K HumanMethylation array [62]. These authors found subtle but statistically significant differences in methylation patterns between the two groups, consistently throughout the genome. These changes correlated with transcriptional and translation changes for specific genes, including *BCL2L2*, *NDRG1*, *LGMN*, and *CTSZ*, and in different functional categories that could be related to an increased susceptibility of oligodendrocytes to damage in MS.

Histone modifications have also been examined in relation to MS, but the few studies have been solely focused on expression analyses and targeted gene-specific ChIP assays [60]. In contrast, miRNA profiling has been carried extensively out by a number of different groups using different technologic platforms. In particular, several genome-wide studies have been performed to identify possible diagnostic biomarkers and biomarkers associated with disease activity and treatment response using whole-blood, blood-derived cells, including CSF, and blood-derived cells, including peripheral blood mononuclear cells, T cells, and B lymphocytes [60,63].

miRNA profiling in brain tissue from MS patients has also been carried out, and altered miRNA profiles have been detected in the demyelinating lesions. Junker et al. [64] compared the miRNA profiles of active ($n = 16$) and inactive white matter brain lesions ($n = 5$) from MS patients with 9 control white matter specimens by using Taqman Human MiRNA arrays (v1.0). They identified 28 miRNAs in active lesions and 35 miRNAs in inactive lesions that were differentially expressed when compared with controls. Three miRNAs were identified, which putatively target CD47, a molecule that inhibits macrophage activity directed at resident brain cells. In another study, miRNA profiles of a small cohort of postmortem brain white matter tissue from 4 controls and 4 MS patients were compared using the Exiqon miRCURY arrays, which showed differential expression for multiple miRNAs, several of which target neurosteroidogenic pathways [65]. Similarly, miRNA profiling of myelinated and demyelinated hippocampi from postmortem MS patients was also carried out using the Exiqon miRCURY arrays [66]. The authors found deregulation of several miRNAs, of which miR-124 was particularly interesting, since it targets mRNAs that encode 26 neuronal proteins, which are deregulated in demyelinated hippocampus.

10.2.5   Major Psychosis

Schizophrenia (SZ) (OMIM 181500) and bipolar disorder (BD) (OMIM 125480) are seemingly etiologically related psychiatric conditions that are collectively called "major psychosis." SZ is characterized by disturbances in thinking and reasoning that can lead to psychotic thoughts (delusions and hallucinations), erratic behavior, and withdrawal from social activity. BD is characterized by recurrent episodes of mania or hypomania interspersed with depressive episodes. Both SZ and BD are complex diseases, and it is generally assumed that they are caused by a combination of genetic and environmental factors [67]. Several new lines of evidence suggest that epigenetic factors could play a role in mediating susceptibility to major psychosis.

The first genome-wide methylome study of major psychosis by Mill et al. [68], who interrogated methylation levels in a relatively strong cohort of SZ patients ($n = 35$), BD ($n = 32$) and controls ($n = 28$) using a microarray-based assay. They enriched the unmethylated fraction of genomic DNA, obtained from postmortem frontal cortex brain tissue, and this was followed by hybridization to CpG-island microarrays, which covered about 12,000 sites in the genome. These authors found significant differences between the SZ and BD patients and controls at a number of genes, several of which had previously been shown to be functionally related to psychosis. These included genes involved in neuronal development and loci implicated in GABAergic and glutamatergic pathways.

Several methylome studies have also been performed using blood samples from patients with psychosis. Nishioka et al. [69] used the Illumina Infinium HumanMethylation27 BeadArray to interrogate DNA methylation profiles in peripheral blood cells from 18 patients with SZ and 15 normal controls. They found 603 differentially methylated sites in SZ patients, and these sites were particularly abundant within CpG Islands. In another study, in which the Illumina 450K HumanMethylation array was used to interrogate DNA methylation levels in peripheral blood cells from a cohort of 65 SZ patients and 42 control individuals, the authors found about 2500 differentially methylated loci in the SZ patients [70]. Finally, using MBD-based enrichment and NGS, Aberg et al. [71] recently analyzed the DNA methylome of peripheral leukocytes in a large cohort of about 1500 SZ patients and controls. They found evidence of differential methylation in the loci of several genes linked to SZ pathogenesis.

The genome-wide study of DNA methylation in monozygotic twins discordant for psychosis has also been a powerful tool to assess the importance of this epigenetic mark in the disease. Using the Illumina Infinium HumanMethylation27 BeadArray, DNA methylation was examined in peripheral blood cells from a collection of 22 twin pairs discordant for SZ or BD [72]. The authors found considerable disease-associated differences in DNA methylation at specific loci across the genome. Pathway analysis showed a significant enrichment of categories related to psychiatric disease and neurodevelopment.

Despite evidence for a potential role of aberrant histone modifications in psychosis [73], at present no genome-wide studies are being performed, according to the current literature. miRNA profiling, however, has been extensively carried out using different technologic platforms. Perkins et al. [74] used a custom-made miRNA microarray, which detects 265 miRNAs, to show the differential expression of 16 miRNAs in postmortem prefrontal cortex tissue from 13 SZ patients and 21 controls. In two subsequent studies based on TaqMan real-time polymerase chain reaction (PCR) arrays, significant differences in miRNA expression was found between patients with psychosis and the controls [75,76]. Santerelli et al. [77] used an Illumina microarray platform, which detects 470 annotated miRNA sequences in the miRBase database, to show significant deregulation of 28 miRNAs in the postmortem prefrontal cortex of a large cohort of SZ patients compared with that in controls ($n = 37$). They found that pathway analysis of predicted targets of the deregulated miRNAs was enriched for categories related to axon guidance and long-term potentiation. Finally, miRNAs were obtained from exosomes isolated from postmortem prefrontal cortices a group of 13 controls, 8 SZ patients, and 9 BD patients, and their expression was examined by Luminex FLEXMAP 3D microfluidic device that assays for 312 miRNAs. The authors found differential expression of several miRNAs in the cohort with psychosis compared with the controls.

10.2.6 Epilepsy

Epilepsy refers to a heterogeneous group of genetic or acquired disease states, characterized by recurrent unprovoked seizures. It is affects 1−2% of the population globally and is the second most prevalent neurologic disease in the world. It is associated with a wide range pathogenetic mechanisms, including cortical damage and stroke and comorbid conditions (e.g., autism and autism spectrum disorders). There is growing evidence which shows that epigenetic mechanisms are important in the pathogenesis of epilepsy. For example, a number of mutations in genes that encode components of the epigenetic machinery have been described, which are responsible for causing human epileptic disorders. Rett syndrome, which is characterized by seizures in 80% of affected children, is the most salient example. This disorder is typically caused by loss-of-function mutations in the *MECP2* gene [78,79].

Genome-wide epigenetic studies also support the conclusion that aberrant epigenetic patterns play a role in the pathogenesis of epilepsy. Genome-wide DNA methylome profiling was done in the hippocampal tissue of a chronic epileptic rat model by using methyl-CpG capture and NGS (Methyl-Seq) [80]. The authors found differential methylation at 2573 loci with genes showing concurrent methylation and expression profile changes enriched for epilepsy-associated pathways, including cytoskeletal organization and synaptic transmission. Another study based on MeDIP-chip analyses of hippocampal tissue from a mouse model of status epilepticus also showed DNA methylation changes at 321 gene loci, associated with nuclear function, such as transcriptional regulation [81]. Finally, one of the most persuasive evidence comes from the study by Williams-Karnesky et al. [82], which showed increased DNA methylation in hippocampal tissue from a rat model of temporal lobe epilepsy, using MeDIP-chip technology, as above. Strikingly, they showed that the adenosine, an endogenous anticonvulsant and antiepileptic factor, could reverse the DNA hypermethylation of the epileptic brains and prevent the progression of epilepsy.

There is also evidence that histone modifications can play a role in the pathogenesis of epilepsy [78], although genome-wide studies have not been performed according to the current literature. Several miRNA profiling studies have been performed in experimental epilepsy different microarray technology, and changes in expression of over 100 miRNAs have been identified [83]. miRNA profiling has also been performed in hippocampal tissue from temporal lobe epilepsy patients. In the study by Kan et al. [84], miRNA expression was examined in resected hippocampal tissue from a cohort of 10 controls and 20 patients with epilepsy by using Exiqon miRCURY arrays.

They found significant deregulation in expression of 165 miRNAs with genes involved in immune response, one of several cellular processes implicated in epilepsy, being the most prominent targets of the deregulated miRNAs. In another study, in which miRNA profiles were examined in postmortem hippocampal tissue from 11 patients with epilepsy and 5 control individuals by TaqMan low-density arrays, the authors found a large-scale reduction in miRNA expression in the disease group [85].

10.2.7   Diabetic Neuropathy

Diabetic neuropathy (DN) is the most common and debilitating complication of diabetes, which can affect about half of the patients with diabetes. The vast majority of patients develop distal polyneuropathy that initiates in the feet and progresses proximally. Typical symptoms of DN include pain, numbness, tingling, and weakness, and DN is associated with substantial morbidity, including depression, susceptibility to foot or ankle fractures, ulceration, and lower-limb amputations [86]. The development of DN is clearly associated with reduced glycemic control, but how this leads to local nerve damage is not well understood.

Gene expression profiling has shown extensive changes in a mouse model of DN and in patients with diabetes [87,88]. In a recent study, Varela-Rey et al. [89] showed extensive differences in the DNA methylome of a mouse model of DN. Using RRBS, they showed that the sciatic nerves from diabetic mice had an aberrantly demethylated DNA profile and found that genes with concomitant changes in DNA methylation and transcript levels were enriched for functional categories that were associated with the transcriptomic signature of these mice. However, so far, the roles of histone modifications and ncRNAs have not been examined in the control of gene expression in this disorder.

10.3   CONCLUSION

The advent of genetics and genomics has transformed our understanding of human diseases, and we have now entered the next stage of this revolution because of the burgeoning field of epigenetics. Epigenetic mechanisms play a key role in many fundamental neurobiologic processes, ranging from neural stem cell maintenance and differentiation, to learning and memory. It is now being increasingly recognized that epigenetics is intrinsically linked to neurologic diseases, ranging from neurodegenerative and neurodevelopmental disorders to neuropsychiatric diseases.

It is now clear that complex neurologic diseases arise as a result of interactions among several factors, including genetic, environmental, stochastic, and lifestyle consequences, and epigenetic mechanisms may present a crucial interface between the genetic and nongenetic risk factors. Over the past few years, technologic advances and increased accessibility to genome-wide approaches have significantly shifted our focus from a gene-centric view to more global analyses in our effort to understand the pathogenesis of neurologic disorders. Major international initiatives, such as the NIH Roadmap Epigenomics Mapping Consortium, have been launched for the production and public dissemination of large-scale epigenomic maps of different systems related to basic biology and disease states [4].

This chapter has provided an up-to-date, although likely not exhaustive, account of recent epigenomic studies performed in different neurologic diseases, focused mainly on DNA methylation (Table 10.2), histone modifications, and miRNAs (Table 10.3). These studies, encompassing both large-scale efforts and more modest ones, have invariably increased our knowledge of the pathogenesis of these disorders. Two recent seminal studies stand out: De Jager et al. [21] and Lunnon et al. [20] have presented the first epigenome-wide association studies (EWAS) in AD, which involved DNA methylome profiling of a notably large cohort of samples, which included hundreds of cases and control individuals, with results replicated in independent cohorts. Importantly, they have provided persuasive evidence that there are significant genome-wide associations between differentially methylated regions and AD, as well as evidence that these changes occur early in the disease process, suggesting that these could be causative events in the disease.

However, despite the thoroughness of these studies and the important conclusions drawn, they share some limitations with all the other studies mentioned earlier. Many of these studies have employed array technologies that target few sites in the genome. For example, the Illumina HumanMethylation450 Beadset, used in the two above-mentioned studies, targets only about 2% of CpG sites in the human genome. Secondly, theses arrays cannot distinguish between DNA methylation and hydoxymethylation, which can have opposing effects on gene regulation [90]. Substantial technologic advances in NGS have made this technologic platform the way forward in epigenetics research, with, by far, its greatest advantage being in its ability to interrogate the entire genome in an comprehensive and unbiased way [4,5].

Another major challenge in the epigenomic mapping of neurologic disease states is the enormous complexity of the human brain. The human brain consists of many different types of neurons, glia, peripheral immune cells, and endothelial cells, and the epigenomic maps being produced do not accurately reflect the epigenomic state of a specific cell type. For example, neurons and glial from the human frontal cortex do not share the same DNA methylation profile [91], and brain regions affected differentially in AD have different transcriptomic profiles [92]. Improvement in techniques to isolate different cell populations from different brain regions, such as fluorescence-activated cell sorting, would invariably help to produce more meaningful epigenomic maps, from which appropriate conclusions can be drawn for their relative importance in disease pathogenesis.

TABLE 10.2 Genome-Wide DNA Methylation Studies in Neurologic Disease

Disease	Platform	Model	References
Alzheimer disease	Illumina Infinium HumanMethylation27 BeadArray	Cerebral cortex from human postmortem AD patients (12 LOAD and 12 controls)	[15]
	Custom-designed Illumina's GoldenGate DNA methylation assay	Frontal cortex samples from two transgenic models of AD, namely APP/PSEN1 and 3xTg-AD mice	[16]
	Illumina 450K HumanMethylation array	Dorsolateral prefrontal cortex from 708 human individuals, post mortem. Results independently verified in another 2 cohorts of 117 subjects and 399 subjects	[21][a]
	Illumina 450K HumanMethylation array	Different brain regions from 122 human individuals, post mortem. Results independently verified in another 2 cohorts of 144 subjects and 62 subjects	[20][a]
Parkinson disease	Illumina 450K HumanMethylation array	Postmortem prefrontal cortex and peripheral blood leukocytes from 5 PD patients and 6 healthy individuals	[36]
	Illumina Infinium HumanMethylation27 BeadArray	Postmortem putamen and cortex from 12 individuals	[37]
Huntington's disease	RRBS	Mouse striatal lines expressing wild-type or mutant HTT	[47]
	5-hmC enrichment and NGS	Striatum and cortex of the YAC128 HD mouse model	[48]
Multiple sclerosis	Custom-designed DNA methylation array (MethDet-56)	Blood plasma from 30 controls and 30 MS patients	[61]

(Continued)

TABLE 10.2 (Continued)

Disease	Platform	Model	References
	Illumina 450K HumanMethylation array	Postmortem normal-appearing white matter brain tissue from 28 MS-affected and 19 control individuals	[62]
Major psychosis	CpG-island microarrays	Postmortem prefrontal cortex from 35 SZ patients, 32 BD patients, and 28 controls	[68]
	Illumina Infinium HumanMethylation27 BeadArray	Peripheral blood cells from 18 patients with SZ and 15 normal controls	[69]
	Illumina 450K HumanMethylation array	Peripheral blood cells from a cohort of 65 SZ patients and 42 control individuals	[70]
	MBD-based enrichment and NGS	Peripheral leukocytes in a large cohort of about 1500 SZ patients and controls	[71]
	Illumina Infinium HumanMethylation27 BeadArray	Peripheral blood cells from a collection of 22 twin pairs discordant for SZ or BD	[72]
Epilepsy	Methyl-Seq	Hippocampal tissue of a chronic epileptic rat model	[80]
	MeDIP-chip	Hippocampal tissue from a mouse model of status epilepticus	[81]
	MeDIP-chip	Hippocampal tissue from a rat model of temporal lobe epilepsy	[82]
Diabetic neuropathy	RRBS	Sciatic nerves from a mouse model of DN (db/db)	[89]

^a*Of special interest.*

TABLE 10.3 Genome-Wide miRNA Profiling in Neurologic Disease

Disease	Platform	Model	References
Alzheimer disease	nCounter Human miRNA Assay kit	Prefontal cortex from 42 Braak-stratified AD patients and 7 controls	[25]
		Hippocampus from 41 LOAD AD patients and 23 controls	
	NGS (Illumina)	Prefontal cortex from 8 AD patients and 4 controls	[25]
	miRCURY LNA array (Exiqon)	Gray and white matter from the superior and middle temporal gyri across a range of early AD pathology ($n = 10$)	[26]
	Microarray (μParaflo)	Parietal lobe of 4 AD patients and 4 controls	[27]
	NGS (Illumina)	Tissue from the superior and middle temporal gyri in 5 AD patients and 2 controls	[28]
	Microarray (MMChip)	Mononuclear blood cells from 16 AD patients and 16 controls	[29]
	miRCURY LNA array (Exiqon)	CSF of 8 controls and 8 patients with AD, still alive	[30]
	NGS (Illumina)	Peripheral blood from 22 healthy controls and 48 AD patients	[31]
Parkinson disease	NCode Multispecies miRNA Microarray V2-arrays (Invitrogen)	*Caenorhabditis elegans* PD models	[40]
	LC Sciences microfluidic chip technology	Brainstem tissue of the α-synuclein (A30P)-transgenic mouse model for PD	[41]
	miRCURY LNA array (Exiqon)	Different brain regions, including the substantia nigra, from 6 control individuals and 11 PD patients	[42]
	miRCURY LNA array (Exiqon)	Peripheral blood mononuclear cells from 13 controls and 19 PD patients	[43]

(Continued)

TABLE 10.3 (Continued)

Disease	Platform	Model	References
	SOLiD sequencing	Blood leukocytes from 6 controls and 7 PD patients	[44]
Huntington disease	Microarray (Agilent)	Striata of two transgenic models of HD (YAC128 and R6/2)	[53]
	Illumina BeadChip Array	Cerebral cortex of the N171-82Q mouse HD model	[54]
	TaqMan Array (Human miRNA Panel)	Cortex of early stage HD patients (*n* = 13) compared with control individuals (*n* = 7)	[55]
	NGS (Illumina)	Frontal cortex and striatum of 2 HD patients and 2 control individuals	[56]
	Microarray (Agilent)	Frontal cortex and striatum of 4 HD patients and 4 control individuals	[56]
	NGS (Illumina)	Prefrontal cortex of 12 HD and 9 control individuals	[57]
Multiple sclerosis	TaqMan Array (Human miRNA Panel)	Active (*n* = 16) and inactive white matter brain lesions (*n* = 5) from MS patients, and 9 control white matter specimens	[64]
	miRCURY LNA array (Exiqon)	Postmortem brain white matter tissue from 4 controls and 4 MS patients	[65]
	miRCURY LNA array (Exiqon)	Myelinated and demyelinated hippocampi from MS patients, post mortem	[66]
Major psychosis	Microarray (custom)	Postmortem prefrontal cortex tissue from 13 SZ patients and 21 controls	[74]
	TaqMan qPCR arrays	Brain tissue from controls and patients with psychosis	[75,76]
	Illumina BeadChip Array	Postmortem prefrontal cortex of 37 SZ patients and 37 controls	[77]

(*Continued*)

TABLE 10.3 (Continued)

Disease	Platform	Model	References
	Luminex FLEXMAP 3D microfluidic assays	Exosomes isolated from postmortem prefrontal cortices from a group of 13 controls, 8 SZ patients, and 9 BP patients	[78]
Epilepsy	Various	Various experimental epilepsy models	[83]
	miRCURY LNA array (Exiqon)	Resected hippocampal tissue from a cohort of 10 controls and 20 patients with epilepsy	[84]
	TaqMan low density arrays	Postmortem hippocampal tissue from 11 patients with epilepsy and 5 control individuals	[85]

In summary, over the recent years, a plethora of epigenomic maps have been produced to further the understanding of pathogenetic mechanisms in neurologic disease. Despite their limitations, these studies have helped us better understand the relative importance of epigenetic mechanisms in these diseases. With continued technologic improvements in the field, we can anticipate that epigenomics would revolutionize our understanding of important complex neurologic disorders.

REFERENCES

[1] Jakovcevski M, Akbarian S. Epigenetic mechanisms in neurological disease. Nat Med 2012; 18(8):1194–204.

[2] Sweatt JD. The emerging field of neuroepigenetics. Neuron 2013;80(3):624–32.

[3] Qureshi IA, Mehler MF. Advances in epigenetics and epigenomics for neurodegenerative diseases. Curr Neurol Neurosci Rep 2011;11(5):464–73.

[4] Rivera CM, Ren B. Mapping human epigenomes. Cell 2013;155(1):39–55.

[5] Telese F, Gamliel A, Skowronska-Krawczyk D, Garcia-Bassets I, Rosenfeld MG. "Seqing" insights into the epigenetics of neuronal gene regulation. Neuron 2013;77(4):606–23.

[6] Urdinguio RG, Sanchez-Mut JV, Esteller M. Epigenetic mechanisms in neurological diseases: genes, syndromes, and therapies. Lancet Neurol 2009;8(11):1056–72.

[7] Qureshi IA, Mehler MF. Understanding neurological disease mechanisms in the era of epigenetics. JAMA Neurol 2013;70(6):703–10.

[8] Zoghbi HY. Rett syndrome: what do we know for sure? Nat Neurosci 2009;12(3):239–40.

[9] International Stroke Genetics C, Wellcome Trust Case Control C, Bellenguez C, Bevan S, Gschwendtner A, Spencer CC, et al. Genome-wide association study identifies a variant in HDAC9 associated with large vessel ischemic stroke. Nat Genet 2012;44(3):328–33.

[10] Li J, Chen J, Ricupero CL, Hart RP, Schwartz MS, Kusnecov A, et al. Nuclear accumulation of HDAC4 in ATM deficiency promotes neurodegeneration in ataxia telangiectasia. Nat Med 2012;18(5):783–90.

[11] Brookmeyer R, Johnson E, Ziegler-Graham K, Arrighi HM. Forecasting the global burden of Alzheimer's disease. Alzheimer's Dement 2007;3(3):186–91.

[12] Ballard C, Gauthier S, Corbett A, Brayne C, Aarsland D, Jones E. Alzheimer's disease. Lancet 2011;377(9770):1019–31.

[13] Mastroeni D, McKee A, Grover A, Rogers J, Coleman PD. Epigenetic differences in cortical neurons from a pair of monozygotic twins discordant for Alzheimer's disease. PLoS One 2009;4(8):e6617.

[14] Lu H, Liu X, Deng Y, Qing H. DNA methylation, a hand behind neurodegenerative diseases. Front Aging Neurosci 2013;5:85.

[15] Bakulski KM, Dolinoy DC, Sartor MA, Paulson HL, Konen JR, Lieberman AP, et al. Genome-wide DNA methylation differences between late-onset Alzheimer's disease and cognitively normal controls in human frontal cortex. J Alzheimer's Dis: JAD 2012;29(3):571–88.

[16] Sanchez-Mut JV, Aso E, Panayotis N, Lott I, Dierssen M, Rabano A, et al. DNA methylation map of mouse and human brain identifies target genes in Alzheimer's disease. Brain 2013;136(Pt 10):3018–27.

[17] Ullrich S, Munch A, Neumann S, Kremmer E, Tatzelt J, Lichtenthaler SF. The novel membrane protein TMEM59 modulates complex glycosylation, cell surface expression, and secretion of the amyloid precursor protein. J Biol Chem 2010;285(27):20664–74.

[18] Siegmund KD, Connor CM, Campan M, Long TI, Weisenberger DJ, Biniszkiewicz D, et al. DNA methylation in the human cerebral cortex is dynamically regulated throughout the life span and involves differentiated neurons. PLoS One 2007;2(9):e895.

[19] Ito H, Usuda N, Atsuzawa K, Iwamoto I, Sudo K, Katoh-Semba R, et al. Phosphorylation by extracellular signal-regulated kinase of a multidomain adaptor protein, vinexin, at synapses. J Neurochem 2007;100(2):545–54.

[20] Lunnon K, Smith R, Hannon E, De Jager PL, Srivastava G, Volta M, et al. Methylomic profiling implicates cortical deregulation of ANK1 in Alzheimer's disease. Nat Neurosci 2014;17(9):1164–70.

[21] De Jager PL, Srivastava G, Lunnon K, Burgess J, Schalkwyk LC, Yu L, et al. Alzheimer's disease: early alterations in brain DNA methylation at ANK1, BIN1, RHBDF2 and other loci. Nat Neurosci 2014;17(9):1156–63.

[22] Cacabelos R, Torrellas C, López-Muñoz F. Epigenomics of Alzheimer's Disease. J Exp Clin Med 2014;6(3):75–82.

[23] Graff J, Rei D, Guan JS, Wang WY, Seo J, Hennig KM, et al. An epigenetic blockade of cognitive functions in the neurodegenerating brain. Nature 2012;483(7388):222–6.

[24] Bennett DA, Yu L, Yang J, Srivastava GP, Aubin C, De Jager PL. Epigenomics of Alzheimer's disease. Trans Res 2015;165(1):200–20.

[25] Lau P, Bossers K, Janky R, Salta E, Frigerio CS, Barbash S, et al. Alteration of the microRNA network during the progression of Alzheimer's disease. EMBO Mol Med 2013;5(10):1613–34.

[26] Wang WX, Huang Q, Hu Y, Stromberg AJ, Nelson PT. Patterns of microRNA expression in normal and early Alzheimer's disease human temporal cortex: white matter versus gray matter. Acta Neuropathol 2011;121(2):193–205.

[27] Nunez-Iglesias J, Liu CC, Morgan TE, Finch CE, Zhou XJ. Joint genome-wide profiling of miRNA and mRNA expression in Alzheimer's disease cortex reveals altered miRNA regulation. PLoS One 2010;5(2):e8898.

[28] Hebert SS, Wang WX, Zhu Q, Nelson PT. A study of small RNAs from cerebral neocortex of pathology-verified Alzheimer's disease, dementia with lewy bodies, hippocampal sclerosis, frontotemporal lobar dementia, and non-demented human controls. J Alzheimer's Dis: JAD 2013;35(2):335−48.

[29] Schipper HM, Maes OC, Chertkow HM, Wang E. MicroRNA expression in Alzheimer blood mononuclear cells. Gene Regul Syst Biol 2007;1:263−74.

[30] Sala Frigerio C, Lau P, Salta E, Tournoy J, Bossers K, Vandenberghe R, et al. Reduced expression of hsa-miR-27a-3p in CSF of patients with Alzheimer disease. Neurology 2013;81(24):2103−6.

[31] Leidinger P, Backes C, Deutscher S, Schmitt K, Mueller SC, Frese K, et al. A blood based 12-miRNA signature of Alzheimer disease patients. Genome Biol 2013;14(7):R78.

[32] de Lau LM, Breteler MM. Epidemiology of Parkinson's disease. Lancet Neurol 2006; 5(6):525−35.

[33] Ammal Kaidery N, Tarannum S, Thomas B. Epigenetic landscape of Parkinson's disease: emerging role in disease mechanisms and therapeutic modalities. Neurotherapeutics 2013;10(4):698−708.

[34] Matsumoto L, Takuma H, Tamaoka A, Kurisaki H, Date H, Tsuji S, et al. CpG demethylation enhances alpha-synuclein expression and affects the pathogenesis of Parkinson's disease. PLoS One 2010;5(11):e15522.

[35] Desplats P, Spencer B, Coffee E, Patel P, Michael S, Patrick C, et al. Alpha-synuclein sequesters Dnmt1 from the nucleus: a novel mechanism for epigenetic alterations in Lewy body diseases. J Biol Chem 2011;286(11):9031−7.

[36] Masliah E, Dumaop W, Galasko D, Desplats P. Distinctive patterns of DNA methylation associated with Parkinson disease: identification of concordant epigenetic changes in brain and peripheral blood leukocytes. Epigenetics 2013;8(10):1030−8.

[37] Kaut O, Schmitt I, Wullner U. Genome-scale methylation analysis of Parkinson's disease patients' brains reveals DNA hypomethylation and increased mRNA expression of cytochrome P450 2E1. Neurogenetics 2012;13(1):87−91.

[38] Coppede F. Genetics and epigenetics of Parkinson's disease. ScientificWorldJournal 2012;2012:489830.

[39] Heman-Ackah SM, Hallegger M, Rao MS, Wood MJ. RISC in PD: the impact of microRNAs in Parkinson's disease cellular and molecular pathogenesis. Front Mol Neurosci 2013;6:40.

[40] Asikainen S, Rudgalvyte M, Heikkinen L, Louhiranta K, Lakso M, Wong G, et al. Global microRNA expression profiling of Caenorhabditis elegans Parkinson's disease models. J Mol Neurosci: MN 2010;41(1):210−18.

[41] Gillardon F, Mack M, Rist W, Schnack C, Lenter M, Hildebrandt T, et al. MicroRNA and proteome expression profiling in early-symptomatic alpha-synuclein(A30P)-transgenic mice. Proteomics Clin Appl 2008;2(5):697−705.

[42] Minones-Moyano E, Porta S, Escaramis G, Rabionet R, Iraola S, Kagerbauer B, et al. MicroRNA profiling of Parkinson's disease brains identifies early downregulation of miR-34b/c which modulate mitochondrial function. Hum Mol Genet 2011;20(15): 3067−78.

[43] Martins M, Rosa A, Guedes LC, Fonseca BV, Gotovac K, Violante S, et al. Convergence of miRNA expression profiling, alpha-synuclein interacton and GWAS in Parkinson's disease. PLoS One 2011;6(10):e25443.

[44] Soreq L, Salomonis N, Bronstein M, Greenberg DS, Israel Z, Bergman H, et al. Small RNA sequencing-microarray analyses in Parkinson leukocytes reveal deep brain

stimulation-induced splicing changes that classify brain region transcriptomes. Front Mol Neurosci 2013;6:10.

[45] Moumne L, Betuing S, Caboche J. Multiple Aspects of Gene Dysregulation in Huntington's Disease. Front Neurol 2013;4:127.

[46] Valor LM, Guiretti D. What's wrong with epigenetics in Huntington's disease? Neuropharmacology 2014;80:103−14.

[47] Ng CW, Yildirim F, Yap YS, Dalin S, Matthews BJ, Velez PJ, et al. Extensive changes in DNA methylation are associated with expression of mutant huntingtin. Proc Natl Acad Sci USA 2013;110(6):2354−9.

[48] Wang F, Yang Y, Lin X, Wang JQ, Wu YS, Xie W, et al. Genome-wide loss of 5-hmC is a novel epigenetic feature of Huntington's disease. Hum Mol Genet 2013;22(18):3641−53.

[49] McFarland KN, Das S, Sun TT, Leyfer D, Xia E, Sangrey GR, et al. Genome-wide histone acetylation is altered in a transgenic mouse model of Huntington's disease. PLoS One 2012;7(7):e41423.

[50] Valor LM, Guiretti D, Lopez-Atalaya JP, Barco A. Genomic landscape of transcriptional and epigenetic dysregulation in early onset polyglutamine disease. J Neurosci 2013;33 (25):10471−82.

[51] Vashishtha M, Ng CW, Yildirim F, Gipson TA, Kratter IH, Bodai L, et al. Targeting H3K4 trimethylation in Huntington disease. Proc Natl Acad Sci USA 2013;110(32):E3027−36.

[52] Lee J, Hwang YJ, Shin JY, Lee WC, Wie J, Kim KY, et al. Epigenetic regulation of cholinergic receptor M1 (CHRM1) by histone H3K9me3 impairs Ca(2 +) signaling in Huntington's disease. Acta Neuropathol 2013;125(5):727−39.

[53] Lee ST, Chu K, Im WS, Yoon HJ, Im JY, Park JE, et al. Altered microRNA regulation in Huntington's disease models. Exp Neurol 2011;227(1):172−9.

[54] Jin J, Cheng Y, Zhang Y, Wood W, Peng Q, Hutchison E, et al. Interrogation of brain miRNA and mRNA expression profiles reveals a molecular regulatory network that is perturbed by mutant huntingtin. J Neurochem 2012;123(4):477−90.

[55] Packer AN, Xing Y, Harper SQ, Jones L, Davidson BL. The bifunctional microRNA miR-9/miR-9* regulates REST and CoREST and is downregulated in Huntington's disease. J Neurosci 2008;28(53):14341−6.

[56] Marti E, Pantano L, Banez-Coronel M, Llorens F, Minones-Moyano E, Porta S, et al. A myriad of miRNA variants in control and Huntington's disease brain regions detected by massively parallel sequencing. Nucleic Acids Res 2010;38(20):7219−35.

[57] Hoss AG, Kartha VK, Dong X, Latourelle JC, Dumitriu A, Hadzi TC, et al. MicroRNAs located in the Hox gene clusters are implicated in huntington's disease pathogenesis. PLoS Genet 2014;10(2):e1004188.

[58] Johnson R. Long non-coding RNAs in Huntington's disease neurodegeneration. Neurobiol Dis 2012;46(2):245−54.

[59] Koch MW, Metz LM, Kovalchuk O. Epigenetic changes in patients with multiple sclerosis. Nat Rev Neurol 2013;9(1):35−43.

[60] Huynh JL, Casaccia P. Epigenetic mechanisms in multiple sclerosis: implications for pathogenesis and treatment. Lancet Neurol 2013;12(2):195−206.

[61] Liggett T, Melnikov A, Tilwalli S, Yi Q, Chen H, Replogle C, et al. Methylation patterns of cell-free plasma DNA in relapsing-remitting multiple sclerosis. J Neurol Sci 2010;290 (1-2):16−21.

[62] Huynh JL, Garg P, Thin TH, Yoo S, Dutta R, Trapp BD, et al. Epigenome-wide differences in pathology-free regions of multiple sclerosis-affected brains. Nat Neurosci 2014;17(1):121−30.

[63]   Koch MW, Metz LM, Kovalchuk O. Epigenetics and miRNAs in the diagnosis and treatment of multiple sclerosis. Trends Mol Med 2013;19(1):23−30.

[64]   Junker A, Krumbholz M, Eisele S, Mohan H, Augstein F, Bittner R, et al. MicroRNA profiling of multiple sclerosis lesions identifies modulators of the regulatory protein CD47. Brain 2009;132(Pt 12):3342−52.

[65]   Noorbakhsh F, Ellestad KK, Maingat F, Warren KG, Han MH, Steinman L, et al. Impaired neurosteroid synthesis in multiple sclerosis. Brain 2011;134(Pt 9):2703−21.

[66]   Dutta R, Chomyk AM, Chang A, Ribaudo MV, Deckard SA, Doud MK, et al. Hippocampal demyelination and memory dysfunction are associated with increased levels of the neuronal microRNA miR-124 and reduced AMPA receptors. Ann Neurol 2013;73(5):637−45.

[67]   Labrie V, Pai S, Petronis A. Epigenetics of major psychosis: progress, problems and perspectives. Trends Genet: TIG 2012;28(9):427−35.

[68]   Mill J, Tang T, Kaminsky Z, Khare T, Yazdanpanah S, Bouchard L, et al. Epigenomic profiling reveals DNA-methylation changes associated with major psychosis. Am J Hum Genet 2008;82(3):696−711.

[69]   Nishioka M, Bundo M, Koike S, Takizawa R, Kakiuchi C, Araki T, et al. Comprehensive DNA methylation analysis of peripheral blood cells derived from patients with first-episode schizophrenia. J Hum Genet 2013;58(2):91−7.

[70]   Kinoshita M, Numata S, Tajima A, Ohi K, Hashimoto R, Shimodera S, et al. Aberrant DNA methylation of blood in schizophrenia by adjusting for estimated cellular proportions. Neuromol Med 2014;16(4):697−703.

[71]   Aberg KA, McClay JL, Nerella S, Clark S, Kumar G, Chen W, et al. Methylome-wide association study of schizophrenia: identifying blood biomarker signatures of environmental insults. JAMA Psychiatry 2014;71(3):255−64.

[72]   Dempster EL, Pidsley R, Schalkwyk LC, Owens S, Georgiades A, Kane F, et al. Disease-associated epigenetic changes in monozygotic twins discordant for schizophrenia and bipolar disorder. Hum Mol Genet 2011;20(24):4786−96.

[73]   Pishva E, Kenis G, van den Hove D, Lesch KP, Boks MP, van Os J, et al. The epigenome and postnatal environmental influences in psychotic disorders. Soc Psychiatry Psychiatr Epidemiol 2014;49(3):337−48.

[74]   Perkins DO, Jeffries CD, Jarskog LF, Thomson JM, Woods K, Newman MA, et al. microRNA expression in the prefrontal cortex of individuals with schizophrenia and schizoaffective disorder. Genome Biol 2007;8(2):R27.

[75]   Moreau MP, Bruse SE, David-Rus R, Buyske S, Brzustowicz LM. Altered microRNA expression profiles in postmortem brain samples from individuals with schizophrenia and bipolar disorder. Biol Psychiatry 2011;69(2):188−93.

[76]   Kim AH, Reimers M, Maher B, Williamson V, McMichael O, McClay JL, et al. MicroRNA expression profiling in the prefrontal cortex of individuals affected with schizophrenia and bipolar disorders. Schizophr Res 2010;124(1-3):183−91.

[77]   Santarelli DM, Beveridge NJ, Tooney PA, Cairns MJ. Upregulation of dicer and microRNA expression in the dorsolateral prefrontal cortex Brodmann area 46 in schizophrenia. Biol Psychiatry 2011;69(2):180−7.

[78]   Qureshi IA, Mehler MF. Sex, epilepsy, and epigenetics. Neurobiol Dis 2014.

[79]   Hwang JY, Aromolaran KA, Zukin RS. Epigenetic mechanisms in stroke and epilepsy. Neuropsychopharmacology 2013;38(1):167−82.

[80]   Kobow K, Kaspi A, Harikrishnan KN, Kiese K, Ziemann M, Khurana I, et al. Deep sequencing reveals increased DNA methylation in chronic rat epilepsy. Acta Neuropathol 2013;126(5):741−56.

[81] Miller-Delaney SF, Das S, Sano T, Jimenez-Mateos EM, Bryan K, Buckley PG, et al. Differential DNA methylation patterns define status epilepticus and epileptic tolerance. J Neurosci 2012;32(5):1577−88.

[82] Williams-Karnesky RL, Sandau US, Lusardi TA, Lytle NK, Farrell JM, Pritchard EM, et al. Epigenetic changes induced by adenosine augmentation therapy prevent epileptogenesis. J Clin Investig 2013;123(8):3552−63.

[83] Henshall DC. MicroRNA and epilepsy: profiling, functions and potential clinical applications. Curr Opin Neurol 2014;27(2):199−205.

[84] Kan AA, van Erp S, Derijck AA, de Wit M, Hessel EV, O'Duibhir E, et al. Genome-wide microRNA profiling of human temporal lobe epilepsy identifies modulators of the immune response. Cell Mol Life Sci: CMLS 2012;69(18):3127−45.

[85] McKiernan RC, Jimenez-Mateos EM, Bray I, Engel T, Brennan GP, Sano T, et al. Reduced mature microRNA levels in association with dicer loss in human temporal lobe epilepsy with hippocampal sclerosis. PLoS One 2012;7(5):e35921.

[86] Vincent AM, Callaghan BC, Smith AL, Feldman EL. Diabetic neuropathy: cellular mechanisms as therapeutic targets. Nat Rev Neurol 2011;7(10):573−83.

[87] Hur J, Sullivan KA, Pande M, Hong Y, Sima AA, Jagadish HV, et al. The identification of gene expression profiles associated with progression of human diabetic neuropathy. Brain 2011;134(Pt 11):3222−35.

[88] Pande M, Hur J, Hong Y, Backus C, Hayes JM, Oh SS, et al. Transcriptional profiling of diabetic neuropathy in the BKS db/db mouse: a model of type 2 diabetes. Diabetes 2011;60(7):1981−9.

[89] Varela-Rey M, Iruarrizaga-Lejarreta M, Lozano JJ, Aransay AM, Fernandez AF, Lavin JL, et al. S-adenosylmethionine levels regulate the schwann cell DNA methylome. Neuron 2014;81(5):1024−39.

[90] Coppieters N, Dieriks BV, Lill C, Faull RL, Curtis MA, Dragunow M. Global changes in DNA methylation and hydroxymethylation in Alzheimer's disease human brain. Neurobiol Aging 2014;35(6):1334−44.

[91] Iwamoto K, Bundo M, Ueda J, Oldham MC, Ukai W, Hashimoto E, et al. Neurons show distinctive DNA methylation profile and higher interindividual variations compared with non-neurons. Genome Res 2011;21(5):688−96.

[92] Twine NA, Janitz K, Wilkins MR, Janitz M. Whole transcriptome sequencing reveals gene expression and splicing differences in brain regions affected by Alzheimer's disease. PLoS One 2011;6(1):e16266.

Chapter 11

Epigenetic Deregulation in Autoimmune Disease

Damiana Álvarez-Errico and Esteban Ballestar
Chromatin and Disease Group, Cancer Epigenetics and Biology Programme, Bellvitge Biomedical Research Institute, Barcelona, Spain

Chapter Outline

11.1 THE LOSS OF IMMUNE TOLERANCE: BREAKING BAD

Organisms have evolved immune systems to avoid succumbing to diseases caused by infectious foreign agents, such as bacteria or viruses, or by self-pathogenic circumstances, such as the development of tumors. Vertebrate immune systems are composed of a complex network of cells and soluble factors that act in a coordinated manner to recognize and distinguish dangerous

M. Fraga & A.F. Fernandez (Eds): Epigenomics in Health and Disease.
DOI: http://dx.doi.org/10.1016/B978-0-12-800140-0.00011-X
© 2016 Elsevier Inc. All rights reserved.
235

from innocuous stimuli and to react accordingly. For a long time, the prevailing view in immunology was the so-called self–non-self theory, which holds that the defense system is based on the immune system's ability to recognize and discriminate self-structures from non-self components. Charles Janeway elaborated on this theory, emphasizing that molecular-associated patterns from foreign infectious agents, named *pathogen-associated-molecular patterns*, act as initial triggers of an immune response when recognized by specific *pathogen-related receptors* on the surface of innate immune cells, which, in turn, stimulate and activate clones of B and T lymphocytes, orchestrating an adaptive, antigen-specific immune response after their expansion [1]. Twenty years ago, Polly Matzinger proposed the alternative *Danger Theory*, which postulates that the immune system's function is based on the recognition of molecular patterns mostly of self-tissues (rather than foreign tissues) that become exposed under stress or pathologic conditions and that are detected as threatening, thereby triggering an immune response. In her words, "endogenous cellular alarm signals from distressed or injured cells" that are not overtly exposed in a healthy situation generate the initiating signals that kick off a defensive action by the immune system [2]. The concept of *damage-associated molecular patterns* includes the danger signals in the body's own cells and tissues recognized by antigen-presenting cells, such as dendritic cells, which migrate to secondary lymphoid organs and trigger, through interaction with specific T cells, an effector response when they recognize them [3]. Given these views, yet regardless of the origin of the initial driving force, the main underlying concerns are how tolerance is achieved and maintained and how it is overcome when a defensive response is required. Immune tolerance consists of preventing self-reactive T and B lymphocytes from acting against a healthy body's own tissues while maintaining the ability of the immune system to protect against harmful insults. Central tolerance is achieved before lymphocytes mature and takes place during development at the primary lymphoid organs, namely, the thymus and the bone marrow. In turn, peripheral tolerance occurs after mature B and T cells are released into the circulation and ensures unresponsiveness to self-components in secondary lymphoid organs and peripheral tissues [4]. Autoimmune diseases lie at the heart of the acquisition of tolerance. In these diseases, the tolerance is disrupted, with a hallmark escape of autoreactive B and T cells, and immune responses are mounted against a healthy body's own tissues.

AIDs are a group of heterogeneous and multifactorial conditions with complex etiopathology, variable distribution across ages, gender, and human populations and affect around 5% of the world's population. They represent a major public health problem as well as an economic burden because of the high morbidity associated with them. It is generally believed that AIDs originate from the cross-talk between genetic predisposition, including a number of frequently described, disease-specific susceptibility loci, and internal and external environmental factors, such as viral infections, nutrition, and exposure to chemicals and radiation [5]. Epigenetics has been proposed as providing a physical link between the environment and the regulation of gene expression, and its contribution to AID development and pathogenesis are discussed in this chapter.

11.2 EPIGENETIC REGULATION

The definition of epigenetics has evolved from the one first proposed by Waddington in the 1940s, which referred to "…the interactions of genes with their environment which bring the phenotype into being" [6]. A popular definition is that of Riggs et al., which states that epigenetics is "the study of mitotically and/or meiotically heritable changes in gene function that cannot be explained by changes in DNA sequence" [7]. More recently, Adrian Bird has added that epigenetic regulation involves "the structural adaptation of chromosomal regions to register, signal, or perpetuate altered activity states" [8]. At any rate, epigenetic modifications comprise the covalent addition or removal of chemical groups either to DNA or to DNA-associated proteins, and as a result, gene expression can be regulated. These changes are discussed in detail below.

11.2.1 DNA Methylation

Genetic information must be regulated in time and space during the development of multicellular organisms. Epigenetic mechanisms, including DNA methylation, are essential for organizing such information, thereby orchestrating proper transcriptional programs. In addition, DNA methylation contributes to other biologic processes, such as genomic imprinting and genome stability [9].

DNA methylation occurs in mammals at the 5' position of cytosine residues, generating 5-methylcytosine, in the context of CpG dinucleotides. The study of DNA methylation has long focused on those located within CpG-rich regions, known as CpG islands (CGIs), particularly those that coincide with gene promoters and that are mostly unmethylated in mammals. However, although CGIs exhibit low levels of methylation in many cell types, the greatest variation in DNA methylation levels among cell types occurs primarily at the borders of CGIs, in regions termed "shores." Addition of methyl groups to cytosines is catalyzed by enzymes called "DNA methyltransferases" (DNMTs). This process is generally associated with gene repression, although recent genome-wide studies have revealed that the functional consequences of DNA methylation are diverse and depend on the genomic context. Three members of the DNMT family are known to exhibit catalytic activity. DNMT1 is considered a "maintenance" transferase due to its higher affinity for hemi-methylated DNA, and acts to restore pre-existing methylation patterns during subsequent DNA replication cycles [10]. On the other hand, DNMT3a and DNMT3b are *de novo* methyltransferases that create and shape

DNA methylation patterns during development [11]. Regarding DNA demethylation processes, it has proven more challenging to identify and characterize the molecules that remove methyl groups from DNA, although it is becoming clear that active demethylation requires the TET2 enzyme and DNA repair machinery [12,13]. Active demethylation occurs in the absence of DNA replication, and within the immune response, it has been associated with the transcriptional activation of several interleukins (ILs) and other factors during activation of T cells [14], among other examples. Passive demethylation, on the other hand, occurs in promoter regions in a replication-dependent manner.

11.2.2 Histone Modifications

Histone proteins interact closely with DNA and have a major role not only in its packaging within the cell nucleus but also in regulating its replication, repair, and especially its transcription through the incorporation of a variety of post-translational modifications of specific amino acid residues. The core histones H2A, H2B, H3, and H4 are organized in two H2A-H2B dimers that associate with an H3-H4 tetramer to form the octamer, which is the protein part of each nucleosome core particle. More than 11 types of post-translational modifications affecting all histones are known, including methylation, acetylation, phosphorylation, sumoylation, and ubiquitination of various amino acid residues [15]. Specific positions within histone tails, mainly lysines and arginine residues, are targeted by a great variety of histone-modifying enzymes, depending on the signals and particular conditions of the cell [16]. Histone acetylation was one of the first modifications to be studied and is generally associated with gene expression. Its balance is maintained by histone acetyltransferases (HATs) and histone deacetylases (HDACs). Histone tails are also methylated at different arginine or lysine residues. This is catalyzed by histone methyltransferases and demethylases. The relationship between K and R methylation with transcriptional activity will depend on the degree of methylation (monomethylation, dimethylation, or trimethylation [17] and the exact amino acid residue. Thus, elevated acetylation and trimethylation of H3K4, H3K36, and H3K79 are associated with active transcription. Conversely, low levels of acetylation and methylation of H3K9, H3K27, and H4K20 residues are usually a mark of gene repression [18].

11.3 LOCAL AND SYSTEMIC AUTOIMMUNE DISORDERS: IMPORTANCE OF THE ENVIRONMENT

The precise definition of different AIDs is difficult due to the significant overlap of similar entities and syndromes. It has also been a challenge to tease apart the causal factors behind AID development. There is substantial evidence that inherited polymorphisms are the main contributors to the development of an AID [19,20]. Genome-wide association studies have revealed strong associations that constitute the main risk factors, in particular with the human leukocyte antigen (HLA) locus for many AIDs. The major histocompatibility complex (MHC) region has been associated with more diseases (mostly autoimmune and inflammatory ones) than any other region in the genome [21]. Several autoimmune diseases display a degree of overlap with regard to the HLA haplotype, as in the case of HLA-DR3 (the MHC class II molecule), which is strongly correlated with systemic lupus erythematosus (SLE), and Sjögren syndrome [22]. Other susceptibility genes associated with the risk of AID development include those involved in immune processes, such as lymphocyte activation, the complement pathway, apoptosis, clearance of immune complexes, and genes acting as key regulators of epigenetic control. Remarkably, several loci that are strongly associated with autoimmune risk for one particular disorder are often correlated with other AIDs, as in the case of variants of the *PTPN22* gene, which, despite varying degrees of risk, are associated with type 1 diabetes, autoimmune thyroid diseases, and rheumatoid arthritis (RA) [23].

With regard to environmental factors, it is well established that, depending on the individual and possibly the genetic background, different AIDs (e.g., silica-associated RA, systemic sclerosis (SSc), and SLE) can be induced by the same or a similar external agent. Conversely, multiple agents can induce resembling syndromes, such as drug-induced lupus-like diseases [24], and this suggests that molecular pathways may converge to generate common steps toward breaking immune tolerance, thereby giving rise to autoimmune pathology.

Given that epigenetic marks can be influenced by the environment, that such an influence could be the basis of disease, and that epigenetic changes may be reversible, epigenetic modifications are particularly attractive in the design of potential pharmacologic or therapeutic interventions [25] (Figure 11.1). In addition, epigenetic modifications also depend on various factors, including transcription factors and various elements of cell signaling pathways. It is likely that polymorphisms in susceptibility genes for SLE, RA, SSc, and other autoimmune diseases, such as IRF5, and STAT4, among others, could influence locally the deposition of epigenetic modifications (Figure 11.1). Ultimately, both environmentally driven epigenetic alterations and those related to specific susceptibility variants will impact gene expression.

11.4 EPIGENETIC REGULATION IN AUTOIMMUNE DISORDERS

11.4.1 Systemic or Rheumatoid Autoimmune Disorders

Previously known as connective tissue diseases, systemic autoimmune diseases (SADs) are understood today to be a group of chronic inflammatory conditions with an autoimmune etiology whose main common feature is the presence of unspecific autoantibodies in serum. The three main SADs, SLE, RA and SSc,

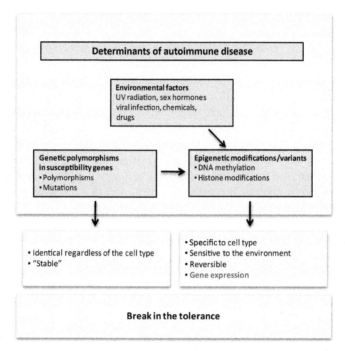

FIGURE 11.1 Scheme depicting the various contributions (genetic, environmental, and epigenetic) related to the loss of immune tolerance, which results in the development of autoimmune diseases (AIDs). The arrows indicate the main direction of influence. Both genetics and environment influence epigenetic modifications, which are ultimately associated with gene expression.

show extensive overlap in their presentation. Other clinical entities and syndromes also overlap extensively with SADs, the most relevant of these being mixed connective tissue disease, Sjögren syndrome, and primary antiphospholipid antibody syndrome. As separate clinical entities, these diseases are not that frequent, but taken together, they affect around 1% of the general population, which makes them a serious biomedical problem.

11.4.1.1 Systemic Lupus Erythematosus

Most epigenetic studies of SADs have focused on SLE. Known as the paradigmatic systemic inflammatory AID, SLE is characterized by its great heterogeneity and high titers of autoantibodies against nuclear antigens, as well as circulating immune complexes that accumulate, forming deposits and causing organ damage [26]. The cross-talk between genetic predisposition and environmental stimuli in this disease is demonstrated by the high women-to-men ratio of those affected, which may be as high as 9:1. There is compelling evidence that environmental factors contribute and modulate the development of these diseases [5]. Some of the strongest evidence about the role of epigenetics in the development of SLE, as a separate component opposed to genetic determinants, comes from a study of monozygotic twins discordant for SLE. In this research, Javierre et al. compared a cohort of identical twins discordant for SLE and found differential DNA methylation patterns between SLE-affected and healthy siblings. This approach offers a useful strategy for analyzing environmental epigenetics without interference from the genetic background. Importantly, a group of genes associated with immune response, including *IFNGR2*, *MMP14*, *LCN2*, *CSF3R*, *PECAM1*, *CD9*, and *AIM2*, were less methylated in SLE twins than in healthy siblings, suggesting potential pathways associated with the triggering of the disease [27]. These are different from those identified in genetic studies, including several HLA genes [28], including *PTPN22*, *IRF5*, *STAT4*, *BLK*, *BANK1*, *PDCD1*, *MECP2*, and *TNSF4*, which are associated with the development of SLE [20]. Epigenetic alterations in SLE may also be the result of the altered or less efficient behavior of some of the susceptibility gene products that interact with elements of the epigenetic machinery. For example, it was recently observed that demethylation of the ERα proximal promoter in patients with SLE is associated with enhanced expression relative to healthy individuals [29]. Concordant with this, another study showed that 17-β-estradiol contributes to global hypomethylation in the CD4 cells of SLE patients through ERa-mediated DNMT1 downregulation [30]. A previous study by the same research group showed that DNMT1 is downregulated and that such a decrease contributes to the hypomethylation observed in CD4 T cells from SLE patients relative to healthy donors [31]. Taken together, these studies suggest a possible contributing mechanism behind the gender bias in the prevalence of SLE.

Over 20 years ago, fundamental work by Bruce Richardson's group demonstrated global hypomethylation in the T cells of SLE patients; these show a 15−20% decrease in the level of methylation [32]. Genes of importance in T-cell function are more strongly expressed due to hypomethylation in the SLE T cells, which, therefore, suggests its dysregulated expression may underlie SLE triggering. *CD11a/ITGAL* [33], *CD70* [34], and *PRF1* [35] are well-characterized examples of genes of this type. Similarly, loss of methylation of the X chromosome gene CD40 ligand (CD154) in female SLE patients has been described [36]. Photosensitivity is a very well-documented trait of SLE, and interestingly enough, peripheral blood mononuclear cells (PBMCs) originating in SLE patients show a significant reduction in global DNA methylation levels upon exposure to both moderate and high doses of ultra-violet B (UVB) radiation, whereas controls only experience such a reduction following exposure to high-dose UVB radiation [37] (Figure 11.2).

More recent genome-wide studies of DNA methylation in CD4 T cells from SLE patients analyzing promoter regions of around 15,000 genes have identified 236 hypomethylated and 105 hypermethylated CpGs in SLE CD4 + T cells [38]. Some hypomethylated genes, such as *CD9*, *MMP9*, and

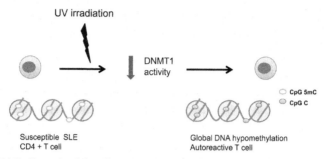

FIGURE 11.2   Example of the effect of medium and high-dose UV irradiation in T cells from SLE genetically susceptible individuals where decreased catalytic activity of DNMT1 impedes proper DNA methylation, and global hypomethylation causes self-reactive T cells. Procainamide and hydralazine have been shown to have similar effects by decreasing DNMT1 levels.

PDGFRA, are particularly relevant in SLE pathogenesis. That is the case for the transcription factor RUNX3, which is associated with T-cell maturation, was significantly hypermethylated in patients relative to healthy donors, in the same study. It is very important to appreciate that global hypomethylation is inversely correlated with disease activity, measured clinically by the SLE Disease Activity Index (SLEDAI). This confirms the role of hypomethylation in SLE pathogenesis [39].

Apart from T cells, autoreactive B cells are key components of SLE and are susceptible to dysregulation, causing disease as a consequence of the unbalanced production of cytokines, hyperproliferation, and exacerbated release of immunogobulins against self-antigens. It is worth noting that B cells from SLE patients show reduced DNA methylation levels upon BCR-mediated B-cell activation due to an impairment in inducing DNMT1 expression [40].

Post-translational modification on histone tails is another widespread epigenetic mechanism of gene expression regulation, in which H3 and H4 acetylation are generally associated with an enhanced transcription capability. Shifts in histone acetylation patterns were further attributed to changes in expression of HDAC and HAT activity. A deviation from homeostasis was observed in SLE, by which T cells overproduced Th2 cytokines, such as IL-6, IL-10 and tumor necrosis factor-α (TNF-α). Changes in histone acetylation could be, at least partially, responsible for this bias, as is suggested by the fact that HDACi trichostatin A reverses this skew as well as the overexpression of CD40L in CD4 T cells freshly isolated from SLE patients [41].

Global H4 acetylation of SLE monocytes, compared with controls, revealed higher acetylation of 179 from a total of 11,321 genes that converged on three main signaling pathways involved in inflammation, namely, interferon type I (IFN-I), nuclear factor-κB (NFκB), and mitogen-activated protein kinase (MAPK). Moreover, 63% of the genes showing greater H4 acetylation are potential regulators of IRF1. Interestingly, treatment of disease-free monocytes with IFN-α resulted in H4ac changes and upregulation of genes similar to those seen in the monocytes of SLE patients. These findings demonstrate the importance of histone H4 changes in the downstream regulation of genes involved in SLE [42].

Chronic inflammation is a major component of SLE pathogenesis. TNF-α levels are usually elevated in patients' serum, and higher levels are associated with greater disease activity. In this respect, higher H4 acetylation levels in the TNF-α, with no changes in the global H3 acetylation locus, have also been demonstrated in SLE patients relative to healthy individuals [43].

Integration of different datasets is also yielding important information about the epigenetic misregulation of immune cells in SLE. For example, pooled DNA methylation, messenger RNA (mRNA), and microRNA (miRNA) expression data from CD4 T cells isolated from patients revealed that an increase in certain immune-related molecules, such as ITGAM1, ITGB2, and ROCK1, may be due to hypomethylation. Similarly, inflammasome genes, such as *NLRP2*, and activating receptor CD300LB are overexpressed and hypomethylated, and this is consistent with the overactivation of CD4 T cells in SLE patients compared with controls [44].

11.4.1.2   Rheumatoid Arthritis

The principal characteristic of RA, a chronic autoimmune inflammatory disorder, is the progressive destruction of the joints. Several types of immune cells, such as various lymphocyte subsets, dendritic cells, and osteoclasts, as well as nonimmune cells, such as synovial fibroblasts (SFs), participate in RA pathogenesis [45]. Twin discordance in monozygotic compared with dizygotic twins suggests that the genetic contribution to RA etiology approaches 65% [46]. This highlights the existence of nongenetic factors, including epigenetic regulation, in the pathogenesis of RA. In this respect, the 3:1 female-to-male gender bias observed in RA may be due to epigenetic escape from X-inactivation of the IRAK1 protein [47]. In a similar way, female RA patients show hypomethylation and upregulation of CD40L. As previously mentioned with regard to SLE, the CD40L gene is located on the X chromosome, and its overexpression in CD4 T cells of female patients may explain the gender-based differences in the incidence of the disease [48]. Even though data are still limited, some studies have demonstrated a role for DNA methylation dysregulation in RA in immune and nonimmune cells. As discussed for SLE, a global loss of methylation has been observed in T cells from RA patients compared with healthy controls [32]. IL-6 is a potent inflammatory mediator and its promoter is hypomethylated in DNA extracted from RA PBMCs compared with controls, which indicates the dysregulation of a potentially important gene [49]. As mentioned above, nonimmune cells in particular fibroblast-like synoviocytes (FLSs) are known to contribute to the inflammatory environment and joint damage in RA. In this context, comparative genome-wide studies of fibroblasts isolated from RA and osteoarthritis (OA) patients revealed more than 1800 differentially methylated loci, around two-thirds of which were hypomethylated, and one-third were hypermethylated [50]. Interestingly, inflammatory effectors, IL-1β in particular, can reduce DNMT levels and catalytic activity in cultured FLSs from RA patients, leading to reduced methylation. With regard to therapeutic intervention, it is remarkable that the effects of IL-1β on DNMT expression are reversible and that the levels return to normal once the inflammatory stimulus is removed [51]. It is therefore difficult to attribute an exclusively causal role to epigenetic regulation, and it is conceivable that its contribution generates feedback and feedforward loops. Multiple levels of regulation have been revealed by the integrated study of high-throughput methylation data, miRNAs and mRNA from synovial fibroblasts of RA and OA patients. This study revealed a group of important genes that were hypomethylated, including *MMP20*, *RASGRF2*, and *TRAF2*, and others that were hypermethylated, such as *ADAMTS2*, *EGF*, and *TIMP2*. De la Rica et al. analyzed DNA methylation in RA synovial fibroblasts and identified changes in novel key target genes, such as *IL6R*, *CAPN8*, and *DPP4*, as well as several *HOX* genes. They also integrated the DNA methylation data with the results of miRNA screening. Integrated analysis identified sets of miRNAs that are controlled by DNA methylation, as well as genes that are regulated by DNA methylation and targeted by miRNAs [52]. Recent epigenome-wide association analysis of blood samples from RA patients has revealed that DNA methylation is a mediator of genetic risk in RA [53]. Specifically, the study identified within the MHC region two clusters whose differential methylation potentially mediates the genetic risk of RA.

Little information is available to link histone modification dysregulation to the development of RA. However, some reports indicate the contribution of such an imbalance to inflammation and joint damage. For instance, Wada et al. have shown an increase in H3 acetylation of IL6 promoter in RA SF compared with OA counterparts, and that *in vitro* treatment of SFs with the HAT inhibitor curcumin, reduced H3 acetylation as well as IL-6 mRNA levels [54].

11.4.1.3 Systemic Sclerosis

Systemic sclerosis (SSc), or scleroderma, is a serious and complex multiorgan autoimmune disorder characterized by immune activation and specific circulating autoantibodies (antinuclear antibodies, anticentromere autoantibodies, antitopoisomerase autoantibodies) affecting the skin, lungs, kidneys, gastrointestinal tract, and vasculature. Organ dysfunction is caused by inflammatory, vascular, and fibrotic processes, with vascular damage affecting mainly the small vessels [55]. Of the SSc patients, 80% are women, and a number of external factors—in particular, occupational exposure to silica dust, vinyl chloride, or drugs (e.g., bleomycin)—have been linked to the disease [56].

As with other systemic AIDs, analysis of DNA methylation in CD4 T cells extracted from SSc patients compared with those from healthy donors revealed a reduction in global 5-methylcytosine levels [57]. As in the case of the specific genes determined in other disorders, such as SLE and RA, which are affected by reduced methylation of their promoters, the CD40L gene on the inactive X chromosome is also correlated with CD40L overexpression in CD4 + T lymphocytes from female SSc patients [58]. The relationship between demethylation and overexpression of this gene and the altered T-lymphocyte signaling in SSc patients is not clear, although it could explain both the gender bias and the T-cell dysfunction seen in this condition. Another key signaling molecule related to T-cell activation is *CD70/ TNFSF7*, which is expressed by activated T lymphocytes, but not by resting T lymphocytes, and whose expression is increased due to hypomethylation of the promoter in CD4 T cells of patients with SSc [59]. As with CD40L, although it is an attractive hypothesis, the role of this signaling pathway in SSc pathogenesis has yet to be demonstrated. The hypermethylation of the T regulatory driving transcription factor FoxP3 in CD4 T cells from SSc patients was the most notable finding. Coincidentally, the level of FoxP3 expression and the number of regulatory T cells were both lower in a group of SSc patients than in healthy controls [60].

The global *in vitro* levels of DNMT1, methyl-CpG binding domain proteins, such as MBD1, MBD2, and MeCP2, are significantly higher in SSc compared with healthy fibroblasts [61].

The analyses of changes in histone modifications and the enzymes involved in their deposition or removal have produced significant results. For instance, there is more HAT p300 in fibroblasts from skin biopsies from SSc patients than in those from healthy controls. p300 upregulation could be associated with the increase in transforming growth factor-β (TGF-β), a profibrotic transcription factor, and enhanced collagen synthesis in SSc [62,63]. Various histone modifications are also more common in SSc fibroblasts than in those of their healthy counterparts, including the total level of histone H3K27me3, although the inhibition of the deposition of this mark seems to be insufficient to block excessive collagen deposition by such fibroblasts [64].

Within the immune compartment, B cells from SSc patients display high levels of acetylation in histone H4, whereas H3K9 has low levels of methylation. However, the mechanistic and clinical relevance remain to be elucidated [65].

11.4.1.4 Sjögren Syndrome

Primary Sjögren syndrome (pSS) is a chronic autoimmune disorder that affects the exocrine (mostly salivary and lachrymal) glands, giving rise to oral and ocular dryness, and some systemic symptoms, such as general pain and fatigue. This disorder is one of the most frequent systemic autoimmune diseases, estimated to affect between 0.01% and 0.1% of the population [66]. The autoantibodies present in the circulation of pSS patients include those against the sicca syndrome (SS)-A/Ro and SSB/La ribonucleoprotein particles. As with other AIDs, pSS is partially heritable, and epigenetics seems to contribute to its development [67]. Analysis of global methylation in cultured salivary gland epithelial cells from pSS patients showed a reduction in 5meC levels measured by enzyme-linked immunosorbent assay (ELISA). However, no differences between pSSs and control T or B lymphocytes were noted. Moreover, the important DNA methylation machinery-related enzymes, DNMT1 and Gadd45-alpha, are downregulated and upregulated, respectively [68]. A genome-wide study performed with naïve CD4 T cells isolated from pSS patients and healthy controls identified 553 hypomethylated and 200 hypermethylated CpG sites. Hypomethylation was prevalent in genes involved in immune activation processes, whereas hypermethylation was prevalent in genes involved in antigen processing and presentation [69].

11.4.2   Tissue-Specific AIDs

11.4.2.1   Type I Diabetes

Type 1 diabetes (T1D) is a prototypic autoimmune disease in which pancreatic B cells, located in the insulin-producing islets of Langerhans, are destroyed, leading to unbalanced glucose homeostasis and insulin deficiency, among other complications. The prevalence of T1D varies considerably between populations, from 0.1 per 100,000 per year in China and Venezuela to 40.9 per 100,000 per year in Finland. Twin studies suggest a concordance rate of 13.0−67.7% in monozygotic twins compared with one of 0.0−12.4% in dizygotic twins [70]. The mechanisms of susceptibility to inflammation that contribute to pancreatic cell destruction include T-cell hyperactivation, and the production of cytokines, such as IL-1β, TNF-α, and IFN-γ, which also interfere with the function and viability of pancreatic β-cells via NF-κB, signal transducer and activator of transcription 1 (STAT1), and MAPK signaling [71].

Upon interrogation of 27,458 CpG sites within 14,475 promoter monocytes extracted from T1D-discordant monozygotic twin pairs, 132 differentially methylated CpG sites were identified. These were significantly associated with the diabetic state. It is of particular note that the methylation-specific positions can be detected before self-reactivity becomes apparent [72].

11.4.2.2   Psoriasis

Psoriasis is an organ-specific autoimmune disease characterized by the excessive proliferation and abnormal differentiation of keratinocytes [73]. The pathogenesis of psoriasis relies upon an exacerbated immune response that includes the malfunction of several immune cell types, including T cells and dendritic cells, and the overproduction of inflammatory cytokine.

Several epigenetic modifications have been observed in psoriasis patients. In the case of DNA methylation, analysis of psoriatic skin lesions by high-throughput techniques revealed a number of regions that were differentially methylated (predominantly hypermethylated) in lesions relative to healthy skin [74]. Some genes, such as *p16INK4a* and *p14^{ARF}*, were hypermethylated, although others, such as *p21* and *SHP-1*, had hypomethylated CpGs on their promoters [75,76].

Histone modification patterns from PBMCs revealed global H4 hypoacetylation [77], and the HDAC1 was overexpressed in psoriasis skin biopsies [78].

11.4.2.3   Inflammatory Bowel Diseases

Inflammatory bowel diseases (IBDs), comprising Crohn disease (CD) and ulcerative colitis (UC), are a group of disorders that affect the gastrointestinal tract of genetically susceptible individuals upon exposure to environmental risk factors [79]. IBD involves chronic inflammation characterized either by flare-ups and then periods of remission or by continuous activity. Together, these disorders, whose incidence is increasing, are considered a major public health issue: Currently they affect European adults with a frequency of 12.7 (CD) and 24.3 (UC) per 100,000 person-years, and a prevalence of 0.5−1.0% [80].

As with other AIDs, the etiology of IBD is unknown, although it is clear that the immune response to the intestinal microbiota is dysregulated and mucosal tolerance is disrupted in the context of a susceptible genetic background [81]. It is clear that the interactions of a susceptible host and its immune system with aspects of the environment, such as the intestinal flora, diet, drugs, and so on, can favor the development of IBD.

Analysis of methylation levels of around 30,000 CpG sites from blood samples taken from CD patients revealed 50 sites that were significantly differentially methylated relative to controls. Differentially methylated genes included immune activation genes, such as *MAPK13*, *FASLG*, *PRF1*, *S100A13*, *RIPK3*, and *IL-21R*. In addition, predicted pathways involved in the regulation of the autoimmunity-related cytokine IL-17 (*IL-17A* and *IL-17F* genes) were also regulated by methylation [82]. Similarly, immune-related genes, such as transcription factors *BCL2*, *STAT3*, and *STAT5*, were found to be differentially methylated in B cells extracted from UC and CD patients compared with unaffected siblings [83]. However, the conclusions from this study need to be re-evaluated, since the B cells had been transformed with Epstein-Barr virus, which could also have affected factors related to the immune response.

DNA methylation studies in rectal biopsy specimens from UC and CD patients have also revealed changes that are common to both pathologies, such as *THRAP2, FANCC, GBGT1, DOK2, TNFSF4, TNFSF12,* and *FUT7.* Some other genes coincided with previously susceptible IBD loci identified by GWAS, such as *CARD9, ICAM3* and *IL8RB.* In this study, differentially methylated genes exclusive to one of the diseases were also identified, as is the case for the serine/threonine−protein kinase ULK1 [84].

In addition, a genome-wide study with intestinal biopsy samples from 20 monozygotic twins discordant for UC identified 61 differentially methylated loci, with several containing genes that regulate inflammation (*CFI, SPINKK4, THY1/CD90*). This study included integration of a three-layer functional map, including transcription, methylation-variable positions, and differentially methylated regions [85].

The pattern of histone H4 acetylation and its localization in *in vivo* rat models of inflammation and CD patients revealed that the global levels of histone H4 acetylation are slightly induced in the noninflamed portions of the intestine but significantly elevated in the inflamed parts and in Peyer patches. The specific K8, K12, and K16 residues were significantly acetylated in patient biopsies. K8 and K16 acetylation have previously been associated with inflammatory gene expression [86]. However, molecules secreted by microbiota, such as butyrate, act as HDAC inhibitors, and several anti-inflammatory mechanisms of HDACi have been proposed. In fact, local butyrate treatment has been used in UC patients, but it is not known whether the main mechanism of action is HDAC inhibition, since other effects, such as the maintenance of barrier function and the homeostatic reduction in epithelial cell production of IL-8, are also possibilities [87].

11.5 MOLECULAR CHARACTERIZATION OF COMMON PATHOGENIC ROUTES: AUTOIMMUNITY IN THE TWENTY-FIRST CENTURY

The complexity of AID etiology, its pathogenicity and impact have made it particularly challenging to define and treat these conditions. For example, it has taken over 50 years to approve a novel and successful drug, belimumab, to treat lupus nephritis. Belimumab is a human recombinant Immunoglobulin G-1λ (IgG-1λ) monoclonal antibody that neutralizes soluble B-cell activating factor (BAFF), interfering with the BAFF-R-mediated survival signals of self-reactive B lymphocytes. Its approval has marked an important milestone in biologic therapy for SLE. However, although belimumab might be useful for treating related or overlapping inflammatory conditions, it has so far only been approved for the treatment of lupus nephritis. Therapy based on epigenetic modifications is attractive because of the reversible nature of such changes and because it may yield cost-effective treatments with fewer undesirable side effects. In this context, hypomethylating agents (azacitidine, decitabine, SGI-110) and histone deacetylase inhibitors (vorinostat, pracinostat, panobinostat) represent promising new treatments for hematologic malignances and could be used in the context of AIDs. For instance, the HDAC inhibitor givinostat (currently in phase II clinical trial for some leukemias) has positive results relative to IL-6 production in RA macrophages and synovial cells [88,89].

With the development of novel strategies for the high-throughput analysis of genomic sequences, a transition from candidate genes studies to epigenomics has already taken place. The lower costs enable the analysis of large multinational patient cohorts, and a great deal of knowledge regarding the identification of differential or aberrant epigenetic patterns is being generated, and this could define reliable novel biomarkers for autoimmune disease prediction and diagnosis. Similarly, the reversible nature of epigenetic modifications, in contrast to the irreversible mutations of DNA sequence, make therapeutic interventions with epigenetic drugs an attractive option, so definition of novel therapeutic targets is also desirable. The latter is particularly important in cases where the toxicity of current treatment or steroid-resistance restricts, such therapeutic options.

11.5.1   Toward Accurate Diagnosis

Different AIDs have common or similar clinical traits and symptoms and share physiopathologic mechanisms and genetic factors, the number of common mechanisms, leading to the term "autoimmune tautology" [90]. For these reasons, there is a need to develop a novel taxonomy of AIDs, based on the identification of specific biomarkers that more accurately define overlapping entities with common biologic pathways that may be dysregulated. In this respect, we need to integrate data from omics and high-throughput technologies, bioinformatics, and systems biology, for which purpose there should be unhindered dialogue between basic researchers and clinicians. It would be better for diagnoses to be based more on particular molecular pathophysiologic characteristics that define clusters of patients who would otherwise be segregated on the basis of a purely clinical classification. This is the case for groups of RA patients showing the typical SLE IFN signal in their peripheral blood cells and the condition RUPUS, which describes SLE patients with severe and extended joint damage.

11.5.2   Toward Individual-Based Treatment

The road to individual-based medicine requires a multifactorial approach, including the integration of omics big data and genome-wide profiling, with which accurate personalized treatments can be designed on the basis of novel molecular taxonomy rather than on purely clinical definitions. The widespread use and lower costs of deep-sequencing techniques are enabling the construction of comprehensive novel patient classifications, leading to ad hoc treatment design.

11.6   CONCLUSIONS AND PERSPECTIVES

AIDs are a group of complex, multifactorial entities with overlapping clinical and molecular traits that often render standard classification meaningless when therapeutic decisions have to be made. Molecular taxonomy based on the integration of big data is an alternative to fine-tuned diagnosis and treatment. Epigenetics is a rapidly growing field, and identifying the misregulated epigenetic mechanisms that contribute to the development of AIDs will enable us to address the environmental contributions to autoimmune pathology.

REFERENCES

[1]   Janeway Jr. CA. The immune system evolved to discriminate infectious nonself from non-infectious self. Immunol Today 1992;13:11−16.

[2]   Matzinger P. Tolerance, danger, and the extended family. Annu Rev Immunol 1994;12:991−1045.

[3]   Seong SY, Matzinger P. Hydrophobicity: an ancient damage-associated molecular pattern that initiates innate immune responses. Nat Rev Immunol 2004;4:469−78.

[4]   Klinman NR. The "clonal selection hypothesis" and current concepts of B cell tolerance. Immunity 1996;5:189−95.

[5]   Gourley M, Miller FW. Mechanisms of disease: Environmental factors in the pathogenesis of rheumatic disease. Nat Clin Pract Rheumatol 2007;3:172−80.

[6]   Waddington CH. The epigenotype. Int J Epidemiol 1942;41:10−13.

[7]   Riggs AD, Russo VEA. Epigenetic mechanisms of gene regulation. Cold Spring Harbor Laboratory Press; 1996.

[8]   Bird A. Perceptions of epigenetics. Nature 2007;447:396−8.

[9]   Jaenisch R, Bird A. Epigenetic regulation of gene expression: how the genome integrates intrinsic and environmental signals. Nat Genet 2003;33(Suppl):245−54.

[10]   Bestor TH. Activation of mammalian DNA methyltransferase by cleavage of a Zn binding regulatory domain. Embo J 1992;11:2611−17.

[11]   Okano M, Bell DW, Haber DA, Li E. DNA methyltransferases Dnmt3a and Dnmt3b are essential for de novo methylation and mammalian development. Cell 1999;99:247−57.

[12]   Ito S, Shen L, Dai Q, Wu SC, Collins LB, Swenberg JA, et al. Tet proteins can convert 5-methylcytosine to 5-formylcytosine and 5-carboxylcytosine. Science 2011;333:1300−3.

[13]   He YF, Li BZ, Li Z, Liu P, Wang Y, Tang Q, et al. Tet-mediated formation of 5-carboxylcytosine and its excision by TDG in mammalian DNA. Science 2011;333:1303−7.

[14]   Bruniquel D, Schwartz RH. Selective, stable demethylation of the interleukin-2 gene enhances transcription by an active process. Nat Immunol 2003;4:235−40.

[15]   Bernstein BE, Meissner A, Lander ES. The mammalian epigenome. Cell 2007;128:669−81.

[16]   Kouzarides T. Chromatin modifications and their function. Cell 2007;128:693−705.

[17]   Allfrey VG, Faulkner R, Mirsky AE. Acetylation and methylation of histones and their possible role in the regulation of rna synthesis. Proc Natl Acad Sci USA 1964;51:786−94.

[18]   Li B, Carey M, Workman JL. The role of chromatin during transcription. Cell 2007;128:707−19.

[19]   Costenbader KH, Gay S, Alarcon-Riquelme ME, Iaccarino L, Doria A. Genes, epigenetic regulation and environmental factors: which is the most relevant in developing autoimmune diseases? Autoimmun Rev 2012;11:604−9.

[20] Delgado-Vega A, Sanchez E, Lofgren S, Castillejo-Lopez C, Alarcon-Riquelme ME. Recent findings on genetics of systemic autoimmune diseases. Curr Opin Immunol 2010;22:698−705.

[21] Trowsdale J, Knight JC. Major histocompatibility complex genomics and human disease. Annu Rev Genomics Hum Genet 2013;14:301−23.

[22] Lessard CJ, Li H, Adrianto I, Ice JA, Rasmussen A, Grundahl KM, et al. Variants at multiple loci implicated in both innate and adaptive immune responses are associated with Sjogren's syndrome. Nat Genet 2013;45:1284−92.

[23] Viatte S, Plant D, Raychaudhuri S. Genetics and epigenetics of rheumatoid arthritis. Nat Rev Rheumatol 2013;9:141−53.

[24] Miller FW, Pollard KM, Parks CG, Germolec DR, Leung PS, Selmi C, et al. Criteria for environmentally associated autoimmune diseases. J Autoimmun 2012;39:253−8.

[25] Jirtle RL, Skinner MK. Environmental epigenomics and disease susceptibility. Nat Rev Genet 2007;8:253−62.

[26] Gatto M, Zen M, Ghirardello A, Bettio S, Bassi N, Iaccarino L, et al. Emerging and critical issues in the pathogenesis of lupus. Autoimmun Rev 2013;12:523−36.

[27] Javierre BM, Fernandez AF, Richter J, Al-Shahrour F, Martin-Subero JI, Rodriguez-Ubreva J, et al. Changes in the pattern of DNA methylation associate with twin discordance in systemic lupus erythematosus. Genome Res 2010;20:170−9.

[28] Lie BA, Thorsby E. Several genes in the extended human MHC contribute to predisposition to autoimmune diseases. Curr Opin Immunol 2005;17:526−31.

[29] Liu HW, Lin HL, Yen JH, Tsai WC, Chiou SS, Chang JG, et al. Demethylation within the proximal promoter region of human estrogen receptor alpha gene correlates with its enhanced expression: Implications for female bias in lupus. Mol Immunol 2014;61:28−37.

[30] Wu Z, Sun Y, Mei X, Zhang C, Pan W, Shi W. 17beta-oestradiol enhances global DNA hypo-methylation in CD4-positive T cells from female patients with lupus, through overexpression of oestrogen receptor-alpha-mediated downregulation of DNMT1. Clin Exp Dermatol 2014;39:525−32.

[31] Wu Z, Li X, Qin H, Zhu X, Xu J, Shi W. Ultraviolet B enhances DNA hypomethylation of CD4 + T cells in systemic lupus erythematosus via inhibiting DNMT1 catalytic activity. J Dermatol Sci 2013;71:167−73.

[32] Richardson B, Scheinbart L, Strahler J, Gross L, Hanash S, Johnson M. Evidence for impaired T cell DNA methylation in systemic lupus erythematosus and rheumatoid arthritis. Arthritis Rheum 1990;33:1665−73.

[33] Richardson BC, Strahler JR, Pivirotto TS, Quddus J, Bayliss GE, Gross LA, et al. Phenotypic and functional similarities between 5-azacytidine-treated T cells and a T cell subset in patients with active systemic lupus erythematosus. Arthritis Rheum 1992;35:647−62.

[34] Oelke K, Lu Q, Richardson D, Wu A, Deng C, Hanash S, et al. Overexpression of CD70 and overstimulation of IgG synthesis by lupus T cells and T cells treated with DNA meth-ylation inhibitors. Arthritis Rheum 2004;50:1850−60.

[35] Luo Y, Zhang X, Zhao M, Lu Q. DNA demethylation of the perforin promoter in CD4(+) T cells from patients with subacute cutaneous lupus erythematosus. J Dermatol Sci 2009;56:33−6.

[36] Lu Q, Wu A, Tesmer L, Ray D, Yousif N, Richardson B. Demethylation of CD40LG on the inactive X in T cells from women with lupus. J Immunol 2007;179:6352−8.

[37] Wang GS, Zhang M, Li XP, Zhang H, Chen W, Kan M, et al. Ultraviolet B exposure of peripheral blood mononuclear cells of patients with systemic lupus erythematosus inhibits DNA methylation. Lupus 2009;18:1037−44.

[38] Jeffries MA, Sawalha AH. Epigenetics in systemic lupus erythematosus: leading the way for specific therapeutic agents. Int J Clin Rheumtol 2011;6:423—39.

[39] Zhu X, Liang J, Li F, Yang Y, Xiang L, Xu J. Analysis of associations between the patterns of global DNA hypomethylation and expression of DNA methyltransferase in patients with systemic lupus erythematosus. Int J Dermatol 2011;50:697—704.

[40] Garaud S, Le Dantec C, Jousse-Joulin S, Hanrotel-Saliou C, Saraux A, Mageed RA, et al. IL-6 modulates CD5 expression in B cells from patients with lupus by regulating DNA methylation. J Immunol 2009;182:5623—32.

[41] Mishra N, Brown DR, Olorenshaw IM, Kammer GM. Trichostatin A reverses skewed expression of CD154, interleukin-10, and interferon-gamma gene and protein expression in lupus T cells. Proc Natl Acad Sci USA 2001;98:2628—33.

[42] Zhang Z, Song L, Maurer K, Petri MA, Sullivan KE. Global H4 acetylation analysis by ChIP-chip in systemic lupus erythematosus monocytes. Genes Immun 2010;11:124—33.

[43] Sullivan KE, Suriano A, Dietzmann K, Lin J, Goldman D, Petri MA. The TNFalpha locus is altered in monocytes from patients with systemic lupus erythematosus. Clin Immunol 2007;123:74—81.

[44] Zhao M, Liu S, Luo S, Wu H, Tang M, Cheng W, et al. DNA methylation and mRNA and microRNA expression of SLE CD4+ T cells correlate with disease phenotype. J Autoimmun 2014.

[45] Lefevre S, Knedla A, Tennie C, Kampmann A, Wunrau C, Dinser R, et al. Synovial fibroblasts spread rheumatoid arthritis to unaffected joints. Nat Med 2009;15:1414—20.

[46] MacGregor AJ, Snieder H, Rigby AS, Koskenvuo M, Kaprio J, Aho K, et al. Characterizing the quantitative genetic contribution to rheumatoid arthritis using data from twins. Arthritis Rheum 2000;43:30—7.

[47] Carrel L, Willard HF. X-inactivation profile reveals extensive variability in X-linked gene expression in females. Nature 2005;434:400—4.

[48] Liao J, Liang G, Xie S, Zhao H, Zuo X, Li F, et al. CD40L demethylation in CD4(+) T cells from women with rheumatoid arthritis. Clin Immunol 2012;145:13—18.

[49] Nile CJ, Read RC, Akil M, Duff GW, Wilson AG. Methylation status of a single CpG site in the IL6 promoter is related to IL6 messenger RNA levels and rheumatoid arthritis. Arthritis Rheum 2008;58:2686—93.

[50] Nakano K, Whitaker JW, Boyle DL, Wang W, Firestein GS. DNA methylome signature in rheumatoid arthritis. Ann Rheum Dis 2013;72:110—17.

[51] Nakano K, Boyle DL, Firestein GS. Regulation of DNA methylation in rheumatoid arthritis synoviocytes. J Immunol 2013;190:1297—303.

[52] de la Rica L, Urquiza JM, Gomez-Cabrero D, Islam AB, Lopez-Bigas N, Tegner J, et al. Identification of novel markers in rheumatoid arthritis through integrated analysis of DNA methylation and microRNA expression. J Autoimmun 2013;41:6—16.

[53] Liu Y, Aryee MJ, Padyukov L, Fallin MD, Hesselberg E, Runarsson A, et al. Epigenome-wide association data implicate DNA methylation as an intermediary of genetic risk in rheumatoid arthritis. Nat Biotechnol 2013;31:142—7.

[54] Wada TT, Araki Y, Sato K, Aizaki Y, Yokota K, Kim YT, et al. Aberrant histone acetylation contributes to elevated interleukin-6 production in rheumatoid arthritis synovial fibroblasts. Biochem Biophys Res Commun 2014;444:682—6.

[55] Broen JC, Radstake TR, Rossato M. The role of genetics and epigenetics in the pathogenesis of systemic sclerosis. Nat Rev Rheumatol 2014.

[56] Katsumoto TR, Whitfield ML, Connolly MK. The pathogenesis of systemic sclerosis. Annu Rev Pathol 2011;6:509—37.

[57] Lei W, Luo Y, Lei W, Luo Y, Yan K, Zhao S, et al. Abnormal DNA methylation in CD4 + T cells from patients with systemic lupus erythematosus, systemic sclerosis, and dermatomyositis. Scand J Rheumatol 2009;38:369−74.

[58] Lian X, Xiao R, Hu X, Kanekura T, Jiang H, Li Y, et al. DNA demethylation of CD40l in CD4 + T cells from women with systemic sclerosis: a possible explanation for female susceptibility. Arthritis Rheum 2012;64:2338−45.

[59] Jiang H, Xiao R, Lian X, Kanekura T, Luo Y, Yin Y, et al. Demethylation of TNFSF7 contributes to CD70 overexpression in CD4 + T cells from patients with systemic sclerosis. Clin Immunol 2012;143:39−44.

[60] Wang YY, Wang Q, Sun XH, Liu RZ, Shu Y, Kanekura T, et al. DNA hypermethylation of the forkhead box protein 3 (FOXP3) promoter in CD4 + T cells of patients with systemic sclerosis. Br J Dermatol 2014;171:39−47.

[61] Altorok N, Almeshal N, Wang Y, Kahaleh B. Epigenetics, the holy grail in the pathogenesis of systemic sclerosis. Rheumatology (Oxford) 2014.

[62] Ghosh AK, Bhattacharyya S, Lafyatis R, Farina G, Yu J, Thimmapaya B, et al. p300 is elevated in systemic sclerosis and its expression is positively regulated by TGF-beta: epigenetic feed-forward amplification of fibrosis. J Invest Dermatol 2013;133:1302−10.

[63] Bhattacharyya S, Ghosh AK, Pannu J, Mori Y, Takagawa S, Chen G, et al. Fibroblast expression of the coactivator p300 governs the intensity of profibrotic response to transforming growth factor beta. Arthritis Rheum 2005;52:1248−58.

[64] Kramer M, Dees C, Huang J, Schlottmann I, Palumbo-Zerr K, Zerr P, et al. Inhibition of H3K27 histone trimethylation activates fibroblasts and induces fibrosis. Ann Rheum Dis 2013;72:614−20.

[65] Wang Y, Yang Y, Luo Y, Yin Y, Wang Q, Li Y, et al. Aberrant histone modification in peripheral blood B cells from patients with systemic sclerosis. Clin Immunol 2013;149:46−54.

[66] Mavragani CP, Moutsopoulos HM. Sjogren's syndrome. Annu Rev Pathol 2014;9:273−85.

[67] Le Dantec C, Varin MM, Brooks WH, Pers JO, Youinou P, Renaudineau Y. Epigenetics and Sjogren's syndrome. Curr Pharm Biotechnol 2012;13:2046−53.

[68] Thabet Y, Le Dantec C, Ghedira I, Devauchelle V, Cornec D, Pers JO, et al. Epigenetic dysregulation in salivary glands from patients with primary Sjogren's syndrome may be ascribed to infiltrating B cells. J Autoimmun 2013;41:175−81.

[69] Altorok N, Coit P, Hughes T, Koelsch KA, Stone DU, Rasmussen A, et al. Genome-wide DNA methylation patterns in naive CD4 + T cells from patients with primary Sjogren's syndrome. Arthritis Rheumatol 2014;66:731−9.

[70] Huber A, Menconi F, Corathers S, Jacobson EM, Tomer Y. Joint genetic susceptibility to type 1 diabetes and autoimmune thyroiditis: from epidemiology to mechanisms. Endocr Rev 2008;29:697−725.

[71] Szablewski L. Role of immune system in type 1 diabetes mellitus pathogenesis. Int Immunopharmacol 2014;22:182−91.

[72] Rakyan VK, Beyan H, Down TA, Hawa MI, Maslau S, Aden D, et al. Identification of type 1 diabetes-associated DNA methylation variable positions that precede disease diagnosis. PLoS Genet 2011;7:e1002300.

[73] Rachakonda TD, Schupp CW, Armstrong AW. Psoriasis prevalence among adults in the United States. J Am Acad Dermatol 2014;70:512−16.

[74] Zhang P, Zhao M, Liang G, Yin G, Huang D, Su F, et al. Whole-genome DNA methylation in skin lesions from patients with psoriasis vulgaris. J Autoimmun 2013;41:17−24.

[75] Zhang K, Zhang R, Li X, Yin G, Niu X. Promoter methylation status of p15 and p21 genes in HPP-CFCs of bone marrow of patients with psoriasis. Eur J Dermatol 2009;19:141−6.

[76] Ruchusatsawat K, Wongpiyabovorn J, Shuangshoti S, Hirankarn N, Mutirangura A. SHP-1 promoter 2 methylation in normal epithelial tissues and demethylation in psoriasis. J Mol Med (Berl) 2006;84:175−82.

[77] Zhang P, Su Y, Zhao M, Huang W, Lu Q. Abnormal histone modifications in PBMCs from patients with psoriasis vulgaris. Eur J Dermatol 2011;21:552−7.

[78] Tovar-Castillo LE, Cancino-Diaz JC, Garcia-Vazquez F, Cancino-Gomez FG, Leon-Dorantes G, Blancas-Gonzalez F, et al. Under-expression of VHL and over-expression of HDAC-1, HIF-1alpha, LL-37, and IAP-2 in affected skin biopsies of patients with psoriasis. Int J Dermatol 2007;46:239−46.

[79] Owczarek D, Cibor D, Glowacki MK, Rodacki T, Mach T. Inflammatory bowel disease: epidemiology, pathology and risk factors for hypercoagulability. World J Gastroenterol 2014;20:53−63.

[80] Karatzas PS, Gazouli M, Safioleas M, Mantzaris GJ. DNA methylation changes in inflammatory bowel disease. Ann Gastroenterol 2014;27:125−32.

[81] Xavier RJ, Podolsky DK. Unravelling the pathogenesis of inflammatory bowel disease. Nature 2007;448:427−34.

[82] Nimmo ER, Prendergast JG, Aldhous MC, Kennedy NA, Henderson P, Drummond HE, et al. Genome-wide methylation profiling in Crohn's disease identifies altered epigenetic regulation of key host defense mechanisms including the Th17 pathway. Inflamm Bowel Dis 2012;18:889−99.

[83] Lin Z, Hegarty JP, Yu W, Cappel JA, Chen X, Faber PW, et al. Identification of disease-associated DNA methylation in B cells from Crohn's disease and ulcerative colitis patients. Dig Dis Sci 2012;57:3145−53.

[84] Cooke J, Zhang H, Greger L, Silva AL, Massey D, Dawson C, et al. Mucosal genome-wide methylation changes in inflammatory bowel disease. Inflamm Bowel Dis 2012;18:2128−37.

[85] Hasler R, Feng Z, Backdahl L, Spehlmann ME, Franke A, Teschendorff A, et al. A functional methylome map of ulcerative colitis. Genome Res 2012;22:2130−7.

[86] Tsaprouni LG, Ito K, Powell JJ, Adcock IM, Punchard N. Differential patterns of histone acetylation in inflammatory bowel diseases. J Inflamm (Lond) 2011;8:1.

[87] Ventham NT, Kennedy NA, Nimmo ER, Satsangi J. Beyond gene discovery in inflammatory bowel disease: the emerging role of epigenetics. Gastroenterology 2013;145:293−308.

[88] Grabiec AM, Korchynskyi O, Tak PP, Reedquist KA. Histone deacetylase inhibitors suppress rheumatoid arthritis fibroblast-like synoviocyte and macrophage IL-6 production by accelerating mRNA decay. Ann Rheum Dis 2012;71:424−31.

[89] Grabiec AM, Reedquist KA. The ascent of acetylation in the epigenetics of rheumatoid arthritis. Nat Rev Rheumatol 2013;9:311−18.

[90] Anaya JM. The diagnosis and clinical significance of polyautoimmunity. Autoimmun Rev 2014;13:423−6.

Chapter 12

Genome-Wide DNA and Histone Modification Studies in Metabolic Disease

Charlotte Ling and Tina Rönn
Department of Clinical Sciences, Epigenetics and Diabetes, Clinical Research Centre,
Lund University Diabetes Centre, Malmö, Sweden

Chapter Outline

12.1 INTRODUCTION

The prevalence of metabolic diseases, such as type 2 diabetes, obesity, and metabolic syndrome, is rapidly increasing worldwide [1]. A main reason for the increased prevalence of metabolic diseases is the increased number of people who live a sedentary life, including physical inactivity and energy-rich diets. It should also be noted that people generally live longer and that the risk for some metabolic diseases, such as type 2 diabetes, increases with aging. Although environmental factors play a key role in predisposing a person to metabolic diseases, there is also a strong genetic component affecting the risk for disease; for example, the risk for type 2 diabetes is estimated to increase by 40% if one first-degree relative has the disease, and the risk is even higher if both parents have diabetes [2]. Additionally, twin studies have estimated a strong heritability for metabolic diseases [3,4]. Also, recent genome-wide association studies

M. Fraga & A.F. Fernandez (Eds): Epigenomics in Health and Disease.
DOI: http://dx.doi.org/10.1016/B978-0-12-800140-0.00012-1
© 2016 Elsevier Inc. All rights reserved.

255

(GWAS) have identified numerous common genetic variants, single nucleotide polymorphisms (SNPs), associated with an increased risk for type 2 diabetes, obesity, and dyslipidemia [5−13]. Nevertheless, although approximately 80 SNPs have been identified to increase the risk for type 2 diabetes, these variants do only explain a small proportion of the estimated heritability of the disease [14]. Initially, researchers have tried to identify rare variants with a large impact on the risk for metabolic disease [15]. However, emerging data suggest that this is not the solution to the missing heritability of metabolic disease [16]. Instead, one may have to look elsewhere, for example, beyond the genome, and the solution could be the epigenome. Importantly, although the epigenome may be dynamic and change upon exposure to environmental factors, such as diet and physical activity, once epigenetic modifications are introduced, they may remain stable and be inherited through cell divisions [17]. This renders epigenetic alterations a potentially important pathogenetic mechanism in complex multifactorial diseases, such as type 2 diabetes and obesity (Figure 12.1).

The epigenome includes DNA methylation and modifications to histone tails as well as noncoding RNA (ncRNA) and has the ability to affect gene regulation during development and in a tissue-specific manner. In differentiated mammalian cells, DNA methylation mainly takes place on a cytosine in the so-called CpG sites. Additionally, there are specific enzymes responsible for

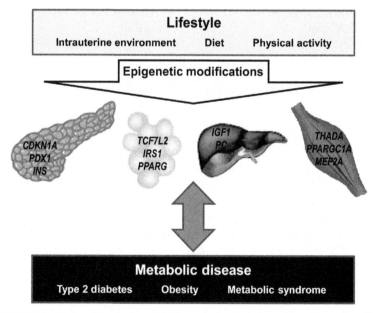

FIGURE 12.1 Cartoon showing how lifestyle may alter the epigenome and hence the risk for metabolic disease, together with examples of candidate genes that exhibit altered DNA methylation in respective tissues either in response to lifestyle interventions and/or in relation to metabolic disease.

adding methyl groups to the DNA, that is, DNMT3A and DNMT3B, which are responsible for *de novo* methylation, as well as DNMT1, which maintains the methylation pattern between cell replication. Demethylation or decreased methylation can take place when the activity of DNMT1 is decreased during replication (passive demethylation) or due to specific enzymes removing methyl groups from DNA (active demethylation). Recent data suggest a role for Ten-Eleven translocation (TET) enzymes in the demethylation process [18]. The amount of available methyl donors in the body does also play a role in the DNA methylation process. Here, one needs to consider the amount of methyl donors in the diet. DNA methylation was initially considered a silencing mark of gene expression. However, emerging data have shown that the role of DNA methylation is more complex than first thought, and its role has been found to depend on the genomic location [19]. DNA methylation has now been associated with gene silencing, activation, splicing, and replication, and it may affect overall genomic stability. With regard to epigenetic modifications of histone tails, multiple different modifications take place at different amino acids. Subsequently, many different enzymes are responsible for adding and removing histone modifications, and the regulation and function of these modifications are thus a network of complex interactions. Histone modifications further interact with DNA methylation and are linked to the chromatin structure [20].

12.2 TYPE 2 DIABETES AND EPIGENETIC MODIFICATIONS

Type 2 diabetes is characterized by chronic hyperglycemia due to impaired insulin secretion from pancreatic islets, together with insulin resistance in target tissues, such as the liver, skeletal muscle, and adipose tissue. Although the search for genetic variants explaining the risk for metabolic disease only requires the analysis of "one genome" per person, the search for epigenetic variation is more complex because numerous tissues and cell types affect the disease etiology at the same time, since the epigenome is cell type specific. It should also be noted that epigenetic modifications may be either the cause or the consequence of disease. Dissection of the impact of epigenetic mechanisms in the pathogenesis of type 2 diabetes subsequently requires analysis of multiple tissues in both case-control studies and in predictive studies. However, if epigenetic differences cannot be identified in subjects with type 2 diabetes compared with normal controls, then it is unlikely that epigenetic mechanisms play a role for the disease. This may therefore be the starting point when dissecting if epigenetics affect type 2 diabetes. In 2008, it was shown for the first time that pancreatic islets from subjects with type 2 diabetes have epigenetic alterations compared with normal controls [21]. Here, Ling et al. found increased DNA methylation of *PPARGC1A* in islets from diabetic versus normal human donors. *PPARGC1A* encodes the transcriptional co-activator PGC1α, which regulates the expression of numerous genes involved in oxidative phosphorylation and thereby adenosine triphosphate (ATP) production. This study also showed that human diabetic islets had decreased *PPARGC1A* expression and lower glucose-stimulated insulin secretion. When the expression of *PPARGC1A* was silenced in human pancreatic islets by using synthetic RNA (siRNA), insulin secretion was reduced, which further supports the role of this gene in the development of type 2 diabetes. Additional studies in human skeletal muscle also found elevated DNA methylation and decreased gene expression of *PPARGC1A* in subjects with type 2 diabetes (Figure 12.1) [22]. Initial epigenetic studies were time consuming, required large amount of DNA and, thus, tissue, and could only study some genomic regions at a time. During the following years, only a few genes were analyzed and discovered to be differentially methylated in subjects with type 2 diabetes. These include key genes for pancreatic islet function and glucose homeostasis, that is, the insulin gene (*INS*), *PDX1*, and *GLP1R*, all studied in pancreatic islets from subjects with type 2 diabetes and normal controls (Figure 12.1) [23–25]. Here, human diabetic islets showed elevated DNA methylation and decreased gene expression of *INS*, *PDX1*, and *GLP1R*. These studies support the role of epigenetic mechanisms in type 2 diabetes. However, genome-wide methods were needed to further improve our understanding of the role of epigenetics in type 2 diabetes. A breakthrough came with the development of the Infinium array from Illumina. First, an array covering approximately 27,000 CpG sites, which corresponds to nearly 0.1% of all CpG sites in the human genome and approximately 14,000 genes, was released. This method was then applied to human pancreatic islets from five human donors with type 2 diabetes and 11 controls, as well as to a number of monozygotic twin pairs discordant for type 2 diabetes, that is, one twin in the pair having diabetes and the other being healthy [26,27]. Although these studies may seem underpowered, they found some sites that were differentially methylated in subjects with diabetes. Interestingly, in human pancreatic islets, the majority of the differentially methylated sites showed decreased methylation in the subjects with diabetes versus normal subjects. Further advances came after the release of the Infinium HumanMethylation450 BeadChip in late 2011 [28]. This array analyzes DNA methylation of approximately 480,000 CpG sites that target approximately 99% of all RefSeq genes in the human genome. Currently, this array has been used to discover nearly 800 genes with altered DNA methylation pattern in pancreatic islets from 15 donors with diabetes versus 34 normal donors [29]. Interestingly, the identified genes include both known candidate genes for type 2 diabetes, for example, *TCF7L2*, *FTO*, and *KCNQ1*, as well as novel candidates, such as *CDKN1A*, *PDE7B*, and *EXOC3L2* (Figure 12.1). In agreement with data from the previous study that used the 27k Infinium array [26], the majority of the differentially methylated sites showed decreased methylation in islets form subjects with type 2 diabetes compared with normal subjects. Among the genes with differential DNA methylation in diabetic islets, 102 also showed altered gene expression, and most of these (97%) exhibited increased messenger RNA (mRNA) expression in parallel with decreased DNA methylation. The direct impact of DNA methylation on the transcriptional activity of some of these genes was further studied using luciferase assays. Here, Dayeh et al. showed that increased DNA methylation of the *CDKN1A* and *PDE7B* promoters resulted in decreased transcriptional activity. To mimic the situation in human islets and functionally dissect the role of the identified genes in the pathogenesis of type 2 diabetes, *CDKN1A* and *PDE7B* were overexpressed in clonal β-cells and α-cells *in vitro*. These cells secrete insulin and glucagon, respectively. Although *CDKN1A* (also known as *p21*) encodes cyclin-dependent kinase inhibitor 1A, which functions as a regulator of cell cycle progression at G1, *PDE7B* encodes phosphodiesterase 7B, which regulated cyclic adenosine monophosphate (cAMP) signaling. Interestingly, overexpression of both of these genes resulted in decreased glucose-stimulated insulin secretion and elevated glucagon secretion in clonal β-cells and α-cells cultured *in vitro*, respectively. Indeed, this is also seen in patients with type 2 diabetes with decrease in insulin secretion and increase in glucagon secretion, both contributing to hyperglycemia. Further analysis of β-cells overexpressing *CDKN1A* showed decreased cell proliferation compared with control cells. In contrast, *EXOC3L2* belongs to the few genes that show increased DNA methylation and decreased gene expression in human pancreatic islets from subjects with type 2 diabetes. This gene is part of the exocyst complex and may thereby play a role in the exocytosis of insulin in pancreatic β-cells. In fact, when the expression of Exoc3l was silenced in clonal β-cells using siRNA, β-cell exocytosis was decreased. In particular, the first-phase insulin secretion seemed to be impaired. To further test if the epigenetic alterations that were identified in pancreatic islets from subjects with type 2 diabetes may occur before development of the disease, the impact of age, hemoglobin A1c (HbA1c), and BMI were related to DNA methylation in pancreatic islets from 87 normal donors with a large span in the studied phenotypes; that is, they were between 26 and 74 years old, had a BMI spanning between 17.6 and 40.1 kg/m^2, and HbA1c between 4.3 and 6.4% (34−56 mmol/mol). All these three factors are known to increase the risk for type 2 diabetes, and they may predispose a person to diabetes by altering the DNA methylation pattern in pancreatic islets. Indeed, these risk factors were associated with differential DNA methylation of some sites that also showed altered DNA methylation in diabetic islets. For example, increased age was associated with decreased DNA methylation of *CDKN1A* and increased methylation of *EXOC3L2* in nondiabetic islets, which is in line with what is also seen in subjects with diabetes. A potential problem when performing epigenome-wide association studies (EWAS) is the fact that differences in cell composition can contribute to identified epigenetic differences. It should therefore be noted that no difference in α-cell and β-cell content was found in the human islets from subjects with type 2 diabetes compared with the normal controls. Additionally, DNA methylation was analyzed in FACS sorted α-cells and β-cells from a small number of normal donors. Still, the genes exhibiting differential islet DNA methylation between subjects with diabetes and controls did not show differences in methylation between α-cells and β-cells.

The Infinium HumanMethylation450 BeadChip has also been used to analyze DNA methylation in adipose tissue from subjects with type 2 diabetes and normal controls. Here, DNA methylation and gene expression were used to analyze in adipose tissue from monozygotic twin pairs discordant for type 2 diabetes [30]. Interestingly, as monozygotic twin pairs are genetically identical, different phenotypes between them may potentially be due to epigenetic alterations. However, although the expression of nearly 200 genes was altered between the twins in the discordant pairs—for example, genes involved in metabolism, such as *ELOVL6, GYS2*, and *FADS1*, which showed decreased expression and genes involved in inflammation, such as *SPP1 (OPN), CCL18*, and *IL1RN*, which showed increased expression—no significant differences in DNA methylation could be identified between the twins after correction for multiple testing. Possible reasons for this result could be lack of power, the fact that the DNA methylation pattern between monozygotic twins is highly heritable, or the fact that type 2 diabetes is not associated with epigenetic differences in human adipose tissue. To test these hypotheses, DNA methylation was further analyzed in adipose tissue from a case-control cohort, including unrelated subjects with or without type 2 diabetes, as well as tissue from monozygotic and dizygotic twin pairs without the disease [30]. First, the heritability of the DNA methylation pattern in human adipose tissue was quite strong, and this was further supported by data from Grundberg et al. [31]. Second, it was possible to identify epigenetic differences in adipose tissue between unrelated cases and controls; that is, 15,627 sites annotated to 7046 genes, such as *PPARG, IRS1, TCF7L2*, and *KCNQ1* were differentially methylated in subjects with diabetes compared with controls (Figure 12.1). Less genome-wide epigenetic information is available from skeletal muscle from subjects with type 2 diabetes. However, MeDIP-chip has been used to identify altered DNA methylation patterns in skeletal muscle from subjects with a family history of type 2 diabetes compared with controls [32]. Some of the identified alterations were then also found in skeletal muscle from monozygotic twin pairs discordant for type 2 diabetes [32]. These data suggest that epigenetic modification in skeletal muscle due to a family history may predispose to the disease later in life. A genome-wide study of DNA methylation in blood samples from type 2 diabetes and control subjects detected differentially methylated regions in genetic loci previously associated with type 2 diabetes (*CENTD2, FTO, KCNJ11, TCF7L2*, and *WFS1*) [33]. Using a longitudinal study, where some of the participants, all healthy at inclusion, progressed to impaired glucose metabolism, these authors also showed that altered DNA methylation may be detected prior to disease development. Together, these data show that type 2 diabetes is associated with epigenetic alterations in pancreatic islets, adipose tissue, skeletal muscle, and blood. However, further EWAS in additional target tissues, such as the liver, are needed to fully dissect the impact of epigenetic modification in the pathogenesis of type 2 diabetes.

It is also relevant to study the genome-wide pattern of histone modifications in subjects with type 2 diabetes and normal controls. However, as many different histone modifications exist and current methods for analyses of histone modifications, such as chromatin immunoprecipitation followed by deep sequencing (ChIP-seq), require a large number of cells, advances in genome-wide histone modification studies of type 2 diabetes case-control cohorts are modest. That being said, some studies have used ChIP-seq to analyze histone modifications in target tissues for type 2 diabetes, that is, human pancreatic islets in a small number of normal donors [34−38]. These studies suggest that epigenetic modifications in enhancer regions play a key role in gene regulations. Additionally, histone modifications of target genes have been analyzed in β-cells exposed to palmitate. The reason for this experiment was that subjects with type 2 diabetes and obese subjects have elevated circulating levels of saturated fatty acids in parallel with impaired islet function. Palmitate does also decrease insulin secretion in β-cells cultured *in vitro*, and it alters the degree of histone modifications of genes showing differential gene expression following palmitate treatment. For example, *Insig1*, *Lss*, *Peci*, *Idi1*, *Hmgcs1*, and *Casr* were subject to epigenetic regulation in β-cells following lipid exposure [39].

Additionally, linking DNA methylation and expression of ncRNA could provide some clues for disease etiology. Indeed, some studies propose a role for differential DNA methylation and altered expression of ncRNA on insulin secretion in human pancreatic islets [40,41].

Furthermore, genetic epidemiologists have proposed that gene environment interactions affect the risk for type 2 diabetes. These gene environment interactions could potentially be explained by chemical modifications of the genome due to environmental exposures, such as epigenetics. Importantly, nearly 25% of all SNPs in the human genome either introduce or remove a CpG site. These so-called CpG-SNPs affect whether a certain site in the genome can be methylated or not, and this could be one explanation for how gene environment interactions affect the risk for disease. For example, close to 50% of all SNPs identified by GWAS to increase the risk for type 2 diabetes either introduce or remove CpG site, and these diabetes SNPs are hence CpG-SNPs [42]. Importantly, this is significantly more CpG-SNPs than expected by chance and double the amount compared with what is found in the overall genome. These CpG-SNP loci associated with type 2 diabetes represent *TCF7L2*, *KCNQ1*, *PPARG*, *HHEX*, *CDKN2A*, *SLC30A8*, *DUSP9*, *CDKAL1*, *ADCY5*, *SRR*, *WFS1*, *IRS1*, *DUSP8*, *HMGA2*, *TSPAN8*, and *CHCHD9*. Notably, all analyzed CpG-SNPs were associated with altered degree of DNA methylation in the CpG-SNP site in human islets. Differential DNA methylation of CpG-SNP sites was also associated with gene expression, alternative splice index, and hormonal secretion in the human islets. In a more recent study, a genome-wide DNA methylation quantitative trait locus (mQTL) analysis was performed in human pancreatic islets [43]. Here, 574,533 SNPs were related to genome-wide DNA methylation data of 468,787 CpG sites targeting 99% of RefSeq genes in human pancreatic islets. The mQTL analysis was performed in both *cis* and *trans*. After correction for multiple testing, 67,438 SNP-CpG pairs in *cis*, corresponding to 36,783 SNPs and 11,735 CpG sites, showed significant associations. It should be noted that this analysis identified significant mQTLs that overlapped with previously reported diabetes loci, such as *KCNJ11*, *ADCY5*, *PDX1*, and *GRB10*. This study tested if identified SNPs associated with altered DNA methylation also significantly affect mRNA expression in human pancreatic islets. Here, an expression quantitative trait locus (eQTL) follow-up analysis was performed, and 302 SNP-mRNA transcript pairs representing 243 unique SNPs and 46 unique mRNA transcripts were identified, showing significant associations between genotypes and mRNA expression levels after correction for multiple testing. It is possible to test if an SNP potentially mediates its effect on gene expression via altered DNA methylation by a causal inference test (CIT). Indeed, in human pancreatic islets, some of the significant mQTL SNPs seemed to mediate their effects on gene expression via alteration of DNA methylation, since they were significant on the basis of the CIT. Additionally, some of the genes which were significant in both the mQTL and eQTL analyses were followed up by functional experiments in clonal β-cells. Here, three selected candidate genes (*GPX7*, *GSTT1*, and *SNX19*) were silenced by siRNA. *GPX7* and *GSTT1* encode proteins that are known to protect against oxidative stress, and *SNX19* encodes sortin nexin 19, which has been found to affect apoptosis. Silencing of *Gpx7* or *Gstt1* increased caspase-3/7 activities in β-cells, whereas knockdown of *Snx19* resulted in increased cell number. Finally, this study found direct correlations between DNA methylation of 22,773 CpGs with gene expression of 4876 genes, and this shows the importance of integrating genetics and epigenetics when studying mechanism contributing to metabolic disease.

12.3   OBESITY AND EPIGENETIC MODIFICATIONS

Obesity is characterized by a BMI above $30 \, kg/m^2$, and subjects with a BMI between 25 and $30 \, kg/m^2$ are considered overweight. Although there are monogenic forms of obesity and common genetic variants associated with BMI, there is a strong environmental component explaining a large proportion of the risk for the disease. For example, it is possible that environmental exposures, such as energy-rich diets early in life, may result in epigenetic modifications that increase the risk for obesity. Indeed, interesting animal studies show that "overeating" early in life contributes to epigenetic alterations in the brain and an altered set-point for food intake, which may contribute to obesity later in life. Additionally, obesity has been shown to affect the epigenome in humans [44,45]. For example, increased BMI was associated with altered DNA methylation in human pancreatic islets [29]. A recent study showed further evidence for obesity accelerating the effect of aging on the DNA methylation pattern in the human liver [46]. Also, nonalcoholic fatty liver disease (NASH) has been associated with an altered DNA methylation pattern in the human liver [47]. Here, expression and methylation differences were seen for genes encoding for key enzymes in metabolism and insulin/insulin-like signaling (Figure 12.1). NAFLD-associated alterations in DNA methylation in the human liver seemed to be partially reversible by bariatric surgery.

There is a growing interest in identifying blood-based epigenetic biomarkers that can mimic epigenetic signatures in target tissues for metabolic disease. Interestingly, Dick et al. recently identified altered DNA methylation of *HIF3A* in both adipose tissue and blood cells due to increased BMI, supporting the use of blood-based epigenetic biomarkers in obesity [48].

12.4   DO DIET AND EXERCISE INTERVENTIONS ALTER THE EPIGENETIC PATTERN AND POTENTIALLY RISK FOR METABOLIC DISEASE?

Lifestyle interventions, including healthy diet and increased physical activity, reduce the risk for metabolic diseases in high-risk groups [49,50]. A lifestyle of physical inactivity and energy-rich diets contributes to insulin resistance and impaired metabolism [51−54]. It is possible that these different lifestyles mediate their effects on metabolism and human metabolic disease via alteration of the epigenetic pattern in key target tissues. Indeed, a 6-month supervised exercise intervention performed in sedentary middle-aged men support this hypothesis. Here, researchers studied the genome-wide DNA methylation pattern as well the transcriptome in human skeletal muscle and adipose tissue before and after the intervention [32,55,56]. Importantly, although these men only exercised twice a week for 1 h during each session, they improved their maximal oxygen uptake (VO$_2$max) and high-density lipoprotein (HDL) levels and decreased their pulse rate, blood pressure, and waistline after the intervention. Additionally, genome-wide changes in the methylation pattern were found in both skeletal muscle and adipose tissue after regular exercise. In human muscle, after Bonferroni corrections, 134 individual genes were found to change the degree of methylation after exercise [32]. Out of these, 115 genes had reduced methylation, and 19 had increased methylation. These include *THADA*, *MEF2A*, *RUNX1*, and *NDUFC2*, which all show decreased methylation after exercise (Figure 12.1). Although *THADA* is a known candidate gene for type 2 diabetes previously identified by GWAS, *MEF2A* encodes a transcription factor known to regulate *GLUT4* expression, and also *RUNX1* encodes a transcription factor known to be regulated by exercise. Moreover, *NDUFC2* is encoding a protein that is part of complex I of the respiratory chain and hence affects mitochondrial ATP production. Next, functional luciferase experiments showed that increased methylation of the promoter of these four genes resulted in reduced transcriptional activity, supporting a key role for exercise in regulating gene transcription via altering the epigenetic pattern. It should also be noted that inverse correlations between the degree of DNA methylation and mRNA expression in skeletal muscle were found for these four genes. In human adipose tissue, the DNA methylation pattern of approximately one-third of all genes changed after 6 months of exercise [55]. These include known candidate genes for both type 2 diabetes and obesity, for example, *TCF7L2*, *KCNQ1*, *CDKAL1*, *THADA*, *FTO*, and *TUB*. Some of the identified genes showed both differential DNA methylation and mRNA expression after exercise. Two of them, *HDAC4* and *NCOR2*, were selected for functional follow-up experiments in adipocytes *in vitro*. Both *HDAC4* and *NCOR2* had increased DNA methylation and decreased mRNA expression in adipose tissue after exercise. Therefore, to mimic the *in vivo* situation, the expression of these two genes was silenced in an adipocyte cell line cultured *in vitro*. *HDAC4* encodes a histone deacetylase, and *NCOR2* encodes a transcriptional co-repressor known to play a role in the SMART complex. Both these proteins have previously been suggested to affect glucose metabolism in adipose tissue. Here, Rönn et al. found that decreased expression of both HDAC4 and NCOR2 was associated with increased insulin-stimulated lipogenesis, suggesting that the epigenetic changes taking place after exercise contribute to improved insulin sensitivity in adipose tissue.

In a different intervention study, Alibegovic et al. studied the impact of physical inactivity on metabolism, gene expression, and DNA methylation in the skeletal muscle of young healthy men [51,52]. These men were exposed to 9 days of bed rest, which resulted in impaired glucose metabolism, differential expression, and increased methylation of *PPARGC1A*.

A diet intervention, including a 5-day high-fat diet, has also been performed in young healthy men, and here the Infinium 27K array was used to study the DNA methylation pattern in skeletal muscle [53,57,58]. This study was based on a randomized crossover setting, where half of the included men were first exposed to a high-fat diet followed by a control diet, and half of the subjects were first exposed to a control diet and then to the high-fat diet. Using this design, it was possible to study if the impact of the high-fat diet was reversible. Interestingly, the high-fat diet changed the degree of methylation of 6508 genes after correction for multiple testing. These genes were enriched in pathways involved in inflammation, reproduction, and cancer. However, these epigenetic changes were only partly reversed after the control diet, suggesting that diet-induced epigenetic changes are difficult to "wash away."

Additionally, to mimic the situation in obesity and type 2 diabetes, human pancreatic islets were exposed to palmitate for 48 h and the genome-wide DNA methylation pattern was studied with the Infinium 450K array. Here, Hall et al. found a global increase in DNA methylation, and CpG sites two times more than expected showed altered DNA methylation after palmitate exposure. Interestingly, these also included known candidate genes for type 2 diabetes, such as *TCF7L2* and *GLIS3*.

Overall, environmental exposures and lifestyle interventions alter the epigenome in multiple human tissues and may thereby contribute to the risk for metabolic disease (Figure 12.1).

12.5   DOES THE INTRAUTERINE ENVIRONMENT ALTER THE EPIGENETIC PATTERN AND POTENTIALLY RISK FOR METABOLIC DISEASE?

One of the first studies pointing to an important role for the intrauterine environment in epigenetic programming and metabolic disease in postnatal life came from the research group led by Professor Simmons [59]. Here, Park et al. showed that restricted blood flow to the fetus causes epigenetic changes to the *Pdx1* gene in pancreatic islets. These changes further contribute to altered *Pdx1* expression and diabetes in postnatal life. *PDX1* encodes a transcription factor that has a key role in the fetal development of pancreatic β-cells. This transcription factor also regulates transcription of the insulin gene in the mature pancreas. Mutations in *PDX1* causes a monogenetic form of diabetes and silencing of *Pdx1* in rodent islets causes diabetes [60,61]. A different study in rodents came from a study by Sandovoci et al., in which they showed that a protein-restricted diet fed to pregnant mothers can alter the epigenetic pattern of *Hnf4a* in pancreatic islets [62]. Again, this resulted in altered islet Hnf4a expression and hyperglycemia in adult life. Another study identified epigenetic changes in islets from offspring of rodent fathers exposed to a high-fat diet [63]. Together, it appears that the intrauterine environment can alter the epigenome and thereby the risk for metabolic disease later in life. Human support, to what is also known as the *Barker hypothesis*, or *thrifty phenotype hypothesis* [64−67], came from the Dutch famine study, in which Heijmans et al. found differential DNA methylation in the offspring of mothers exposed to the Dutch famine during World War II compared with offspring of the same mothers when plenty of food was available [68,69]. Offspring exposed to famine *in utero* were also shown to have an increased risk for metabolic disease later in life. Also, the impact of low birth weight, as a measure of an impaired intrauterine environment, on DNA methylation has been studied in humans [53,57]. These subjects exhibit metabolic impairments, such as increased glucose and insulin levels, and increased waist circumference in parallel with altered DNA methylation in skeletal muscle even when they are young and healthy.

12.6   CONCLUSIONS

Genome-wide epigenetic alterations can be found in multiple target tissues from subjects with metabolic disease compared with healthy subjects. These epigenetic variations in combination with genetic variants are likely to interact and thereby predispose a person to disease. Indeed, further support for this hypothesis comes from studies demonstrating that environmental risk factors for metabolic disease, such as physical inactivity and diet, can alter the epigenetic pattern of known candidate genes for both type 2 diabetes and obesity. Additionally, quantitative trait locus studies have shown that the genome-wide SNP pattern affect the genome-wide DNA methylation pattern and thereby gene expression and metabolic phenotypes. In conclusion, recent genome-wide studies highlight how epigenetics plays a key role in metabolic disease in humans.

REFERENCES

[1] Chen L, Magliano DJ, Zimmet PZ. The worldwide epidemiology of type 2 diabetes mellitus—present and future perspectives. Nat Rev Endocrinol 2012;8(4):228–36.

[2] Kobberling J, Tillil H. Genetic and nutritional factors in the etiology and pathogenesis of diabetes mellitus. World Rev Nutr Diet 1990;63:102–15.

[3] Kaprio J, Tuomilehto J, Koskenvuo M, Romanov K, Reunanen A, Eriksson J, et al. Concordance for type 1 (insulin-dependent) and type 2 (non-insulin-dependent) diabetes mellitus in a population-based cohort of twins in Finland. Diabetologia 1992;35(11):1060–7.

[4] Newman B, Selby JV, King MC, Slemenda C, Fabsitz R, Friedman GD. Concordance for type 2 (non-insulin-dependent) diabetes mellitus in male twins. Diabetologia 1987;30 (10):763–8.

[5] Sladek R, Rocheleau G, Rung J, Dina C, Shen L, Serre D, et al. A genome-wide association study identifies novel risk loci for type 2 diabetes. Nature 2007;445(7130):881–5.

[6] McCarthy M. Genomics, type 2 diabetes, and obesity. N Eng J Med 2010;363 (24):2339–50.

[7] Kathiresan S, Melander O, Guiducci C, Surti A, Burtt NP, Rieder MJ, et al. Six new loci associated with blood low-density lipoprotein cholesterol, high-density lipoprotein cholesterol or triglycerides in humans. Nat Genet 2008;40(2):189–97.

[8] Speliotes EK, Willer CJ, Berndt SI, Monda KL, Thorleifsson G, Jackson AU, et al. Association analyses of 249,796 individuals reveal 18 new loci associated with body mass index. Nat Genet 2010;42(11):937–48.

[9] Saxena R, Voight BF, Lyssenko V, Burtt NP, de Bakker PI, Chen H, et al. Genome-wide association analysis identifies loci for type 2 diabetes and triglyceride levels. Science 2007;316(5829):1331–6.

[10] Frayling TM, Timpson NJ, Weedon MN, Zeggini E, Freathy RM, Lindgren CM, et al. A common variant in the FTO gene is associated with body mass index and predisposes to childhood and adult obesity. Science 2007;316(5826):889–94.

[11] Dupuis J, Langenberg C, Prokopenko I, Saxena R, Soranzo N, Jackson AU, et al. New genetic loci implicated in fasting glucose homeostasis and their impact on type 2 diabetes risk. Nat Genet 2010;42(2):105–16.

[12] Zeggini E, Scott LJ, Saxena R, Voight BF, Marchini JL, Hu T, et al. Meta-analysis of genome-wide association data and large-scale replication identifies additional susceptibility loci for type 2 diabetes. Nat Genet 2008;40(5):638–45.

[13] Zeggini E, Weedon MN, Lindgren CM, Frayling TM, Elliott KS, Lango H, et al. Replication of genome-wide association signals in UK samples reveals risk loci for type 2 diabetes. Science 2007;316(5829):1336–41.

[14] Manolio TA, Collins FS, Cox NJ, Goldstein DB, Hindorff LA, Hunter DJ, et al. Finding the missing heritability of complex diseases. Nature 2009;461(7265):747−53.

[15] Grarup N, Sandholt CH, Hansen T, Pedersen O. Genetic susceptibility to type 2 diabetes and obesity: from genome-wide association studies to rare variants and beyond. Diabetologia 2014;57(8):1528−41.

[16] Steinthorsdottir V, Thorleifsson G, Sulem P, Helgason H, Grarup N, Sigurdsson A, et al. Identification of low-frequency and rare sequence variants associated with elevated or reduced risk of type 2 diabetes. Nat Genet 2014;46(3):294−8.

[17] Anway MD, Cupp AS, Uzumcu M, Skinner MK. Epigenetic transgenerational actions of endocrine disruptors and male fertility. Science 2005;308(5727):1466−9.

[18] Chen H, Kazemier HG, de Groote ML, Ruiters MH, Xu GL, Rots MG. Induced DNA demethylation by targeting Ten-Eleven translocation 2 to the human ICAM-1 promoter. Nucleic Acids Res 2014;42(3):1563−74.

[19] Jones PA. Functions of DNA methylation: islands, start sites, gene bodies and beyond. Nat Rev Genet 2012;13(7):484−92.

[20] Hashimoto H, Vertino PM, Cheng X. Molecular coupling of DNA methylation and histone methylation. Epigenomics 2010;2(5):657−69.

[21] Ling C, Del Guerra S, Lupi R, Ronn T, Granhall C, Luthman H, et al. Epigenetic regulation of *PPARGC1A* in human type 2 diabetic islets and effect on insulin secretion. Diabetologia 2008;51(4):615−22.

[22] Barres R, Osler ME, Yan J, Rune A, Fritz T, Caidahl K, et al. Non-CpG methylation of the PGC-1alpha promoter through DNMT3B controls mitochondrial density. Cell Metab 2009;10(3):189−98.

[23] Yang BT, Dayeh TA, Kirkpatrick CL, Taneera J, Kumar R, Groop L, et al. Insulin promoter DNA methylation correlates negatively with insulin gene expression and positively with HbA(1c) levels in human pancreatic islets. Diabetologia 2011;54(2):360−7.

[24] Yang BT, Dayeh TA, Volkov PA, Kirkpatrick CL, Malmgren S, Jing X, et al. Increased DNA methylation and decreased expression of PDX-1 in pancreatic islets from patients with type 2 diabetes. Mol Endocrinol 2012;26(7):1203−12.

[25] Hall E, Dayeh T, Kirkpatrick CL, Wollheim CB, Dekker Nitert M, Ling C. DNA methylation of the glucagon-like peptide 1 receptor (GLP1R) in human pancreatic islets. BMC Med Genet 2013;14:76.

[26] Volkmar M, Dedeurwaerder S, Cunha DA, Ndlovu MN, Defrance M, Deplus R, et al. DNA methylation profiling identifies epigenetic dysregulation in pancreatic islets from type 2 diabetic patients. EMBO J 2012;31(6):1405−26.

[27] Ribel-Madsen R, Fraga MF, Jacobsen S, Bork-Jensen J, Lara E, Calvanese V, et al. Genome-wide analysis of DNA methylation differences in muscle and fat from monozygotic twins discordant for type 2 diabetes. PLoS One 2012;7(12):e51302.

[28] Bibikova M, Barnes B, Tsan C, Ho V, Klotzle B, Le JM, et al. High density DNA methylation array with single CpG site resolution. Genomics 2011;98(4):288−95.

[29] Dayeh T, Volkov P, Salo S, Hall E, Nilsson E, Olsson AH, et al. Genome-wide DNA methylation analysis of human pancreatic islets from type 2 diabetic and non-diabetic donors identifies candidate genes that influence insulin secretion. PLoS Genet 2014;10(3): e1004160.

[30] Nilsson E, Jansson PA, Perfilyev A, Volkov P, Pedersen M, Svensson MK, et al. Altered DNA methylation and differential expression of genes influencing metabolism and inflammation in adipose tissue from subjects with type 2 diabetes. Diabetes 2014;63 (9):2962−76.

[31] Grundberg E, Meduri E, Sandling JK, Hedman AK, Keildson S, Buil A, et al. Global analysis of DNA methylation variation in adipose tissue from twins reveals links to disease-associated variants in distal regulatory elements. Am J Hum Genet 2013;93(5):876−90.

[32] Nitert MD, Dayeh T, Volkov P, Elgzyri T, Hall E, Nilsson E, et al. Impact of an exercise intervention on DNA methylation in skeletal muscle from first-degree relatives of patients with type 2 diabetes. Diabetes 2012;61(12):3322−32.

[33] Toperoff G, Aran D, Kark JD, Rosenberg M, Dubnikov T, Nissan B, et al. Genome-wide survey reveals predisposing diabetes type 2-related DNA methylation variations in human peripheral blood. Hum Mol Genet 2012;21(2):371−83.

[34] Parker SC, Stitzel ML, Taylor DL, Orozco JM, Erdos MR, Akiyama JA, et al. Chromatin stretch enhancer states drive cell-specific gene regulation and harbor human disease risk variants. Proc Natl Acad Sci USA 2013;110(44):17921−6.

[35] Pasquali L, Gaulton KJ, Rodriguez-Segui SA, Mularoni L, Miguel-Escalada I, Akerman I, et al. Pancreatic islet enhancer clusters enriched in type 2 diabetes risk-associated variants. Nat Genet 2014;46(2):136−43.

[36] Stitzel ML, Sethupathy P, Pearson DS, Chines PS, Song L, Erdos MR, et al. Global epigenomic analysis of primary human pancreatic islets provides insights into type 2 diabetes susceptibility loci. Cell Metab 2010;12(5):443−55.

[37] Bhandare R, Schug J, Le Lay J, Fox A, Smirnova O, Liu C, et al. Genome-wide analysis of histone modifications in human pancreatic islets. Genome Res 2010;20(4):428−33.

[38] Bramswig NC, Everett LJ, Schug J, Dorrell C, Liu C, Luo Y, et al. Epigenomic plasticity enables human pancreatic alpha to beta cell reprogramming. J Clin Invest 2013;123 (3):1275−84.

[39] Malmgren S, Spegel P, Danielsson AP, Nagorny CL, Andersson LE, Dekker Nitert M, et al. Coordinate changes in histone modifications, mRNA levels and metabolite pro-files in clonal INS-1 832/13 beta-cells accompany functional adaptations to lipotoxicity. J Biol Chem 2013;288(17):11973−87.

[40] Kameswaran V, Bramswig NC, McKenna LB, Penn M, Schug J, Hand NJ, et al. Epigenetic regulation of the DLK1-MEG3 microRNA cluster in human type 2 diabetic islets. Cell Metab 2014;19(1):135−45.

[41] Hall E, Volkov E, Dayeh T, Esguerra J, Salö S, Eliasson L, et al. Sex differences in the genome-wide DNA methylation pattern and impact on gene expression, microRNA levels and insulin secretion in human pancreatic islets. Genome Biol 2014;15(12):522.

[42] Dayeh TA, Olsson AH, Volkov P, Almgren P, Ronn T, Ling C. Identification of CpG-SNPs associated with type 2 diabetes and differential DNA methylation in human pancreatic islets. Diabetologia 2013;56(5):1036−46.

[43] Olsson AH, Volkov P, Bacos K, Dayeh T, Hall E, Nilsson EA, et al. Genome-wide associations between genetic and epigenetic variation influence mRNA expression and insulin secretion in human pancreatic islets. PLoS Genet 2014;10(11):e1004735.

[44] Franks PW, Ling C. Epigenetics and obesity: the devil is in the details. BMC Med 2010;8:88.

[45] Milagro FI, Campion J, Cordero P, Goyenechea E, Gomez-Uriz AM, Abete I, et al. A dual epigenomic approach for the search of obesity biomarkers: DNA methylation in relation to diet-induced weight loss. FASEB J 2011;25(4):1378−89.

[46] Horvath S, Erhart W, Brosch M, Ammerpohl O, von Schonfels W, Ahrens M, et al. Obesity accelerates epigenetic aging of human liver. Proc Natl Acad Sci USA 2014;111 (43):15538−43.

[47]   Ahrens M, Ammerpohl O, von Schonfels W, Kolarova J, Bens S, Itzel T, et al. DNA methylation analysis in nonalcoholic fatty liver disease suggests distinct disease-specific and remodeling signatures after bariatric surgery. Cell Metab 2013;18(2):296−302.

[48]   Dick KJ, Nelson CP, Tsaprouni L, Sandling JK, Aissi D, Wahl S, et al. DNA methylation and body-mass index: a genome-wide analysis. Lancet 2014;383(9933):1990−8.

[49]   Knowler WC, Barrett-Connor E, Fowler SE, Hamman RF, Lachin JM, Walker EA, et al. Reduction in the incidence of type 2 diabetes with lifestyle intervention or metformin. N Engl J Med 2002;346(6):393−403.

[50]   Tuomilehto J, Lindstrom J, Eriksson JG, Valle TT, Hamalainen H, Ilanne-Parikka P, et al. Prevention of type 2 diabetes mellitus by changes in lifestyle among subjects with impaired glucose tolerance. N Engl J Med 2001;344(18):1343−50.

[51]   Alibegovic AC, Hojbjerre L, Sonne MP, van Hall G, Stallknecht B, Dela F, et al. Impact of 9 days of bed rest on hepatic and peripheral insulin action, insulin secretion, and whole-body lipolysis in healthy young male offspring of patients with type 2 diabetes. Diabetes 2009;58(12):2749−56.

[52]   Alibegovic AC, Sonne MP, Hojbjerre L, Bork-Jensen J, Jacobsen S, Nilsson E, et al. Insulin resistance induced by physical inactivity is associated with multiple transcriptional changes in skeletal muscle in young men. Am J Physiol Endocrinol Metab 2010;299(5): E752−63.

[53]   Brons C, Jacobsen S, Nilsson E, Ronn T, Jensen CB, Storgaard H, et al. Deoxyribonucleic acid methylation and gene expression of PPARGC1A in human muscle is influenced by high-fat overfeeding in a birth-weight-dependent manner. J Clin Endocrinol Metab 2010;95(6):3048−56.

[54]   Brons C, Jacobsen S, Hiscock N, White A, Nilsson E, Dunger D, et al. Effects of high-fat overfeeding on mitochondrial function, glucose and fat metabolism, and adipokine levels in low-birth-weight subjects. Am J Physiol Endocrinol Metab 2012;302(1):E43−51.

[55]   Ronn T, Volkov P, Davegardh C, Dayeh T, Hall E, Olsson AH, et al. A six months exercise intervention influences the genome-wide DNA methylation pattern in human adipose tissue. PLoS Genet 2013;9(6):e1003572.

[56]   Ronn T, Volkov P, Tornberg A, Elgzyri T, Hansson O, Eriksson KF, et al. Extensive changes in the transcriptional profile of human adipose tissue including genes involved in oxidative phosphorylation after a 6-month exercise intervention. Acta Physiol (Oxf) 2014;211(1):188−200.

[57]   Jacobsen SC, Gillberg L, Bork-Jensen J, Ribel-Madsen R, Lara E, Calvanese V, et al. Young men with low birthweight exhibit decreased plasticity of genome-wide muscle DNA methylation by high-fat overfeeding. Diabetologia 2014;57(6):1154−8.

[58]   Jacobsen SC, Brons C, Bork-Jensen J, Ribel-Madsen R, Yang B, Lara E, et al. Effects of short-term high-fat overfeeding on genome-wide DNA methylation in the skeletal muscle of healthy young men. Diabetologia 2012;55(12):3341−9.

[59]   Park JH, Stoffers DA, Nicholls RD, Simmons RA. Development of type 2 diabetes following intrauterine growth retardation in rats is associated with progressive epigenetic silencing of Pdx1. J Clin Invest 2008;118(6):2316−24.

[60]   Stoffers DA, Ferrer J, Clarke WL, Habener JF. Early-onset type-II diabetes mellitus (MODY4) linked to IPF1. Nat Genet 1997;17(2):138−9.

[61]   Ahlgren U, Jonsson J, Jonsson L, Simu K, Edlund H. beta-cell-specific inactivation of the mouse Ipf1/Pdx1 gene results in loss of the beta-cell phenotype and maturity onset diabetes. Genes Dev 1998;12(12):1763−8.

[62] Sandovici I, Smith NH, Nitert MD, Ackers-Johnson M, Uribe-Lewis S, Ito Y, et al. Maternal diet and aging alter the epigenetic control of a promoter-enhancer interaction at the Hnf4a gene in rat pancreatic islets. Proc Natl Acad Sci USA 2011;108(13):5449−54.

[63] Ng SF, Lin RC, Laybutt DR, Barres R, Owens JA, Morris MJ. Chronic high-fat diet in fathers programs beta-cell dysfunction in female rat offspring. Nature 2010;467(7318): 963−6.

[64] Barker DJ. The developmental origins of insulin resistance. Horm Res 2005;64(Suppl 3):2−7.

[65] Hales CN, Barker DJ, Clark PM, Cox LJ, Fall C, Osmond C, et al. Fetal and infant growth and impaired glucose tolerance at age 64. BMJ 1991;303(6809):1019−22.

[66] Barker DJ. The intrauterine environment and adult cardiovascular disease. Ciba Found Symp 1991;156:3−10, discussion 6.

[67] Hales CN, Barker DJ. Type 2 (non-insulin-dependent) diabetes mellitus: the thrifty phenotype hypothesis. Diabetologia 1992;35(7):595−601.

[68] Heijmans BT, Tobi EW, Stein AD, Putter H, Blauw GJ, Susser ES, et al. Persistent epigenetic differences associated with prenatal exposure to famine in humans. Proc Natl Acad Sci USA 2008;105(44):17046−9.

[69] Tobi EW, Lumey LH, Talens RP, Kremer D, Putter H, Stein AD, et al. DNA methylation differences after exposure to prenatal famine are common and timing- and sex-specific. Hum Mol Genet 2009.

Chapter 13

Clinical Applications of Epigenomics

Michael A. McDevitt[1,2]
[1]*Division of Hematology, Johns Hopkins University School of Medicine, Baltimore, MD, USA,*
[2]*Division of Hematological Malignancy, Johns Hopkins University School of Medicine, Baltimore, MD, USA*

Chapter Outline

M. Fraga & A.F. Fernandez (Eds): Epigenomics in Health and Disease.
DOI: http://dx.doi.org/10.1016/B978-0-12-800140-0.00013-3
© 2016 Elsevier Inc. All rights reserved.

13.1 INTRODUCTION

As described in previous chapters of this book, epigenetic programs are critical to normal cellular development and differentiation programs. These facts are supported by evolving knowledge of epigenetic regulatory controls that go awry in disease states, including cancer, perhaps the most vigorously studied to date. If we consider that the field of epigenomics is at its scientific infancy, then by that analogy, clinical applications are at a neonatal developmental stage. The quest for knowledge of the complete set of epigenetic modifications of the cellular material of normal and diseased tissues of an individual, provides the same anticipation for breakthrough diagnostic understanding and new therapeutic opportunities projected for genetic advances when automated, large capacity genomic technologies first came on line. For this chapter, I primarily utilize investigations of cancer epigenomes, particularly in the myeloid malignancies to illustrate the potential clinical applications of epigenomics. Not far behind however, will be applications influencing potential discovery and treatment of immunological, neurological, obstetrical, and metabolic conditions to name just a few. Selected examples of potential future developments are outlined in this chapter.

13.2 CANCER IS AN EPIGENETIC DISEASE

Epigenetic changes result in alterations in gene expression or phenotype through mechanisms other than changes in the DNA sequence. Epigenetic mechanisms regulate normal development, and evolving data are elucidating roles of epigenetic factors in the control of stem cell function, lineage selection, and hematopoietic cell differentiation and other normal cellular functions. Normal blood cell production coincides with widespread developmental changes in genome-wide patterns of DNA methylation [1–3]. Similar widespread changes have been recently reported in studies of embryonic stem (ES) cell and other differentiation systems as well [4]. Histone modifications, including the histone code [5], provide additional dynamic epigenetic regulation, and microRNAs (miRNAs) and other RNAs provide an even additional layer of potential epigenetic regulation of gene expression and thus cellular physiology. Cancer cells under selection to survive and proliferate alter these normal developmental and cellular processes at each of these epigenetic levels to their advantage. Historically, the relative roles of genetic versus epigenetic abnormalities in the pathogenesis of human cancers have been debated. It is clear now that genetic factors and epigenetic factors cooperate at all stages of cancer development [6,7]. Tumor suppressor gene silencing in association with promoter hypermethylation in cancer cells is one particularly prominent mechanism and will be the first paradigm reviewed [8]. Based on potential sensitivity and specificity considerations and access to clinical samples (aberrant promoter methylation has been detected in urine, ejaculate, saliva, sputum, breast, and blood, and bone marrow), assays of DNA methylation have significant clinical application potential.

13.2.1 Methylation as Qualitative Disease Markers: Illuminating Disease Biology and Facilitating Accurate Diagnosis

13.2.1.1 Colorectal Cancer

Epigenetic studies of colorectal cancer (CRC) have provided many landmark discoveries increasing our knowledge of cancer biology and treatment. For example:

- CRC was one of the earliest cancers evaluated for epigenetic alterations.
- Investigations of CRC have identified general critical cancer development pathways.
- CRC has established that tumor suppressor gene silencing can occur through epigenetic mechanisms without mutation.
- CRC has provided a model of how epigenetic alterations can influence treatment response and drug resistance.
- CRC serves a model for clinical biomarker development.

CRC is the third most common and second most fatal malignancy worldwide, with a 6% lifetime risk [9]. Early biologic investigations of colorectal carcinogenesis have led to a more broad general understanding of the genetic basis of carcinogenesis. The classic adenoma progressing to invasive cancer pathway was proposed by Fearon and Vogelstein in 1990 [10]. The first experiments on DNA methylation in human cancer compared samples of human CRC with matched normal mucosa isolated from the same patients. Widespread hypomethylation involving approximately one-third of single-copy genes was observed [11]. Silencing of tumor suppressor genes through promoter CpG methylation is also well established in CRC pathogenesis. The biologic consequences of loss of gene function associated with a coding-region mutations or epigenetic silencing by promoter CpG island methylation appear to be similar in studied cases [12,13].

Among the numerous CRC genes that are hypermethylated and silenced are *hMLH1* and *O6-MGMT*, which encode DNA-repair proteins. When the function of these genes is lost, the cells are susceptible to mutations because of faulty DNA repair. When *O6-MGMT* function is lost, the cells have a diminished capacity to repair alkylation damage to the base guanosine, and they become susceptible to guanosine to adenine mutations. Such loss of function sensitizes cells to the effects of chemotherapy that depend on alkylating mechanisms. This has been studied in a number of clinical trials and is now being clinically utilized in conjunction with selected glioblastoma treatment planning [14].

Silencing of the *MLH1* gene leads to a form of genomic instability termed *microsatellite instability*. This has potential diagnostic and treatment implications, including potential sensitivity to chemotherapeutic agents utilized in therapies for CRC and other cancers [15,16]. *MLH1* is one of a number of commonly aberrantly hypermethylated and silenced genes in CRC [17]. Some are more typically methylated early in the disease process, and others later, including at the metastatic stage.

Epigenetic disease biomarkers that detect acquired aberrant tumor-associated methylation offer the potential to provide sensitive, noninvasive, cost-effective early detection of CRC, before tumors metastasize and become deadly. Commercially available DNA methylation test kits for cancer genes, including *VIM*, *SEPT9*, *SHOX2*, and *MGMT*, have been recently reviewed [18].

13.2.1.1.1 Gene Panels

Combinations of methylated markers with and without additional clinical parameters have been explored, with a goal of increased sensitivity and specificity [18]. In 2014, the U.S. Food and Drug Administration (FDA) approved a stool DNA test (Cologuard) to screen average-risk adults 50 years or older for CRC [19]. The stool test includes molecular assays for DNA mutation (KRAS) and methylation biomarkers (NDRG4 and BMP3 methylation) and a non-DNA immunochemical assay for human hemoglobin. In a clinical study, the test was 92.3% sensitive in detecting cases of CRC in asymptomatic average-risk persons but detected less than half of advanced precancerous lesions and produced a substantial number of false-positive results [20]. In early stage lung cancer, methylation of the promoter region of four genes in patients with stage I non−small cell lung cancer (NSCLC) treated with curative intent by means of surgery were associated with early disease recurrence [21]. A panel of three markers hypermethylated and hypomethylated in urine sediments accurately predicts bladder cancer recurrence [22]. See McDevitt [23] for review of additional examples.

13.2.1.1.2 Global (Epigenomic) Analyses for Disease Classification

The potential clinical application of epigenetic markers for diagnosis and treatment choices has been actively investigated in hematologic malignancies, in part because of the easy access to blood samples or bone marrow. Aberrant DNA methylation in the 5′ regulatory region of the calcitonin gene was reported in primary acute leukemia samples in 1987 [24]. Myeloid malignancies, including myelodysplastic syndrome (MDS), myeloprolifera-tive neoplasms (MPNs), and acute myelogenous leukemia (AML), are clonal stem cell malignancies with tremendous biologic and clinical heterogeneity at presentation. Current diagnosis relies on clinical variables, cell surface immunophenotyping, and histomorphologic pathological criteria. Even with this information, diagnostic assignment is often imprecise. Cytogenetic and mutational data can provide useful clonal markers and offer prognostic value but are typically not disease specific. Even within clinically defined AML subsets, there are significant variations in response to standard multidrug induction chemotherapy, ranging from complete remission to primary refrac-tory disease. Whole genome array-based gene expression profiling has been applied to this problem seeking to improve diagnosis, classification, and prognosis. Elegant studies by Valk et al. [25] showed that surveying the expression levels of thousands of genes with DNA microarray technology identify recurrent patterns. Some clustered had known diagnostic cytogenetic abnormalities, such as Inv [16], or specific mutations. Novel AML subtypes were identified by gene expression array profiling. More recently, a set of 24 prognostic genes using similar approaches was identified, and the prognostic impact of this gene signature was confirmed in two independent validation cohorts [26]. These studies are an example of an increasing number of reports across tumor types that seek to use RNA expression differences as biomarkers for diagnosis, prognosis, or treatment planning. There are signifi-cant practical limitations to such approaches, however, and epigenomic derived biomarkers may offer significant practical advantages.

As described above for colon cancer, DNA methylation assays offer a number of potential technical advantages relative to RNA expression analy-ses. Figueroa et al. [27] applied large-scale genome-wide DNA methylation profiling to a cohort of 344 AML patient samples. Clustering these patients by methylation data segregated the patients into 16 groups. Five of the groups identified define new AML subtypes that share no features with stan-dard genetic, cytogenetic, or histomorphologic features. In addition, the anal-ysis was successful in subdividing known genetically classified AML groups. Finally, this research group derived a 15-gene methylation classifier predic-tive of overall survival in an independent patient cohort. Other groups using separate technologies and samples have reported similar findings [28]. This is just one of now many examples where epigenomic profiling may improve the molecular profiling of complex human malignancies.

13.3 EPIGENOMICS AS A TOOL TO UNRAVEL CANCER MECHANISMS

A significant advantage of the application of epigenomic studies to poorly understood malignancies relates to the potential to unravel undiscovered can-cer mechanisms. Abnormal expression of the *EVI1* gene, as a result of inv(3) (q21q26.2)/t(3,3)(q21;q26.2) or through other mechanisms, is associated with very unfavorable AML outcomes through unknown mechanisms [29]. EVI1 is a sequence-specific transcriptional repressor protein. Lugthart studied potential aberrant epigenetic programming in EVI1 overexpressing AML cells [30] with a large-scale genome wide DNA methylation profiling experiment. A specific promoter DNA methylation signature was uncovered in the EVI1 overexpressing AMLs, suggesting that EVI1 may be directing aberrant promoter DNA methylation patterning as a key pathogenic mechanism.

Another recent example of the intertwining of genetic mutations and epigenetic mechanisms in AML relates to recurrent FLT3 and TET2 mutations. Shih et al. showed that combined mutations in Tet2 and Flt3 alter DNA methylation and gene expression to an extent not seen with either mutation alone [31]. Together, a fully penetrant, lethal AML in mice is observed, resistant to FLT3 pharmacologic inhibitors. The mechanism of the cooperative synergistic gain-of-function effects on DNA methylation and gene expression are currently unknown.

In summary, methylation profiling demonstrates the power of epigenomic investigations when applied to carefully collected and characterized clinical samples. When integrated with clinical, chromosomal, and clinical data, this approach offers the potential for improved risk stratification and outcome predictions. These studies have identified previously unrecognized AML disease subtypes. DNA methylation profiling has also been successful in other hematologic malignancies, for example, in chronic lymphocytic leukemia (CLL), multiple myeloma (MM), and splenic marginal zone lymphoma [32−34].

13.4 METHYLATION PROFILING IN OTHER CANCERS

A pivotal study demonstrating the power of epigenetic characterization of cancer phenotypes was reported by Noushmehr et al., who investigated glioblastoma tumors [35]. In this study, the investigators profiled promoter DNA methylation alterations in 272 glioblastoma tumors in the context of The Cancer Genome Atlas (TCGA). They found that a distinct subset of samples display concerted hypermethylation at a large number of loci, indicating the existence of a glioma-CpG island methylator phenotype (G-CIMP). The CIMP was first identified and has been most extensively studied in CRC but has also been identified in many other tumor types, including bladder, breast, endometrial, and gastric cancers, glioblastoma, and hepatocellular, lung, ovarian, pancreatic, renal cell, prostate, leukemia, and other cancers [36]. In the Noushmehr study, G-CIMP tumors were found to belong to a specific clinical subgroup and were tightly associated with somatic IDH1 mutations. Patients with G-CIMP tumors are younger at the time of diagnosis and experience significantly improved outcome. Thus, this epigenomic investigation has identified G-CIMP as a distinct subset of human gliomas on molecular and clinical grounds. Epigenomic studies are ideally suited to identify and characterize hypermethylator phenotypes. However, considerable variation exists on identifying cases with CIMP from study to study. Sanchez-Vega performed a comprehensive computational study of CIMP that reveals pan-cancer commonalities and tissue-specific differences underlying current hypermethylation of CpG islands across cancers. The results have the potential to refine the molecular subtypes of cancer into more homogeneous subgroups, with the potential to improved stratification for clinical trials and improved clinical outcomes [37].

Another study that demonstrates the power of TCGA cooperative group investigations relates to findings from the Gastric Cancer Team [38]. Gastric cancer is a leading cause of cancer deaths, but analysis of its molecular and clinical characteristics has been complicated by histologic heterogeneity and poor understanding of the underlying pathogenesis. A comprehensive molecular evaluation of 295 primary gastric adenocarcinomas was described as part of TCGA. On the basis of their findings, a novel molecular classification schema has been proposed, dividing gastric cancer into four subtypes, including, again, a subset with extreme DNA hypermethylation, and another with microsatellite unstable tumors. Identification of these subtypes provides a roadmap for patient stratification and novel trials of targeted therapies. A number of other tumor types have been studied or are under study by TCGA and other investigative groups. These results have the potential to ultimately revolutionize how we diagnose and classify malignant diseases and provide unique insights into the overlap of genomic copy number changes, genetic mutations, and epigenetic alterations.

13.5 METHYLATION MARKERS AS CLINICAL PREDICTORS OF DISEASE PROGRESSION AND THE POTENTIAL POWER OF QUANTITATIVE BIOMARKERS

Early analyses of CpG promoter methylation of single genes and multigene panels have identified an association of CpG promoter methylation and MDS disease progression in children and adults [23]. The promoter region of the cyclin-dependent kinase inhibitor p15(INK4B), which contains a CpG island that is hypermethylated in many hematologic malignancies, has been a case study. Cameron et al. showed that the entire CpG island region of p15 is largely devoid of methylation in normal lymphocytes, but methylation of varying density is found in primary acute leukemia samples [39]. Methylation density is generally conserved between the alleles from each sample, but marked heterogeneity for the specific CpG sites methylated has been observed. When patterns of methylation and expression were compared, the density of methylation within the CpG Island, and not any specific location, correlated best with transcriptional loss. Leukemias with methylation of approximately 40% of the CpG dinucleotides on each allele had complete gene silencing, with variable, but diminished, expression and less dense CpG island methylation. These results were an early model of the transcriptional silencing of a cell cycle regulatory tumor suppressor gene, in conjunction with aberrant hypermethylation, suggesting that silencing is best understood as an evolutionary process that involves progressively increasing methylation of the entire p15 CpG Island.

The application of epigenomic assays evaluating global methylation levels across a spectrum of early versus late disease states has provided additional insight into the biology of myeloid disease progression. Jiang et al. identified a significant increase in methylation in the transition from MDS to AML [40]. Using a different technical approach, Figueroa et al. evaluated the extent of promoter hypermethylation of 14,000 loci/promoters in patients with MDS and secondary AML compared with normal CD34 cells and *de novo* AML samples [41]. The patients with MDS and secondary AML displayed more extensive aberrant DNA methylation, involving thousands of genes. Investigation of the regions of aberrant methylation also provides clues to potential oncogenic pathways for these diseases (WNT, MAPK).

In summary, epigenomic investigation of quantitative in addition to qualitative measures of epigenetic alterations is likely to provide an additional level of biologic characterization and biomarker development. Further development, with a focus on epigenetic marker thresholds, disease specificity, and relationships among response versus resistance to therapies, will likely significantly advance the field.

13.6 GENETIC MUTATION IN EPIGENETIC REGULATORS

One clinical area where epigenetics and human disease study intersect are genetic diseases induced by germline mutation in epigenetic regulators. Disease-causing genes have been identified in enzymes that add or remove histone or DNA modifications, and those that remodel nucleosome arrays affecting gene expression patterns. Other mutations affect chromatin reader proteins and regulators of higher order chromatin structure, which also leads to unique disease states. These have been the subject of an excellent recent review [42]. Examples include MECP2 (Rett syndrome), EZH2 (Weaver syndrome), and CREBBP (Rubinstein-Taybi syndrome). In monogenic syndromes, the causative mutation can be diagnosed by mutation analysis from any clinically accessible tissue, such as blood lymphocytes or skin fibroblasts.

The heritable genetic diseases caused by mutations in epigenetic regulator genes present with a variety of clinical symptoms and pathologic characteristics. There are, however, common features and concepts. Many of the diseases linked to mutations in epigenetic regulators present with autosomal dominant or X-linked patterns of inheritance. Many of epigenetic regulator mutations are associated with cognitive defects in affected individuals in spite of the fact that the mutations are present in every cell of the body. These findings support the concept of a unique and important role for epigenetic regulation in normal brain development and function and suggest exploration of epigenetic biomarkers in the neurocognitive arena.

Another finding going back to the rationale for epigenetic investigations of cancer, relate to the frequency of mutations in epigenetic regulatory genes associated with developmental genetic disorders, which, when mutated in the germline configuration, are often also present as recurrent oncogenic or tumor suppressor mutations in the context of somatically acquired mutations in human cancer cells.

For example, Piazza et al. identified mutations of SETBP1 through whole-exome sequencing of samples of the mixed myelodysplastic/myelo-proliferative disorder atypical chronic myeloid leukemia (aCML) [43]. aCML shares clinical and laboratory features with CML, but it lacks the BCR-ABL1 fusion. Additional targeted sequencing identified SETBP1 in 24% of a larger aCML cohort and other myeloid malignancies. For the most part, the mutations were identical to changes seen in individuals with Schinzel-Giedion syndrome. A separate study by Makishima et al. [44] identified somatic SETBP1 mutations again matching germline mutations in Schinzel-Giedion syndrome in 17% of secondary acute myeloid leukemias (sAML), and 15% of chronic myelomonocytic leukemia cases. Somatic mutations of SETBP1 appear to cause a gain of function and are associated with a poor prognosis.

In fact, recent whole-genome sequencing of thousands of human cancers has uncovered an unexpectedly large number of mutations in genes that control the epigenome. Many of these recurrent mutations are in epigenetic writers, readers, erasers, and chromatin remodeling enzymes, including polycomb complex members [45]. These represent potential cancer drivers for both pediatric as well as adult cancers. Recent analysis of 633 genes in over 1000 pediatric tumors performed to explore the landscape of somatic mutations in epigenetic regulators by Huether et al. [46]. Results demonstrated a significant variation in the frequency of gene mutations across 21 different pediatric cancer subtypes. A particularly high frequency of mutations were identified in high-grade gliomas, T-cell ALL, and medulloblastoma, with a paucity of mutations seen in low-grade glioma and retinoblastoma. Taken together, these studies highlight the close collaboration between genetic and epigenetic pathways in cancer formation.

13.7   OTHER EPIGENETIC METHYLATION PATTERNS IN CANCER

13.7.1   Gene Body

Approximately 40% of human genes do not have bona fide CpG islands in their promoters [47], although there may be methylation of non-CpG Islands [48]. The transcribed regions of most genes are heavily methylated, and the level of methylation has been positively correlated with the level of expression [49]. Potential mechanisms to explain this correlation have included potential silencing of alternative promoters, retrotransposon elements, and other functional elements to maintain the efficiency of transcription. Yang et al. investigated gene body methylation after treatment with a methyltransferase inhibitor to investigate altered gene expression and potential therapeutic targets in cancer [50]. In this study, decitabine treatment not only reactivated genes but decreased the overexpression of other genes, many of which were involved in metabolic processes regulated by the *c-MYC* oncogene. Detailed studies of chromatin state and epigenetic chromatin marks, nucleosome positioning, and other epigenomic features identified in this study illustrate how detailed epigenomic studies have the potential to increase our understanding of cancer biology and to identify future therapeutic targets.

13.7.2   Hypomethylation Regions, LOCKs, and LADs

Most epigenetic investigations have generally focused on promoter CpG Island and other genomic loci hypermethylation. The earliest observation in cancer epigenetics, however, was identification of widespread hypomethylation of genes in cancer [11]. This observation has been confirmed with modern technologies, where large heterochromatin regions termed LOCKs (large organized chromatin lysine modifications) and LADs (lamin-associated domains) have been identified with whole-genome bisulfite sequencing studies [51,52]. When evaluated across a panel of a number of solid human cancers, these elements appear to be universal epigenetic alterations in human cancers. Careful correlations with gene expression profiles and additional investigations have resulted in the presentation of provocative general models of epigenetic dysregulation, which have important implications in cancer diagnostics, therapeutics, and general disease pathways [7].

Interestingly, within the classical myeloproliferative disorder subset of hematopoietic malignancies—essential thrombocytosis (ET), polycythemia vera (PV), primary myelofibrosis (PMF), and secondary myelofibrosis (SMF)—epigenomic studies have also been informative with regard to hyper versus hypomethylation patterns. A study utilizing the HELP (HpaII tiny fragment enriched by ligation-mediated polymerase chain reaction [LM-PCR]) assay was utilized to study genome-wide methylation in PV, essential thrombocytosis, and PMF samples compared with healthy controls [53]. PV and essential thrombocytosis were found to be characterized by aberrant promoter hypermethylation, whereas PMF is an epigenetically distinct subgroup characterized by both aberrant hypermethylation and hypomethylation. Aberrant hypomethylation in PMF was seen to occur in non-CpG island loci, showing further qualitative differences between the disease subgroups. Within the PMF subgroup, cases with ASXL1 disruptions formed an epigenetically distinct subgroup with relatively increased methylation. These results show epigenetic differences between PMF and PV/essential thrombocytosis and revealed methylomic signatures of ASXL1 and TET2 mutations. High-resolution methylome analysis also revealed widespread functional hypomethylation during normal adult human erythropoiesis [54].

13.7.3   Shores

Taking a comprehensive genome-wide approach examining normal tissues and CRCs with matched normal mucosa controls, Irizarry et al. showed that most methylation alterations in colon cancer occur not in promoters and CpG islands but in sequences up to 2 kb distant, which they termed "CpG island shores" [55]. CpG island shore methylation was strongly related to gene expression and was highly conserved in the mouse, discriminating tissue types. Strikingly, there was a notable overlap of the locations of colon cancer–related methylation changes with those that distinguished normal tissues. This group has hypothesized that epigenetic plasticity and mechanisms that are utilized in normal development and responses to tissue injury become constitutively activated in cancer, causing epigenetic heterogeneity, which leads to most of the classical cancer hallmarks. A "tissue-injury" pathway example is presented below in the context of nonmethylation epigenetic changes.

13.8   OTHER RECURRENT EPIGENOMIC PATTERNS IN CANCER (NONMETHYLATION)

13.8.1   Genome-Scale Epigenetic Reprogramming During Epithelial-to-Mesenchymal Transition

Epithelial-to-mesenchymal transition (EMT) is an extreme example of cell plasticity, which is utilized in normal development, injury repair, and cancer malignancy progression. McDonald et al. [56] studied this process and did not observe changes in DNA methylation. They did, however, find a global reduction of the heterochromatin mark H3 Lys9 dimethylation (H3K9Me2), an increase in euchromatin mark H3 Lys4 trimethylation (H3K4Me3), and an increase in the transcriptional mark H3 Lys36 trimethylation (H3K36Me3) during EMT-related epigenetic reprograming induced by transforming growth factor-β (TGF-β). These changes were largely dependent on the lysine-specific demethylase-1 (Lsd1), influencing EMT-driven cell migration and chemotherapy resistance. Genome-scale mapping showed that chromatin changes were mainly specific to large organized heterochromatin K9 modifications (LOCKs), suggesting that EMT is characterized by reprogramming of specific chromatin domains across the genome. These experiments are part of a series of studies that have led Feinberg et al. to propose a model in which the epigenome can modulate cellular plasticity in development and disease by regulating the effects of noise [57,58]. In this model, the epigenome facilitates phase transitions in development and mediates robustness during cell fate commitment. They hypothesize that distinct chromatin domains, which they identified as being dysregulated in diseases, including cancer, and remodeled during development, might underlie cellular plasticity more generally.

13.8.2   Polycomb Complexes and Bivalent Chromatin

Mechanisms by which other chromatin modifications play a role in cancer, including nucleosome remodeling and histone modifications in addition to DNA methylation, have been increasingly studied. These are interrelated and, again, involve developmental pathways. Epigenomic studies again of colon cancer have revealed that approximately 50% of genes with cancer-specific promoter CpG island hypermethylation are among the approximately 10% of genes that are controlled by the polycomb group complex (PcG) in ES cells [48]. PcG-mediated transcriptional repression, often in the setting of bivalent chromatin, appears to mediate a poised transcriptional state of CpG island promoters from genes in ES cells that are important for regulating lineage determination. A working model hypothesizes that a molecular progression during tumorigenesis starts with an abnormally expanding adult stem or progenitor cell compartment in which vulnerable genes with promoter CpG islands undergo quantitative replacement of flexible PcG-mediated gene silencing with the more stable silencing that is associated with DNA methylation [59,60].

Support for such concepts and rationale for developing clinically applied biomarkers include studies of the myeloid candidate chromosome 5 tumor suppressor gene *CTNNA1* [61]. The gene encoding a-catenin (*CTNNA1*) was found to be expressed at a much lower level in leukemia-initiating stem cells from individuals with AML or MDS with a 5q deletion than in individuals with MDS or AML lacking a 5q deletion or in normal hematopoietic stem cells. Analysis in a myeloid leukemia line with deletion of the 5q31 region showed that the *CTNNA1* promoter of the retained allele was suppressed by both methylation and histone deacetylation. In a separate study [62], CpG promoter methylation and repressive chromatin marks (H3K27me3) were identified in *CTNNA1* repressed AML cell lines and primary leukemia cells, with the most repressive state correlating with DNA methylation. These results suggest a progressive, acquired epigenetic inactivation of *CTNNA1*, including histone modifications as well as promoter CpG methylation, as a component of leukemia progression in patients with both 5q-myeloid and non-5q-myeloid malignancies.

13.8.3   Post-Translational Histone Modifications and Nucleosomes

Histones are chemically modified by different enzymes at external N-terminal and C-terminal tails as well as internally. A large number of post-translational modifications include acetylation, phosphorylation, methylation, carbonylation, and others, which set up the "histone code" [5]. The chemical modifications on the histones affect physicochemical properties of chromatin, thus activating or silencing genes, and/or are read by other proteins to bring about distinct downstream events.

Several recent studies have identified recurrent mutations in critical residues of the H3 histone tail in pediatric or young adults with aggressive glioma brain tumors [63,64]. It has been reported that the K27M mutant pediatric high-grade gliomas display a global decrease of the repressive post-translational histone modification H3K27me3 [65,66], which under physiologic conditions is mainly established by the H3K27-specific histone methyltransferase enhancer of zeste 2 (EZH2) within the polycomb repressive complex 2 (PRC2) [67]. Mechanistically, reduction of H3K27me3 levels is caused by an inhibitory effect of the K27M mutant H3.3 protein [66]. Bender et al. [68] showed that the mutations lead to aberrant recruitment of the PRC2 complex to K27M mutant H3.3 and enzymatic inhibition of the H3K27me3-establishing methyltransferase EZH2. By performing chromatin immunoprecipitation followed by next-generation sequencing (NGS) and whole-genome bisulfite sequencing in primary pHGGs, these authors showed that reduced H3K27me3 levels and DNA hypomethylation act in concert to activate aberrant gene expression programs in the K27M mutant tumors.

Histone acetylation is also dysregulated in the genetic disease Runstein-Taybi syndrome, and affected individuals have a mutation in *CREBBP* that encodes the CREB-binding protein with histone acetylase activity [69]. A number of recent discoveries have linked several neurobiologic disorders to genes whose products actively regulate histone acetylation. Disordered nucleosome remodeling represents an important potential epigenetic mechanism in human developmental and intellectual disability disorders [70]. Nucleosome remodeling is driven primarily through nucleosome remodeling complexes with specialized adenosine triphosphate (ATP)-dependent enzymes. These enzymes directly interact with the DNA or chromatin structure, as well as with histone subunits, to restructure the shape and organization of nucleosome positioning to ultimately regulate gene expression. Taken together, these reports indicate that further studies investigating nucleosome density and positioning will be important to unravel this important biologic layer of epigenetic regulation [71].

13.8.4   RNA-Mediated Epigenetic Regulation

The full spectrum of epigenetic silencing of tumor suppressor genes (TSGs) whose normal function is to inhibit normal cellular growth is unknown. DNA methylation may be a final end point, yet as described above, the early steps in the process remain enigmatic. One possible mechanism for gene regulation involves noncoding RNAs (ncRNAs), such as miRNA and antisense RNAs. Widespread sense−antisense transcripts have been systematically identified in mammalian cells. Yu et al. [72] evaluated the potential role of one antisense RNA in the silencing p15, a cyclin-dependent kinase inhibitor implicated in leukemia. They found an inverse relation between p15 anti-sense (p15AS) and p15 sense expression in leukemia. Functional studies suggested that the natural antisense RNA may be a trigger for heterochromatin formation and DNA methylation and TSG silencing in tumorigenesis.

13.8.5 miRNA

miRNAs are a class of small noncoding RNAs of 19−23 nucleotides that can positively (through silencing of transcriptional repressors) or negatively regulate gene expression and influence epigenetic mechanisms, including DNA methylation. It is estimated that as many as 30% of mammalian genes may be regulated by miRNAs [73]. miRNA profiles, similar to epigenomic DNA methylation arrays, have promising applications in terms of disease-specific diagnostic, prognostic, and therapeutic choices. There is increasing evidence for the direct roles of pathogenic miRNAs in human cancers. Loss of tumor suppressor type miRNAs or amplified onco-miRNAs through cancer genome deletions or amplifications have been described. For example, miRNAs constitute the first class of ncRNAs identified to have widely dysregulated expression in AML and associated with clinical features, including outcome [74,75]. High levels of mir-155 and miR-3151 expression have been associated with poor outcome in older patients with AML and normal cytogenetics [76,77]. Like chromatin patterns and mutations in some epigenetic regulators, several onco-miRNA expressions have been associated with chemotherapy drug resistance [78].

13.8.6 LncRNA

Long noncoding RNAs (lncRNAs) are transcripts longer than 200 nucleo-tides, located within the intergenic stretches or overlapping antisense transcripts of protein coding genes [79,80]. LncRNAs have numerous biologic roles, including imprinting and epigenetic regulation. The lncRNA HOX transcript antisense RNA (HOTAIR) has been found to be upregulated in multiple cancers, including colon cancer [81−83]. Garzon et al. examined expression of a large number of lncRNAs in 148 untreated older AML patients by using a custom microarray platform [84]. This study identified expression relationships of specific lncRNAs with specific mutations associated with myeloid malignancies in primary AML samples, including mutations in epigenetic regulators (ASXL1, DNMT3A, TET2, and IDH1). The group was able to develop an lncRNA prognosis score that was informative for complete response to standard AML induction therapy, disease-free survival, and overall survival. This study is an example of how epigenomic investigations can identify a small subset of biomarkers that provide potential insight into disease mechanisms and possible future treatment response and survival predictors.

In summary, the cancer epigenome is extraordinarily complex, with multilayered levels of epigenetic regulation. There is a critical need for future epigenomic studies to unravel this complexity and identify driver epigenetic changes, possible coordinate, combinatorial, synergistic, and antagonistic contributions to human disease states.

13.9 EPIGENOMICS AND EPIGENETIC THERAPY

Although somatically heritable, the reversibility of DNA methylation by pharmacologic interventions makes DNA methyltransferases an attractive therapeutic target. Increasing knowledge from epigenomic studies are rapidly providing additional novel potential epigenetic targets for therapeutic intervention. A complete review of epigenetic therapy is beyond the scope of this review, but epigenomic studies will continue to be critical to direct and optimize current approaches.

13.10 POTENTIAL APPLICATIONS OF EPIGENETIC AND GENETIC BIOMARKERS WITH EPIGENETIC THERAPIES

The azanucleotides azacitidine and decitabine have been shown to induce hematologic response and potentially prolong survival in higher-risk myelodysplastic syndromes. They are inhibitors of DNA methyltransferase-1 and induce DNA hypomethylation. Induction of apoptosis is also clinically relevant, in particular during the first treatment cycles, when cytopenia is a frequent side effect. Since the hypomethylating effect is reversible and the malignant clone has been shown to persist in most responding patients, few to many cycles (quite variable between patients) are necessary to achieve and maintain responses, whereas treatment interruption is associated with rapid relapse [85]. Methylation studies have shown global and gene-specific hypermethylation in myelodysplastic syndromes, but there is a poor relationship between the degree of demethylation following hypomethylating treatment and hematologic response [86−88].

Changes in azanucleotide metabolism genes and development of acquired drug resistance likely plays a significant role. Finally, we may not have applied the correct assays to identify key predictive epigenomic biomarkers of response to methyltransferase inhibitors (MTIs). In the future, methylation analysis that concentrates not only on promoters but also on gene bodies and intergenic regions may identify key genes in patients with the highest probability of response to azanucleotides and allow a patient-tailored approach. With the discovery of mutations of epigenetic factors in myelodysplastic syndromes and AML, where MTI therapy is utilized, opportunities to look at genetic mutation profiles in relation to clinical responses or resistance to MTI has been investigated.

13.10.1 TET2 Mutations and AZA Responses

Itzykson et al. investigated the impact of ten-eleven-translocation 2 (TET2) mutations on the clinical response rate to azacitidine in myelodysplastic syndromes and low-blast-count AMLs [89]. The impact of TET2 mutations on response to azacitidine (AZA) were evaluated in 86 MDS and AML cases with low blast counts. Fifteen percent of the patients were identified to have TET2 mutations. The response rate (including hematologic improvement) was significantly improved in patients with TET2 mutations relative to patients without TET2 mutations. Several additional clinical studies support a finding of increased sensitivity to hypomethylating agents in patients with TET2 mutations [90,91], which may reflect an ability to reverse the effects of TET2 mutations on DNA methylation in these cases.

A large scale epigenomic evaluation has been recently applied to this problem by Meldi et al. [92]. Chronic myelomonocytic leukemia (CMML) is a mixed myelodysplastic/myeloproliferative syndrome that is characterized by very frequent mutations in genes encoding epigenetic modifiers and aberrant DNA methylation [93]. DNA methyltransferase inhibitors (DMTIs) are also used to treat these disorders formerly classified as a subtype of myelodysplastic syndrome; however, similar to the results for myelodysplastic syndrome, patient response is highly variable, with few means to predict who might benefit. Baseline differences in mutations, DNA methylation, and gene expression in 40 CMML patients who were responsive or resistant to decitabine were evaluated in order to develop a molecular means of predicting response at diagnosis. Although somatic mutations did not differentiate responders from nonresponders in this small sample cohort, 167 differentially methylated regions (DMRs) of DNA at baseline did. These DMRs were primarily localized to nonpromoter regions and overlapped with distal regulatory enhancers. Using the methylation profiles, the group developed an epigenetic classifier that accurately predicted DAC response at the time of diagnosis.

The MTIs are FDA approved for MDS treatment. They have been tested in early-phase clinical trials in other cancers. Azacitidine and decitabine have been used to treat patients with lymphoma, CLL, and myeloproliferative disorders but have shown minimal to modest clinical benefit at best [94–96]. It is possible that improvements in the optimal dose and schedule, and combination therapy with pretherapy selection of patients with unique disease parameters could improve responses. For example, the use of decitabine in patients with splenic marginal zone lymphoma, who were identified to have a high methylation profile by a recent epigenomic study [97], may offer hope for good clinical outcomes.

13.10.2 MTIs and Solid Tumors

MTIs applied at low doses, which preserve on-target effects and minimize off-target toxicity, offer some promise in the treatment of solid tumors [98].

Of particular interest is the potential integration with conventional therapies, allowing potential reversal of drug resistance mechanisms and increasing therapeutic sensitivity [99]. In preclinical studies, low-dose azacitidine has been shown to alter many key pathways associated with tumorigenesis, but a dominant finding is the upregulation of immunomodulatory pathways [100,101]. It is possible that such treatments could sensitize the tumors and allow for more effective application of cancer immunotherapies. Induction of the expression of immune regulatory molecules, such as PD1 and PDL1, might make immune checkpoint inhibitor therapies more effective and potentially even curative for some patients [100,101].

Similarly, the development of novel dosing strategies or combination with other epigenetic therapies holds promise to exploit the activities of MTI compounds for an improved therapeutic effect in hematologic cancers or solid tumors [102,103]. There are many scientific and clinical rationales for combining epigenetic therapeutics with drug combinations. Current MTI and histone deacetylase inhibitors (HDACs) as single agents generally do not have a high rate of complete responses and do not offer a curative approach. There is considerable preclinical support for the synergy of DNA methylation and HDAC inhibitors [104]. Clinical trials of combination studies are not yet mature, but evolving data indicate that significant challenges remain in translating the preclinical data to clinically effective therapies [105]. Correlative epigenomic investigations will be critical to facilitating successful development of epigenetic therapy at all clinical developmental stages.

13.11   EPIGENOME-WIDE ASSOCIATION STUDIES FOR COMMON HUMAN DISEASES

Epigenetic changes are observed in many disease states, and chromatin, DNA, miRNA, and other molecules may be biomarkers, disease drivers, or therapeutic targets. Elucidating the nongenetic determinants of complex human diseases represents a particularly critical biomedical challenge, one that epigenomics can help solve.

Genome-wide association studies (GWAS) have identified genetic loci that have associations or are causative in human diseases. Sensitivity is generally a problem, and most of GWAS hits do not affect protein coding sequences. Understanding disease mechanisms from the results is often limited as well. A powerful integrated genetic and epigenetic mapping study evaluating causal variants in a number of different autoimmune diseases was recently reported by Kai-How Farh et al. [106]. Using this innovative and integrative approach, the group found that 90% of disease associated variants are noncoding, with approximately 60% mapping to immune-cell enhancers, many of which gain histone acetylation and transcribe enhancer-associated RNA upon immune stimulation. This study demonstrated the potential power of integrated genetic and epigenomic studies across the entire human disease spectrum, and the results have broad implications for understanding disease biology and treatment. The challenges and promises of epigenome-wide association studies (EWAS) studies have been recently reviewed [107].

Mouse modeling, where available, provides considerable potential advantages to discovery approaches evaluating the contribution of variation in the methylation levels contributing to complex traits. Orozco et al. [108] examined the heritable epigenetic factors can contribute to complex disease etiology of complex traits that are precursors to heart disease, diabetes, and osteoporosis. They profiled DNA methylation in the liver of 90 inbred mouse strains and looked at genome-wide expression levels, proteomics, metabolomics, and 68 clinical traits as part of their EWAS. They found epigenetic associations with numerous clinical traits, including bone density, insulin resistance, expression, and protein and metabolite levels. Their results indicate that natural variation in methylation levels can contribute to the etiology of complex clinical traits.

The study by Multhaup et al. [109] is another example of a study that utilizes the power of complementing conventional human genetic studies with cross-species epigenomics and clinical epidemiology. The group explored the epigenetics of type 2 diabetes. First, they performed dietary manipulation of genetically homogeneous mice and identified differentially DNA-methylated genomic regions. Next, they searched for epigenetic conservation in humans in adipose samples from lean patients and from obese patients before and after Roux-en-Y gastric bypass, identifying regions where both the location and direction of methylation change were conserved. Functional analysis of genes associated with these regions revealed four genes with roles in insulin resistance, demonstrating the potential general utility of this approach.

These and related studies are examples of ongoing work in the field and illustrate the potential for epigenomic studies to improve our understanding and clinical approach to complex human diseases and possible transgenerational phenotypes and effects [110].

13.12 SUMMARY AND FUTURE DIRECTIONS RELATED TO CLINICAL APPLICATIONS OF EPIGENOMICS

Epigenomic studies have great potential for advancing preclinical knowledge and impacting clinical outcomes. Epigenomics is an evolving field, and a considerable amount of work remains to be done. With continued technologic and scientific developments, many goals may be achievable in a relatively short time frame. Epigenomics will be utilized as a discovery tool to better define the epigenetic contributions to the pathologic basis of human disease at all levels, including:

- Early stage disease
  - Identification of potential disease predisposition markers, better definition of pre-disease, and early disease states

- Exploring the influence of and interaction with environment
- Interaction with genetic factors
- Potential Example: "...accurately separating patients with early stage myelodysplastic syndrome with modest cytopenias versus those with nonclonal reactive or infectious processes"
- Classical disease
  - Generation of epigenetic biomarkers that improve diagnostic precision and staging and
  - Direct choice of and monitor response/resistance to therapeutic interventions
  - Minimal residual disease monitoring
  - Potential Example: "...identifying an epigenetic subset of a cancer diagnosis with clinical referral for epigenetic-directed therapy"
- Disease acceleration
  - Improved understanding of clonal selection and tumor heterogeneity
  - Development of disease acceleration biomarkers with improved prognosis and alternate treatment plans
  - Early detection of metastatic disease
  - Potential Example: "...identification of a patient classically diagnosed as having early stage myelodysplastic syndrome, but with epigenetic markers demonstrating an aggressive/poor prognosis disease that would benefit from an early stem cell transplant consultation"

At all levels, epigenomics will facilitate the identification of unique targets for pharmacologic and other therapeutic interventions. In combination with genetic studies, including sequencing, and GWAS studies, we can anticipate improved identification of common disease predisposition determinants and pathways, with a better understanding of environmental causes and contributions to disease and potential "epigenetic memory." Preclinical discoveries will need prospective testing in carefully conducted clinical trials. Having access to high-quality patient sample banks annotated for clinical determinants and clinical outcomes will be critical for epigenomic biomarker development.

REFERENCES

[1] Broske AM, Vockentanz L, Kharazi S, Huska MR, Mancini E, Scheller M, et al. DNA methylation protects hematopoietic stem cell multipotency from myeloerythroid restriction. Nat Genet 2009;41:1207–15.

[2] Trowbridge JJ, Snow JW, Kim J, Orkin SH. DNA methyltransferase 1 is essential for and uniquely regulates hematopoietic stem and progenitor cells. Cell Stem Cell 2009;5: 442–9.

[3] Ji H, Ehrlich LI, Seita J, et al. Comprehensive methylome map of lineage commitment from haematopoietic progenitors. Nature 2010;467:338–42.

[4] Dixon JR, Jung I, Selvaraj S, et al. Chromatin architecture reorganization during stem cell differentiation. Nature 2010;331:331–6.

[5] Strahl BD, Allis CD. The language of covalent histone modifications. Nature 2000;403:41–5.

[6] Baylin SB, Jones PA. A decade of exploring the cancer epigenome - biological and translational implications. Nat Rev Cancer 2011;11(10):726–34.

[7] Timp W, Feinberg AP. Cancer as a dysregulated epigenome allowing cellular growth advantage at the expense of the host. Nat Rev Cancer 2013;13(7):497–510.

[8] Herman JG, Baylin SB. Gene silencing in cancer in association with promoter hypermethylation. N Engl J Med 2003;349(21):2042–54.

[9] Jemal A, Bray F, Center MM, et al. Global cancer statistics. CA Cancer J Clin 2011;61 (2):69–90.

[10] Fearon ER, Vogelstein B. A genetic model for colorectal tumorigenesis. Cell 1990;61:759–67.

[11] Feinberg AP, Vogelstein B. Hypomethylation distinguishes genes of some human cancers from their normal counterparts. Nature 1983;301:89–92.

[12] Jones PA, Baylin SB. The fundamental role of epigenetic events in cancer. Nat Rev Genet 2002;3:415–28.

[13] Jones PA, Laird PW. Cancer epigenetics comes of age. Nat Genet 1999;21:163–7.

[14] Wick W, Weller M, van den Bent M, Sanson M, Weiler M, von Deimling A, et al. MGMT testing--the challenges for biomarker-based glioma treatment. Nat Rev Neurol 2014;10(7):372–85.

[15] Newton K, Jorgensen NM, Wallace AJ, et al. Tumour MLH1 promoter region methylation testing is an effective prescreen for Lynch Syndrome (HNPCC). J Med Genet 2014;51:789–96.

[16] Carethers JM, Chauhan DP, Fink D, et al. Mismatch repair proficiency and *in vitro* response to 5 flurouracil. Gastroenterology 1999;117:123–31.

[17] Lao VV, Grady WM. Epigenetics and colorectal cancer. Nat Rev Gastroenterol Hepatol 2011;8:686–700.

[18] Mikeska T, Craig JM. DNA methylation biomarkers: cancer and beyond. Genes 2014;5:821–64.

[19] [No authors listed]. A stool DNA test (Cologuard) for colorectal cancer screening. JAMA 2014;312(23):2566.

[20] Imperiale TF, Ransohoff DF, Itzkowitz SH, Levin TR, Lavin P, Lidgard GP, et al. Multitarget stool DNA testing for colorectal-cancer screening. N Engl J Med 2014;370 (14):1287–97.

[21] Brock M, Hooker C, Ota-Machida E, et al. DNA methylation markers and early recurrence in stage I lung cancer. NEJM 2008;358(11):1118–28.

[22] Su SF, de Castro Abreu AL, Chihara Y, Tsai Y, Andreu-Vieyra C, Daneshmand S, et al. A panel of three markers hyper- and hypomethylated in urine sediments accurately predicts bladder cancer recurrence. Clin Can Res 2014;20(7):1978–89.

[23] McDevitt MA. Clinical applications of epigenetic markers and epigenetic profiling in myeloid malignancies. Semin Oncol 2012;39(1):109–22.

[24] Baylin SB, Fearon ER, Vogelstein B, de Bustros A, Sharkis SJ, Burke PJ, et al. Hypermethylation of the 5′ region of the calcitonin gene is a property of human lymphoid and acute myeloid malignancies. Blood 1987;70(2):412–17.

[25] Bullinger L, Valk PJ. Gene expression profiling in acute myeloid leukemia. J Clin Oncol 2005;23(26):6296–305.

[26] Li Z, Herold T, He C, Valk PJ, Chen P, Jurinovic V, et al. Identification of a 24-gene prognostic signature that improves the European LeukemiaNet risk classification of acute myeloid leukemia: an international collaborative study. J Clin Oncol 2013;31(9):1172–81.

[27] Figueroa ME, Lugthart S, Li Y, Erpelinck-Verschueren C, Deng X, Christos PJ, et al. DNA methylation signatures identify biologically distinct subtypes in acute myeloid leukemia. Cancer Cell 2010;17(1):13−27.

[28] Bullinger L, Ehrich M, Döhner K, Schlenk RF, Döhner H, Nelson MR, et al. Quantitative DNA methylation predicts survival in adult acute myeloid leukemia. Blood 2010;115(3):636−42.

[29] Lugthart S, Gröschel S, Beverloo HB, Kayser S, Valk PJ, van Zelderen-Bhola SL, et al. Clinical, molecular, and prognostic significance of WHO type inv(3)(q21q26.2)/t(3;3) (q21;q26.2) and various other 3q abnormalities in acute myeloid leukemia. J Clin Oncol 2010;28(24):3890−8.

[30] Lugthart S, Figueroa ME, Bindels E, Skrabanek L, Valk PJ, Li Y, et al. Aberrant DNA hypermethylation signature in acute myeloid leukemia directed by EVI1. Blood 2011;117 (1):234−41.

[31] Shih AH, Jiang Y, Meydan C, Shank K, Pandey S, Barreyro L, et al. Mutational cooperativity linked to combinatorial epigenetic gain of function in acute myeloid leukemia. Cancer Cell 2015;27(4):502−15.

[32] Agirre X, Castellano G, Pascual M, Heath S, Kulis M, Segura V, et al. Whole-epigenome analysis in multiple myeloma reveals DNA hypermethylation of B cell-specific enhancers. Genome Res 2015;25(4):478−87.

[33] Landau DA, Clement K, Ziller MJ, Boyle P, Fan J, Gu H, et al. Locally disordered methylation forms the basis of intratumor methylome variation in chronic lymphocytic leukemia. Cancer Cell 2014;26(6):813−25.

[34] Arribas AJ, Rinaldi A, Mensah AA, et al. DNA methylation profiling identifies two splenic marginal zone lymphoma subgroups. Blood 2015;125(16):2530−43.

[35] Noushmehr H, Weisenberger DJ, Diefes K, Phillips HS, Pujara K, Berman BP, et al. Identification of a CpG island methylator phenotype that defines a distinct subgroup of glioma,; Cancer Genome Atlas Research Network. Cancer Cell 2010;17(5):510−22.

[36] Hughes LA, Melotte V, de Schrijver J, de Maat M, Smit VT, Bovée JV, et al. The CpG island methylator phenotype: what's in a name? Cancer Res 2013;73(19):5858−68.

[37] Sánchez-Vega F, Gotea V, Margolin G, Elnitski L. Pan-cancer stratification of solid human epithelial tumors and cancer cell lines reveals commonalities and tissue-specific features of the CpG island methylator phenotype. Epigenetics Chromatin 2015;8(14):1−24.

[38] Bass AJ, Thorsson V, Shmulevich I, et al. Comprehensive molecular characterization of gastric adenocarcinoma, Cancer Genome Atlas Research Network. Nature 2014;513 (7517):202−9.

[39] Cameron EE, Baylin SB, Herman JG. p15(INK4B) CpG island methylation in primary acute leukemia is heterogeneous and suggests density as a critical factor for transcriptional silencing. Blood 1999;94(7):2445−51.

[40] Jiang Y, Dunbar A, Gondek LP, Mohan S, Rataul M, O'Keefe C, et al. Aberrant DNA methylation is a dominant mechanism in MDS progression to AML. Blood 2009;113(6):1315−25.

[41] Figueroa ME, Skrabanek L, Li Y, Jiemjit A, Fandy TE, Paietta E, et al. MDS and secondary AML display unique patterns and abundance of aberrant DNA methylation. Blood 2009;114(16):3448−58.

[42] Brookes E, Shi Y. Diverse epigenetic mechanisms of human disease. Annu Rev Genet 2014;48:237−68.

[43] Piazza R, Valletta S, Winkelmann N, Redaelli S, Spinelli R, Pirola A, et al. Recurrent SETBP1 mutations in atypical chronic myeloid leukemia. Nat Genet 2013;45(1):18−24.

[44] Makishima H, Yoshida K, Nguyen N, Przychodzen B, Sanada M, Okuno Y, et al. Somatic SETBP1 mutations in myeloid malignancies. Nat Genet 2013;45(8):942−6.

[45] Plass C, Pfister SM, Lindroth AM, Bogatyrova O, Claus R, Lichter P. Mutations in regulators of the epigenome and their connections to global chromatin patterns in cancer. Nat Rev Genet 2013;14(11):765−80.

[46] Huether R, Dong L, Chen X, Wu G, Parker M, Wei L, et al. The landscape of somatic mutations in epigenetic regulators across 1,000 paediatric cancer genomes. Nat Commun 2014;5(3630):1−7.

[47] Takai D, Jones PA. Comprehensive analysis of CpG islands in human chromosomes 21 and 22. Proc Natl Acad Sci USA 2002;99(6):3740−5.

[48] Jones PA, Baylin SB. The epigenomics of cancer. Cell 2007;128(4):683−92.

[49] Bender CM, Gonzalgo ML, Gonzales FA, Nguyen CT, Robertson KD, Jones PA. Roles of cell division and gene transcription in the methylation of CpG islands. Mol Cell Biol 1999;19(10):6690−8.

[50] Yang X, Han H, De Carvalho DD, et al. Gene body methylation can alter gene expression and is a therapeutic target in cancer. Cancer Cell 2014;26:577−90.

[51] Hansen KD, Timp W, Bravo HC, Sabunciyan S, Langmead B, McDonald OG, et al. Increased methylation variation in epigenetic domains across cancer types. Nat Genet 2011;43(8):768−75.

[52] Berman BP, Weisenberger DJ, Aman JF, Hinoue T, Ramjan Z, Liu Y, et al. Regions of focal DNA hypermethylation and long-range hypomethylation in colorectal cancer coincide with nuclear lamina-associated domains. Nat Genet 2011;44(1):40−6.

[53] Nischal S, Bhattacharyya S, Christopeit M, Yu Y, Zhou L, Bhagat TD, et al. Methylome profiling reveals distinct alterations in phenotypic and mutational subgroups of myeloproliferative neoplasms. Cancer Res 2013;73(3):1076−85.

[54] Yu Y, Mo Y, Ebenezer D, Bhattacharyya S, Liu H, Sundaravel S, et al. High resolution methylome analysis reveals widespread functional hypomethylation during adult human erythropoiesis. J Biol Chem 2013;288(13):8805−14.

[55] Irizarry RA, Ladd-Acosta C, Wen B, Wu Z, Montano C, Onyango P, et al. The human colon cancer methylome shows similar hypo- and hypermethylation at conserved tissue-specific CpG island shores. Nat Genet 2009;41(2):178−86.

[56] McDonald OG, Wu H, Timp W, Doi A, Feinberg AP. Genome-scale epigenetic reprogramming during epithelial-to-mesenchymal transition. Nat Struct Mol Biol 2011;18(8):867−74.

[57] Pujadas E, Feinberg AP. Regulated noise in the epigenetic landscape of development and disease. Cell 2012;148(6):1123−31.

[58] Feinberg AP. Epigenetic stochasticity, nuclear structure and cancer: the implications for medicine. J Intern Med 2014;276(1):5−11.

[59] Cedar H, Bergman Y. Linking DNA methylation and histone modification: patterns and paradigms. Nat Rev Genet 2009;10:295−304.

[60] Ohm JE, Baylin SB. Stem cell chromatin patterns: an instructive mechanism for DNA hypermethylation? Cell Cycle 2007;6:1040−3.

[61] Liu TX, Becker MW, Jelinek J, Wu WS, Deng M, Mikhalkevich N, et al. Chromosome 5q deletion and epigenetic suppression of the gene encoding alpha-catenin (CTNNA1) in myeloid cell transformation. Nat Med 2007;13(1):78−83.

[62] Ye Y, McDevitt MA, Guo M, Zhang W, Galm O, Gore SD, et al. Progressive chromatin repression and promoter methylation of CTNNA1 associated with advanced myeloid malignancies. Cancer Res 2009;69(21):8482−90.

[63] Wu G, Broniscer A, McEachron TA, Lu C, Paugh BS, Becksfort J, et al. Somatic histone H3 alterations in pediatric diffuse intrinsic pontine gliomas and non-brainstem glioblastomas. Nat Genet 2012;44(3):251−3.

[64] Schwartzentruber J, Korshunov A, Liu XY, Jones DT, Pfaff E, Jacob K, et al. Driver mutations in histone H3.3 and chromatin remodelling genes in paediatric glioblastoma. Nature 2012;482(7384):226−31.

[65] Chan KM, Fang D, Gan H, Hashizume R, Yu C, Schroeder M, et al. The histone H3.3K27M mutation in pediatric glioma reprograms H3K27 methylation and gene expression. Genes Dev 2013;27:985−90.

[66] Lewis PW, Müller MM, Koletsky MS, Cordero F, Lin S, Banaszynski LA, et al. Inhibition of PRC2 activity by a gain-of-function H3 mutation found in pediatric glioblastoma. Science 2013;340:857−61.

[67] Margueron R, Reinberg D. The Polycomb complex PRC2 and its mark in life. Nature 2011;469:343−9.

[68] Bender S, Tang Y, Lindroth AM, Hovestadt V, Jones DT, Kool M, et al. Reduced H3K27me3 and DNA hypomethylation are major drivers of gene expression in K27M mutant pediatric high-grade gliomas. Cancer Cell 2013;24(5):660−72.

[69] Petrij F, Giles RH, Dauwerse HG, Saris JJ, Hennekam RC, Masuno M, et al. Rubinstein-Taybi syndrome caused by mutations in the transcriptional co-activator CBP. Nature 1995;376(6538):348−51.

[70] López AJ, Wood MA. Role of nucleosome remodeling in neurodevelopmental and intellectual disability disorders. Front Behav Neurosci 2015;9(100):1−10.

[71] Zentner GE, Henikoff S. Regulation of nucleosome dynamics by histone modifications. Nat Struct Mol Biol 2013;20(3):259−66.

[72] Yu W, Gius D, Onyango P, Muldoon-Jacobs K, Karp J, Feinberg AP, et al. Epigenetic silencing of tumour suppressor gene p15 by its antisense RNA. Nature 2008;451 (7175):202−6.

[73] Mack GS. MicroRNA gets down to business. Nat Biotechnol 2007;25(6):631−8.

[74] Marcucci G, Radmacher MD, Maharry K, Mrózek K, Ruppert AS, Paschka P, et al. MicroRNA expression in cytogenetically normal acute myeloid leukemia. N Engl J Med 2008;358(18):1919−28.

[75] Garzon R, Volinia S, Liu CG, Fernandez-Cymering C, Palumbo T, Pichiorri F, et al. MicroRNA signatures associated with cytogenetics and prognosis in acute myeloid leukemia. Blood 2008;111(6):3183−9.

[76] Marcucci G, Maharry KS, Metzeler KH, Volinia S, Wu YZ, Mrózek K, et al. Clinical role of microRNAs in cytogenetically normal acute myeloid leukemia: miR-155 upregulation independently identifies high-risk patients. J Clin Oncol 2013;31(17):2086−93.

[77] Eisfeld AK, Marcucci G, Maharry K, Schwind S, Radmacher MD, Nicolet D, et al. miR-3151 interplays with its host gene BAALC and independently affects outcome of patients with cytogenetically normal acute myeloid leukemia. Blood 2012;120(2):249−58.

[78] MacDonagh L, Gray SG, Finn SP, Cuffe S, O'Byrne KJ, Barr MP. The emerging role of microRNAs in resistance to lung cancer treatments. Cancer Treat Rev 2015;41(2):160−9.

[79] Kapranov P, Cheng J, Dike S, Nix DA, Duttagupta R, Willingham AT, et al. RNA maps reveal new RNA classes and a possible function for pervasive transcription. Science 2007;316(5830):1484−8.

[80] Derrien T, Johnson R, Bussotti G, Tanzer A, Djebali S, Tilgner H, et al. The Gencode v7 catalog of human long noncoding RNAs: analysis of their gene structure, evolution, and expression. Genome Res 2012;22(9):1775−89.

[81] Rinn JL, Kertesz M, Wang JK, Squazzo SL, Xu X, Brugmann SA, et al. Functional demarcation of active and silent chromatin domains in human HOX loci by noncoding RNAs. Cell 2007;129(7):1311−23.

[82] Gupta RA, Shah N, Wang KC, Kim J, Horlings HM, Wong DJ, et al. Long non-coding RNA HOTAIR reprograms chromatin state to promote cancer metastasis. Nature 2010;464(7291):1071−6.

[83] Kogo R, Shimamura T, Mimori K, Kawahara K, Imoto S, Sudo T, et al. Long noncoding RNA HOTAIR regulates polycomb-dependent chromatin modification and is associated with poor prognosis in colorectal cancers. Cancer Res 2011;71(20):6320−6.

[84] Garzon R, Volinia S, Papaioannou D, Nicolet D, Kohlschmidt J, Yan PS, et al. Expression and prognostic impact of lncRNAs in acute myeloid leukemia. PNAS 2014;111(52):18679−84.

[85] Zeidan AM, Linhares Y, Gore SD. Current therapy of myelodysplastic syndromes. Blood Rev 2013;27(5):243−59.

[86] Yan P, Frankhouser D, Murphy M, Tam HH, Rodriguez B, Curfman J, et al. Genome-wide methylation profiling in decitabine-treated patients with acute myeloid leukemia. Blood 2012;120(12):2466−74.

[87] Klco JM, Spencer DH, Lamprecht TL, Sarkaria SM, Wylie T, Magrini V, et al. Genomic impact of transient low-dose decitabine treatment on primary AML cells. Blood 2013;121 (9):1633−43.

[88] Voso MT, Santini V., Fabiani E, Fianchi L, Criscuolo M, Falconi G, et al. Why methylation is not a marker predictive of response to hypomethylating agents. Haematologica 2014;99(4):613−19.

[89] Itzykson R, Kosmider O, Cluzeau T, Mansat-De Mas V, Dreyfus F, Beyne-Rauzy O, et al. Groupe Francophone des Myelodysplasies (GFM) Impact of TET2 mutations on response rate to azacitidine in myelodysplastic syndromes and low blast count acute myeloid leukemias. Leukemia 2011;25(7):1147−52.

[90] Bejar R, Lord A, Stevenson K, Bar-Natan M, Pérez-Ladaga A, Zaneveld J, et al. TET2 mutations predict response to hypomethylating agents in myelodysplastic syndrome patients. Blood 2014;124(17):2705−12.

[91] Traina F, Visconte V, Elson P, Tabarroki A, Jankowska AM, Hasrouni E, et al. Impact of molecular mutations on treatment response to DNMT inhibitors in myelodysplasia and related neoplasms. Leukemia 2014;28(1):78−87.

[92] Meldi K, Qin T, Buchi F, Droin N, Sotzen J, Micol JB, et al. Specific molecular signatures predict decitabine response in chronic myelomonocytic leukemia. J Clin Invest 2015;125(5):1857−72.

[93] Jankowska AM, Makishima H, Tiu RV, Szpurka H, Huang Y, Traina F, et al. Mutational spectrum analysis of chronic myelomonocytic leukemia includes genes associated with epigenetic regulation: UTX, EZH2, and DNMT3A. Blood 2011;118 (14):3932−41.

[94] Blum KA, Liu Z, Lucas DM, Chen P, Xie Z, Baiocchi R, et al. Phase I trial of low dose decitabine targeting DNA hypermethylation in patients with chronic lymphocytic leukaemia and non-Hodgkin lymphoma: dose-limiting myelosuppression without evidence of DNA hypomethylation. Br J Haematol 2010;150(2):189−95.

[95] Mesa RA, Verstovsek S, Rivera C, Pardanani A, Hussein K, Lasho T, et al. 5-Azacitidine has limited therapeutic activity in myelofibrosis. Leukemia 2009;23(1):180−2.

[96] Quintás-Cardama A, Tong W, Kantarjian H, Thomas D, Ravandi F, Kornblau S, et al. A phase II study of 5-azacitidine for patients with primary and post-essential thrombocythemia/polycythemia vera myelofibrosis. Leukemia 2008;22(5):965−70.

[97]   Arribas AJ, Rinaldi A, Mensah AA, Kwee I, Cascione L, Robles EF, et al. DNA methyl-ation profiling identifies two splenic marginal zone lymphoma subgroups with different clinical and genetic features. Blood 2015;125(12):1922−31.

[98]   Tsai HC, Li H, Van Neste L, Cai Y, Robert C, Rassool FV, et al. Transient low doses of DNA-demethylating agents exert durable antitumor effects on hematological and epithe-lial tumor cells. Cancer Cell 2012;21(3):430−46.

[99]   Matei D, Fang F, Shen C, Schilder J, Arnold A, Zeng Y, et al. Epigenetic resensitization to platinum in ovarian cancer. Cancer Res 2012;72(9):2197−205.

[100]   Wrangle J, Wang W, Koch A, Easwaran H, Mohammad HP, Vendetti F, et al. Alterations of immune response of Non-Small Cell Lung Cancer with Azacytidine. Oncotarget 2013;4(11):2067−79.

[101]   Li H, Chiappinelli KB, Guzzetta AA, Easwaran H, Yen RW, Vatapalli R, et al. Immune regulation by low doses of the DNA methyltransferase inhibitor 5-azacitidine in common human epithelial cancers. Oncotarget 2014;5(3):587−98.

[102]   Griffiths EA, Gore SD. DNA methyltransferase and histone deacetylase inhibitors in the treatment of myelodysplastic syndromes. Semin Hematol 2008;45(1):23−30.

[103]   Juergens RA, Wrangle J, Vendetti FP, Murphy SC, Zhao M, Coleman B, et al. Combination epigenetic therapy has efficacy in patients with refractory advanced non-small cell lung cancer. Cancer Discov 2011;1(7):598−607.

[104]   Cameron EE, Bachman KE, Myöhänen S, Herman JG, Baylin SB. Synergy of demethyl-ation and histone deacetylase inhibition in the re-expression of genes silenced in cancer. Nat Genet 1999;21(1):103−7.

[105]   Prebet T, Sun Z, Figueroa ME, Ketterling R, Melnick A, Greenberg PL, et al. Prolonged administration of azacitidine with or without entinostat for myelodysplastic syndrome and acute myeloid leukemia with myelodysplasia-related changes: results of the US Leukemia Intergroup trial E1905. J Clin Oncol 2014;32(12):1242−8.

[106]   Farh KK, Marson A, Zhu J, Kleinewietfeld M, Housley WJ, Beik S, et al. Genetic and epigenetic fine mapping of causal autoimmune disease variants. Nature 2015;518 (7539):337−43.

[107]   Rakyan VK, Down TA, Balding DJ, Beck S. Epigenome-wide association studies for common human diseases. Nat Rev Genet 2011;12(8):529−41.

[108]   Orozco LD, Morselli M, Rubbi L, Guo W, Go J, Shi H, et al. Epigenome-wide associa-tion of liver methylation patterns and complex metabolic traits in mice. Cell Metab 2015;21(6):905−17.

[109]   Multhaup ML, Seldin MM, Jaffe AE, Lei X, Kirchner H, Mondal P, et al. Mouse-human experimental epigenetic analysis unmasks dietary targets and genetic liability for diabetic phenotypes. Cell Metab 2015;21(1):138−49.

[110]   Öst A, Lempradl A, Casas E, Weigert M, Tiko T, Deniz M, et al. Paternal diet defines offspring chromatin state and intergenerational obesity. Cell 2014;159(6):1352−64.

Index

Note: Page numbers followed by "*f*" and "*t*" refer to figures and tables, respectively.

Printed in the United States
By Bookmasters